Metabolomics and Plant Defence

Metabolomics and Plant Defence

Guest Editors

Shitou Xia
Junxing Lu

MDPI

Basel • Beijing • Wuhan • Barcelona • Belgrade • Novi Sad • Cluj • Manchester

Guest Editors

Shitou Xia
College of Bioscience and Biotechnology
Hunan Agricultural University
Changsha
China

Junxing Lu
College of Life Sciences
Chongqing Normal University
Chongqing
China

Editorial Office
MDPI AG
Grosspeteranlage 5
4052 Basel, Switzerland

This is a reprint of the Special Issue, published open access by the journal *Metabolites* (ISSN 2218-1989), freely accessible at: https://www.mdpi.com/journal/metabolites/special_issues/27641OU3B6.

For citation purposes, cite each article independently as indicated on the article page online and as indicated below:

Lastname, A.A.; Lastname, B.B. Article Title. *Journal Name* **Year**, *Volume Number*, Page Range.

ISBN 978-3-7258-6850-6 (Hbk)
ISBN 978-3-7258-6851-3 (PDF)
https://doi.org/10.3390/books978-3-7258-6851-3

Contents

About the Editors

Shitou Xia

Shitou Xia, Ph.D., is a second-level Professor at Hunan Agricultural University, where he serves as the head of the Plant Science doctoral program. He obtained his Ph.D. in Botany through a joint program between Hunan Agricultural University and the University of British Columbia (UBC), Canada. His primary research interests focus on plant physiology and molecular biology, specifically investigating the molecular mechanisms of plant innate immunity and resistance regulated by phytohormones such as salicylic acid, using *Arabidopsis thaliana*, *Brassica napus*, and *Oryza sativa* as materials. Prof. Xia has made significant contributions to understanding plant–pathogen interactions and resistance signaling pathways. His research has been supported by numerous grants, including projects from the National Natural Science Foundation of China. He is a recognized expert in his field, honored as an Outstanding Scientific Worker in Hunan Province and a recipient of the First Prize of the Hunan Provincial Science and Technology Progress Award. His scholarly work includes over 60 publications in reputable international journals such as *Nature Plants*, *Plant Physiology*, and *Molecular Plant*. Actively engaged in academic service, Prof. Xia serves as a Responsible Editorial Board Member for the Chinese Bulletin of Botany, and as a review expert for the National Natural Science Foundation of China. He is also dedicated to higher education, leading nationally top-rated courses such as "Plant Physiology". Prof. Xia contributes to the scientific community as a Guest Editor for Special Issues and as a reviewer for several international journals.

Junxing Lu

Junxing Lu, Ph.D., is an Associate Professor within the College of Life Sciences at Chongqing Normal University. Dr. Lu's research focuses on molecular plant pathology and crop disease resistance. His work is dedicated to elucidating the pathogenic and developmental mechanisms of the destructive fungal pathogen *Sclerotinia sclerotiorum*, as well as deciphering the molecular regulatory networks underlying disease resistance in rapeseed oil and Arabidopsis. Dr. Lu's academic foundation includes a Ph.D. in Biochemistry and Molecular Biology from Southwest University and a B.Sc. from South China Agricultural University, which was strengthened through international research stints as a visiting scholar at the University of British Columbia (UBC) and Sichuan University. He has been a principal or key investigator on more than 10 competitive grants, including projects from the National Natural Science Foundation of China. His contributions to the field are documented in over 50 scholarly publications, appearing in reputable international journals such as *Nature*, *Plant Physiology*, and *Plant Biotechnology Journal*. In service to the broader scientific community, Dr. Lu engages in the peer-review process for several international journals and has served as a Guest Editor for Special Issues.

Preface

We are pleased to present this Reprint, a curated collection of research articles from the Special Issue "Metabolomics and Plant Defence" of *Metabolites*. In an era of escalating environmental challenges, understanding the molecular basis of plant resilience is more critical than ever. This volume is dedicated to exploring the sophisticated metabolic dialogues that plants engage in to survive and thrive under stress.

The studies compiled here employ cutting-edge metabolomic and multi-omics approaches to decode the chemical language of plant defense. From the priming of barley against fungal pathogens to the antioxidant strategies of banana against bacterial wilt, and from the metabolic acclimation of tomato to heat and aluminum stress to the developmental metabolomics of sorghum, this Reprint offers a panoramic view of plant metabolic adaptability. It underscores a central theme: defense is not merely a reaction but a strategic reprogramming of primary and secondary metabolism, balancing resistance with growth.

This Reprint is intended for plant scientists, biochemists, agricultural researchers, and advanced students seeking to comprehend advances in our knowledge of plant–environment interactions. We believe that this collection serves as a valuable resource, highlighting how fundamental discoveries in metabolic pathways can inform practical strategies for crop improvement.

We extend our deepest gratitude to all the contributing authors for sharing their insightful work. Our sincere thanks also go to the numerous expert reviewers whose rigorous evaluations and constructive feedback were indispensable in shaping the quality of this Special Issue. Finally, we acknowledge the dedicated editorial team of *Metabolites* for their professional support and for making this Reprint possible.

Shitou Xia and Junxing Lu
Guest Editors

metabolites

MDPI

Editorial

Metabolomics and Plant Defense

Junxing Lu [1,*] and Shitou Xia [2,*]

1 Chongqing Key Laboratory of Plant Environmental Adaptations, College of Life Science, Chongqing Normal University, Chongqing 401331, China
2 Hunan Provincial Key Laboratory of Phytohormones and Growth Development, College of Bioscience and Biotechnology, Hunan Agricultural University, Changsha 410128, China
* Correspondence: junxlu@163.com (J.L.); xstone0505@hunau.edu.cn (S.X.)

1. Introduction

Plant metabolomics is pivotal in understanding plant defense mechanisms against environmental stresses [1,2], which enables researchers to profile the complex metabolic changes in plants in response to biotic and abiotic factors such as pathogens, drought, and temperature extremes [3,4]. By analyzing metabolic fingerprints, scientists can identify key metabolites that are upregulated or synthesized during stresses, including defense phytohormones like salicylic acid, jasmonic acid, and ethylene, which are crucial for mounting an effective resistance [5,6]. Additionally, metabolomic studies reveal the accumulation of secondary metabolites such as flavonoids and phenolics, which often possess antioxidant properties and can protect plants from oxidative damage caused by various stresses [7]. These are fundamental for understanding plant biology and developing strategies to enhance crop resilience and improve agricultural sustainability in the face of climate change.

The field of metabolomics has become instrumental in deciphering the intricate metabolic responses of plants to biotic and abiotic stresses, particularly in the context of disease resistance and environmental adaptability [8]. This Special Issue of *Metabolites*, entitled "Metabolomics and Plant Defense" contains nine original research articles and one review, focusing on distinct plant–pathogen interactions and plant–environmental stress responses to enhance our understanding of how plants reprogram their metabolism as a defense strategy against pathogens and environmental challenges.

2. Metabolomics Insights into Plant Pathogen Resistance

In recent years, metabolomics has emerged as a powerful tool for understanding the complex metabolic responses of plants to biotic stresses, particularly in the context of disease resistance [1,6]. This Special Issue features four original research articles on distinct plant–pathogen interactions, which collectively contribute to our understanding of how plants employ metabolic reprogramming in resistance against pathogens.

Singh et al. (contribution 1) investigated the metabolic responses of the wild tomato species *Solanum cheesmaniae* to the pathogen *Alternaria solani*, which causes early blight. Untargeted metabolomics was applied to identify significant metabolic biomarkers and pathways associated with plant resistance. The study identified 10,943 metabolite features with 3371 compounds annotated using the KEGG database, implicating secondary metabolites, cofactors, steroids, terpenoids, and fatty acids. A total of 541 upregulated and 485 downregulated metabolite features were identified, which play roles in defense, infection prevention, signaling, plant growth, and homeostasis under stress. Orthogonal partial least squares discriminant analysis (OPLS-DA) revealed 34 upregulated and 41 downregulated biomarker metabolites, which could contribute to disease-resistant metabolic traits and tomato breeding programs.

Barley (*Hordeum vulgare*), a highly versatile cereal crop, is widely cultivated but is susceptible to fungal pathogens, notably the ascomycete *Pyrenophora teres f. teres* (*Ptt*), which causes the 'net blotch net form' (NBNF) disease. Hamany Djande et al. (contribution 2) provided insights into the metabolomic reconfiguration of barley plants primed with 3,5-dichloroanthranilic acid (3,5-DCAA) in response to *Ptt* infection, shedding light on how priming can arm barley with a rapid and robust defense mechanism against necrotrophic fungi, a group known to cause significant crop losses worldwide. Both untargeted and targeted metabolomics analyses were conducted, revealing a complex interplay of metabolites that are reprogrammed in primed barley. Key metabolites such as 5-oxo-proline and citric acid were significantly associated with the priming response. These findings advance our understanding of the regulated and reprogrammed metabolic responses that constitute defense priming in barley against *Ptt*.

The tea industry, a vital sector in global agriculture, faces significant challenges from diseases including gray blight caused by *Pestalotiopsis*-like species. Zheng et al. (contribution 3) provided valuable insights into the metabolic defense mechanisms of tea plants (*Camellia sinensis*) in response to this disease. Using a multi-omics approach, the researchers identified 64 differentially accumulated metabolites, including increased levels of antimicrobial compounds like caffeine and (−)-epigallocatechin 3-gallate, along with a decrease in the synthesis of (+)-catechin and (−)-epicatechin. A dynamic modulation of different branches of the flavonoid biosynthetic pathway was revealed in tea plants under fungal challenge. The findings underscore the strategic reprogramming of tea plant metabolism, prioritizing the production of potent antimicrobial flavanols while reducing the allocation of resources to less bioactive compounds. The metabolic shift is likely crucial for bolstering the plant's defense against fungal pathogens.

Diseases such as Moko, caused by *Ralstonia solanacearum*, also pose significant threats to the global banana industry. Mihai et al. (contribution 4) examined the metabolic response of *Musa* spp. from the Ecuadorian coast to this disease, shedding light on their defense mechanisms. A higher accumulation of secondary metabolites with antioxidant capacity was identified in diseased samples compared to healthy ones, suggesting strategic metabolic reprogramming in response to pathogenic attack. The study employed a multi-omics approach, utilizing LC-MS to analyze active compounds with antioxidant properties in four common *Musa* varieties. The results indicated an upregulation of phenolic compounds and flavonoids such as kaempferol and quercetin glycosides, which are crucial for the plant's defense against *R. solanacearum*.

These four studies collectively illustrate the dynamic and complex nature of plant metabolic responses to pathogen infection and reveal how plants employ a suite of metabolic changes, including the upregulation of secondary metabolites and antioxidants, to combat pathogen invasion. These metabolic shifts not only provide immediate defense but also prime the plant for future encounters, highlighting the plant's ability to "remember" and respond more effectively to subsequent attacks. Metabolomic approaches offer a window into the plant's chemical arsenal, allowing scientists to identify and manipulate key metabolic pathways to enhance disease resistance.

3. Metabolomics Insights into Plant Environmental Stress Adaptation

The intricate interplay between environmental stresses and plant metabolism is akin to a complex dance, where each step, choreographed by nature, carries profound implications for agriculture and food security. Within this Special Issue, four original research articles delve into the metabolic responses of plants under various stressors.

Hamany Djande et al. (contribution 5) explored the metabolic reprogramming in barley plants following the application of dichlorinated functional analogs of salicylic acid, shedding light on how these plants bolster their defenses against environmental stresses. Employing a multi-omics approach, the research revealed significant alterations in both primary and secondary metabolites, with a particular emphasis on the phenylpropanoid pathway—a known marker of induced resistance in plants. The application of 3,5-dichloroanthranilic acid, 2,6-dichloropyridine-4-carboxylic acid, and 3,5-dichlorosalicylic acid to barley at the third leaf stage triggered a metabolic response, underscoring the potential of these dichlorinated compounds as inducers of acquired resistance. The study observed the accumulation of barley-specific metabolites, for instance hordatines and precursors, indicating a strategic metabolic reprogramming that enhances the plant's defense mechanisms.

In the face of environmental adversities, β-cyclocitral (βCC) emerges as a beacon of hope for bolstering the resilience of *Solanum lycopersicum* (tomato). This apocarotenoid, an oxidative product of β-carotene, primes plants for stress acclimation without impeding growth—a delicate balance rarely achieved. Deshpande and Mitra (contribution 6) meticulously documented how βCC induces metabolic changes, enhancing both defense mechanisms and growth in tomato plants. Through a sophisticated LC-MS/MS analysis, the research identified 57 compounds. Among which, amino acids and phytophenols, upregulated by βCC, not only fortify the plant against biotic stresses but also support its growth. This dual action underscores the wisdom of plant biology, where βCC orchestrates a temporal separation in metabolite accumulation, optimizing resource allocation. The findings are a testament to βCC's promise as a metabolic signal, priming crops for a fortified defense against environmental challenges.

Amid the growing challenges posed by acidic soils that hinder global crop production, pyroligneous acid (PA) emerges as a promising ally for tomato plants. Aluminum toxicity, a pervasive issue in such soils, can be mitigated by PA application, according to a recent study by Ofoe et al. (contribution 7). This research uncovers the metabolic nuances of how PA treats tomato plants under aluminum stress, revealing a complex interplay of central carbon metabolism (CCM). Forty-eight differentially expressed metabolites were identified as involved in CCM, indicating PA's role in regulating plant metabolism to adapt to aluminum stress. Notably, PA increased metabolites associated with glycolysis and the tricarboxylic acid (TCA) cycle, suggesting a shift towards enhanced energy production and organic acid biosynthesis, which is crucial for aluminum tolerance.

Understanding plant responses to water stress is critical for agricultural resilience, particularly in light of climate change. Abbey et al. (contribution 8) explored the metabolic shifts in Mexican mint (*Plectranthus amboinicus*) under varying watering regimes. The research reveals that 68 key metabolites of central carbon metabolism are significantly impacted by water stress, highlighting the plant's metabolic plasticity. Under drought conditions, glycolytic metabolites surge, suggesting an energy conservation strategy. Conversely, flooding enhances Calvin cycle metabolites, indicating a photosynthetic adaptation. Notably, the resumption of regular watering prompts a swift metabolic rebound, underscoring the plant's ability to restore normal physiological activities.

Collectively, these studies deepen our understanding of how plants reprogram their metabolism in response to environmental stresses and highlight the complexity of plant metabolic responses and the potential for manipulating these pathways to enhance stress tolerance. By revealing the metabolic underpinnings of plant resilience, these studies may open avenues for developing crops that can maintain productivity under challenging conditions.

4. Metabolomic Insights into Sorghum Seedling Development and Tomato Heat Stress Management

Sorghum, an important crop in arid regions, faces considerable challenges during its early developmental stages, particularly from soil-borne pathogens. Dubery et al. (contribution 9) provide a thorough examination of the metabolic reprogramming in sorghum seedlings, offering insights that are essential for enhancing crop resilience and ensuring food security, as it uncovers the role of specialized metabolites in plant growth, development, and defense mechanisms against environmental stresses, including the common damping-off disease in sorghum seedlings. Using targeted metabolomics, the research identified shifts in metabolite patterns across various growth stages, revealing a significant increase in metabolite content as the plants mature. Notably, the study found a correlation between apigenin flavone derivatives and growth stages, which could play a key role in the development of stress-resilient crops and the improvement of crop quality and yield.

Amidst the relentless rise in global temperatures, the agricultural sector faces significant challenges, particularly for heat-sensitive crops like tomatoes. Khan et al. (contribution 10) offer an exhaustive analysis of the adverse effects of heat stress on tomato plants. The authors detail how elevated temperatures negatively impact tomato growth, nutrient availability, photosynthesis, reproduction, and cellular integrity. Excessive production of reactive oxygen species (ROS) under heat stress is especially alarming, as it can lead to substantial cellular damage. The review emphasizes the need to develop heat-tolerant tomato varieties and highlights the potential of modern functional genomics techniques—transcriptomics, proteomics, and metabolomics—to identify key genes, proteins, and metabolites associated with heat stress tolerance. The integration of these technologies with traditional breeding methods could lead to the creation of robust tomato cultivars capable of withstanding high temperatures.

5. Conclusions

In conclusion, the comprehensive insights provided by the studies in this Special Issue underscore the pivotal role of metabolomics in plant defense mechanisms against both biotic and abiotic stresses. The observed metabolic reprogramming across diverse plant species, including tomatoes, barley, tea, bananas, and others, highlights the dynamic nature of plant responses to pathogens and environmental challenges. While these studies provide valuable insights into plant metabolic adaptations under stress, their focus on single stress factors or model crops might limit our understanding of combinatorial stress responses and their broader applicability in agricultural systems. To address these gaps and further explore the complexity of plant–environment interactions, we are organizing a follow-up Special Issue, "Metabolomics and Plant Defence, 2nd Edition". This new issue will emphasize multi-omics integration, cross-species comparative analyses, and field-based studies under realistic climate conditions. The integration of metabolomics with other omics technologies and traditional breeding methods can facilitate the development of resilient, stress-adapted crop varieties, thereby contributing to the establishment of sustainable and climate-resilient agricultural systems.

Author Contributions: Conceptualization, J.L. and S.X.; methodology, J.L.; software, S.X.; validation, J.L. and S.X.; formal analysis, S.X.; investigation, J.L.; resources, S.X.; data curation, J.L.; writing—original draft preparation, J.L. and S.X.; writing—review and editing, S.X. and J.L.; visualization, J.L.; supervision, J.L. and S.X.; project administration, S.X.; funding acquisition, J.L. All authors have read and agreed to the published version of the manuscript.

Funding: This research was partially supported by the Chongqing Municipal Science and Technology Bureau (CSTB2024NSCQ-MSX0502).

Conflicts of Interest: The authors declare no conflict of interest.

List of Contributions:

1. Singh, D.P.; Bisen, M.S.; Prabha, R.; Maurya, S.; Yerasu, S.R.; Shukla, R.; Tiwari, J.K.; Chaturvedi, K.K.; Farooqi, M.S.; Srivastava, S.; et al. Untargeted Metabolomics of Alternaria Solani-Challenged Wild Tomato Species Solanum Cheesmaniae Revealed Key Metabolite Biomarkers and Insight into Altered Metabolic Pathways. *Metabolites* **2023**, *13*, 585. https://doi.org/10.3390/metabo13050585.
2. Hamany Djande, C.Y.; Tugizimana, F.; Steenkamp, P.A.; Piater, L.A.; Dubery, I.A. Metabolomic Reconfiguration in Primed Barley (Hordeum Vulgare) Plants in Response to Pyrenophora Teres f. Teres Infection. *Metabolites* **2023**, *13*, 997. https://doi.org/10.3390/metabo13090997.
3. Zheng, S.; Du, Z.; Wang, X.; Zheng, C.; Wang, Z.; Yu, X. Metabolic Rewiring in Tea Plants in Response to Gray Blight Disease Unveiled by Multi-Omics Analysis. *Metabolites* **2023**, *13*, 1122. https://doi.org/10.3390/metabo13111122.
4. Mihai, R.A.; Terán-Maza, V.A.; Portilla-Benalcazar, K.A.; Ramos-Guaytarilla, L.E.; Vizuete-Cabezas, M.J.; Melo-Heras, E.J.; Cubi-Insuaste, N.S.; Catana, R.D. Secondary Metabolites and Antioxidant Activity against Moko Disease as a Defense Mechanism of *Musa* spp. from the Ecuadorian Coast Area. *Metabolites* **2024**, *14*, 307. https://doi.org/10.3390/metabo14060307.
5. Hamany Djande, C.Y.; Steenkamp, P.A.; Piater, L.A.; Tugizimana, F.; Dubery, I.A. Metabolic Reprogramming of Barley in Response to Foliar Application of Dichlorinated Functional Analogues of Salicylic Acid as Priming Agents and Inducers of Plant Defence. *Metabolites* **2023**, *13*, 666. https://doi.org/10.3390/metabo13050666.
6. Deshpande, S.; Mitra, S. β-Cyclocitral-Mediated Metabolic Changes Optimize Growth and Defense Responses in *Solanum lycopersicum* L. *Metabolites* **2023**, *13*, 329. https://doi.org/10.3390/metabo13030329.
7. Ofoe, R.; Thomas, R.H.; Abbey, Lord Coordinated Regulation of Central Carbon Metabolism in Pyroligneous Acid-Treated Tomato Plants under Aluminum Stress. *Metabolites* **2023**, *13*, 770. https://doi.org/10.3390/metabo13060770.
8. Abbey, Lord; Ofoe, R.; Wang, Z.; Chada, S. How Central Carbon Metabolites of Mexican Mint (*Plectranthus amboinicus*) Plants Are Impacted under Different Watering Regimes. *Metabolites* **2023**, *13*, 539. https://doi.org/10.3390/metabo13040539.
9. Dubery, I.A.; Nephali, L.P.; Tugizimana, F.; Steenkamp, P.A. Data-Driven Characterization of Metabolome Reprogramming during Early Development of Sorghum Seedlings. *Metabolites* **2024**, *14*, 112. https://doi.org/10.3390/metabo14020112.
10. Khan, Q.; Wang, Y.; Xia, G.; Yang, H.; Luo, Z.; Zhang, Y. Deleterious Effects of Heat Stress on the Tomato, Its Innate Responses, and Potential Preventive Strategies in the Realm of Emerging Technologies. *Metabolites* **2024**, *14*, 283. https://doi.org/10.3390/metabo14050283.

References

1. Patel, M.K.; Pandey, S.; Kumar, M.; Haque, M.I.; Pal, S.; Yadav, N.S. Plants Metabolome Study: Emerging Tools and Techniques. *Plants* **2021**, *10*, 2409. [CrossRef] [PubMed]
2. Muñoz-Hoyos, L.; Stam, R. Metabolomics in Plant Pathogen Defense: From Single Molecules to Large-Scale Analysis. *Phytopathology* **2023**, *113*, 760–770. [CrossRef] [PubMed]
3. Shiade, S.R.G.; Zand-Silakhoor, A.; Fathi, A.; Rahimi, R.; Minkina, T.; Rajput, V.D.; Zulfiqar, U.; Chaudhary, T. Plant Metabolites and Signaling Pathways in Response to Biotic and Abiotic Stresses: Exploring Bio Stimulant Applications. *Plant Stress* **2024**, *12*, 100454. [CrossRef]
4. Yang, Y.; Guo, Y. Elucidating the Molecular Mechanisms Mediating Plant Salt-Stress Responses. *New Phytol.* **2018**, *217*, 523–539. [CrossRef] [PubMed]
5. Kumari, M.; Yagnik, K.N.; Gupta, V.; Singh, I.K.; Gupta, R.; Verma, P.K.; Singh, A. Metabolomics-Driven Investigation of Plant Defense Response against Pest and Pathogen Attack. *Physiol. Plant.* **2024**, *176*, e14270. [CrossRef] [PubMed]
6. Castro-Moretti, F.R.; Gentzel, I.N.; Mackey, D.; Alonso, A.P. Metabolomics as an Emerging Tool for the Study of Plant-Pathogen Interactions. *Metabolites* **2020**, *10*, 52. [CrossRef] [PubMed]

7. Manickam, S.; Rajagopalan, V.R.; Kambale, R.; Rajasekaran, R.; Kanagarajan, S.; Muthurajan, R. Plant Metabolomics: Current Initiatives and Future Prospects. *Curr. Issues Mol. Biol.* **2023**, *45*, 8894–8906. [CrossRef] [PubMed]
8. Shen, S.; Zhan, C.; Yang, C.; Fernie, A.R.; Luo, J. Metabolomics-Centered Mining of Plant Metabolic Diversity and Function: Past Decade and Future Perspectives. *Mol. Plant* **2023**, *16*, 43–63. [CrossRef] [PubMed]

metabolites

MDPI

Article

Untargeted Metabolomics of *Alternaria solani*-Challenged Wild Tomato Species *Solanum cheesmaniae* Revealed Key Metabolite Biomarkers and Insight into Altered Metabolic Pathways

Dhananjaya Pratap Singh [1,*], Mansi Singh Bisen [1], Ratna Prabha [2], Sudarshan Maurya [1], Suresh Reddy Yerasu [1], Renu Shukla [3], Jagesh Kumar Tiwari [1], Krishna Kumar Chaturvedi [2], Md. Samir Farooqi [2], Sudhir Srivastava [2], Anil Rai [2], Birinchi Kumar Sarma [4], Nagendra Rai [1], Prabhakar Mohan Singh [1], Tusar Kanti Behera [1,*] and Mohamed A. Farag [5,*]

1 ICAR-Indian Institute of Vegetable Research, Varanasi 221305, India
2 ICAR-Indian Agricultural Statistics Research Institute, Library Avenue, New Delhi 110012, India
3 Indian Council of Agricultural Research, New Delhi 110012, India
4 Department of Mycology and Plant Pathology, Institute of Agricultural Sciences, Banaras Hindu University, Varanasi 221005, India
5 Pharmacognosy Department, College of Pharmacy, Cairo University, Cairo 11562, Egypt
* Correspondence: dhananjaya.singh@icar.gov.in (D.P.S.); tusar@rediffmail.com (T.K.B.); mohamed.farag@pharma.cu.edu.eg (M.A.F.)

Abstract: Untargeted metabolomics of moderately resistant wild tomato species *Solanum cheesmaniae* revealed an altered metabolite profile in plant leaves in response to *Alternaria solani* pathogen. Leaf metabolites were significantly differentiated in non-stressed versus stressed plants. The samples were discriminated not only by the presence/absence of specific metabolites as distinguished markers of infection, but also on the basis of their relative abundance as important concluding factors. Annotation of metabolite features using the *Arabidopsis thaliana* (KEGG) database revealed 3371 compounds with KEGG identifiers belonging to biosynthetic pathways including secondary metabolites, cofactors, steroids, brassinosteroids, terpernoids, and fatty acids. Annotation using the *Solanum lycopersicum* database in PLANTCYC PMN revealed significantly upregulated (541) and downregulated (485) features distributed in metabolite classes that appeared to play a crucial role in defense, infection prevention, signaling, plant growth, and plant homeostasis to survive under stress conditions. The orthogonal partial least squares discriminant analysis (OPLS-DA), comprising a significant fold change ($\geq$2.0) with VIP score ($\geq$1.0), showed 34 upregulated biomarker metabolites including 5-phosphoribosylamine, kaur-16-en-18-oic acid, pantothenate, and *O*-acetyl-L-homoserine, along with 41 downregulated biomarkers. Downregulated metabolite biomarkers were mapped with pathways specifically known for plant defense, suggesting their prominent role in pathogen resistance. These results hold promise for identifying key biomarker metabolites that contribute to disease resistive metabolic traits/biosynthetic routes. This approach can assist in mQTL development for the stress breeding program in tomato against pathogen interactions.

Keywords: metabolite profiling; secondary metabolites; metabolomics; biotic stress; *Solanum cheesmaniae*; *Alternaria solani*; early blight

1. Introduction

Chemical diversity and metabolite composition of plants exposed to biotic challenges in the environment critically define their standing status under the conditions of the progression of pathogen infection [1]. Following pathogen invasion, pathophysiological and molecular cascades are triggered to sense the stress [2], which simultaneously activate an integrated cellular network of physiological, biochemical, and molecular mechanisms [3,4].

Further development of pathogen invasion and disease progression starts with the exchange of signaling molecules [5] and chemical crosstalk [6] as a ping-pong mechanism in the host plant and pathogen interaction [7]. Accordingly, a complex cellular and metabolic response is activated inside the host plant against pathogen invasion to circumvent disease progression. As the infection progresses, metabolic networks of the pathogens and host become intertwined, thereby leading to mutually influenced metabolism in both organisms [8]. In plants, an array of metabolites and their intermediates are biosynthesized, typically to either accumulate or become degraded to categorically avoid or restrict pathogen invasion [9]. In order to protect themselves from pathogens, plants need to accumulate a diverse array of primary and secondary metabolites that warrant distinct developmental and functional protective roles [10]. These metabolites, if mapped in pathogen-challenged and unchallenged plants in an integrated way, can establish their identity and functional role and provide exclusive clues regarding plant responses toward pathogens.

Solanum cheesmaniae is one of the two main wild tomato species from the Galápagos Islands with enhanced plant performance against stresses [11]. Because of its potential stress-tolerance and disease-resistance traits [12], this species has successfully been used in stress-resistance breeding programs for the improvement of commercial tomato varieties [13]. Early blight-resistant sources have been identified in different wild tomato species, with the accession WIR3928 belonging to *S. cheesmaniae* showing resistance to early blight throughout the vegetative phase and moderate resistance at the flowering stage [14]. Being a wildtype, the plant is supposed to possess stringent metabolic and molecular buildup of metabolic events that strongly activate defense mechanisms. Such metabolic buildup possibly makes plants resistant against invasion by pathogens such as *Alternaria solani* (causing early blight), a devastating pathogenic disease causing >50% damage to cultivated tomato crop under field conditions [15]. For comparatively mapping such chemical buildup inside plant cells and tissues, metabolomics has recently emerged as a practical analytical and informatics tool to offer unbiased qualitative and quantitative screening of metabolites at different status [16]. It was, therefore, hypothesized that the robust intrinsic metabolic fitness of *S. cheesmaniae* can offer the plant overprotection against *A. solani* pathogen. In this study, untargeted metabolomics analysis of *S. cheesmaniae* plants grown under normal and pathogen-challenged conditions was attempted to aid in identifying significantly distinct metabolic biomarkers and biosynthetic clues linked to plant tolerance responses against *A. solani*. The study holds promise for developing disease-diagnostic tools based on key biomarker metabolites, which could lead to the development of mQTLs to target disease-resistance metabolic traits/biosynthetic routes for supporting speed breeding programs in tomato against biotic stresses.

2. Materials and Methods

2.1. Plant Material and Growth Conditions

S. cheesmaniae accession WIR3928 (wild species moderately resistant to early blight) was used in this study to assess the post-inoculation metabolomic response toward *A. solani* (Indian Type Culture Collection (ITCC) 4632, ICAR-IARI, New Delhi, India). Seeds were surface sterilized using 2% sodium hypochlorite solution for 5 min using agitation followed by washing with sterilized water and drying on sterile filter paper. Dried seeds were seeded inside plastic pots with dimensions 20 × 20 × 14 cm containing a sterilized mixture of field soil, cocopeat, perlite, and vermiculite in a ratio of 3:1:1:1 (*w*/*w*), and then left in greenhouse to grow for 3 weeks. Afterward, seedlings were transplanted in 10 pots containing similar soil substrates. At the early flowering stage (60 days), lower leaves of 10 potted plants were spray-inoculated with the pathogen (2.1×10^4 CFU/mL inoculums). Pots were kept under high relative humidity (RH 86–90%) for 24 h, which was further reduced to 70% for the next 5 days to allow for disease development. The disease incidence (%) was calculated from days 1 to 5 as reported by Yerasu et al. [14]. On the fifth day post inoculation, sampling of upper leaves was performed separately from the pathogen-challenged and non-inoculated plants.

2.2. Leaf Sample Preparation and Metabolite Extraction

A comprehensive method was employed to capture both polar and nonpolar metabolites of tomato leaves [17]. Leaf tissues frozen with liquid nitrogen were pulverized with an ice-cold mortar and pestle. Five grams of leaves from pathogen-challenged and unchallenged plants were extracted using ethyl acetate (100%, 20 mL) overnight to capture nonpolar to moderately polar metabolites [18]. The supernatant was separated using centrifugation (6000× *g*, 4 °C, 15 min) and dried on a water bath set at 55 °C to yield 1 mL of ethyl acetate extract. Tissues were further extracted with methanol and water/methanol (10 mL each, 1:1, *v*/*v*) separately using sonication for 15 min and then vortexed for 2 h. The mixture was kept overnight, and tissue debris was then pelleted using a benchtop centrifuge (6000× *g*, 4 °C, 15 min) to yield 1 mL of each extract using a rotary evaporator (55 °C). All three extracts were pooled together and transferred to a 5 mL glass tube to obtain dried extract using rotavapor at 55 °C for 1 h. The samples were finally reconstituted in 1 mL of HPLC-grade 100% methanol as solvent, filtered through a 0.22 µm syringe filter, and transferred into chromatographic vials, which were capped and stored at 4 °C for further analysis. Three biological replications for each sample were prepared and analyzed in triplicate. The overall results represented two independent experiments. A quality check of the samples in aliquots was also performed to monitor sample stability, as well as instrument and analysis deviation. The extraction protocol standards were followed as per the minimum reporting standards explained for chemical analysis [19,20].

2.3. Untargeted LC–MS Analysis

A reconstituted extract of tomato leaves (200 µL) was filtered using 0.22 µm syringe filters and analyzed on a Dionex Ultimate 3000 (Thermo Fisher Scientific, Waltham, MA, USA) HPLC system coupled to a Q Exactive mass spectrometer (Thermo Fisher Scientific, USA). The analysis was performed using a Hypersil Gold C18 (2.1 mm × 100 mm, 1.9 µm) column set at 35 °C throughout the sample run of 31 min. The injection volume was 15 µL per sample. The mobile phase was A (0.05% formic acid in water) and B (0.05% formic acid in acetonitrile), with a flow rate of 350 µL/min. The elution conditions were 5% solvent B for 0 to 2 min, followed by 5–95% solvent B for 2.01–22 min, 95% solvent B for 22.01–27 min, and 5% solvent B for 27.01–31 min.

The Q Exactive MS system was used for data acquisition. The full scan MS range was set at 100–1500 *m*/*z*, and the first-order resolution was set at 140,000. Data acquisition was obtained in positive and negative ionization mode with an AGC target of 1×10^6. The parameters for the ion source were set as follows: sheath gas flow rate at 60 (arbitrary unit, au), aux gas flow rate at 20 (au), and sweep gas flow rate at 10 (au). A capillary voltage of (+) 3.2 kV was applied at a capillary temperature of 275 °C. The S-Lens level was at 55 rf, and the probe and aux gas heat temperatures were at 250 and 350 °C, respectively. Each analysis was performed in three technical replicates. A quality check (QC) sample was run after every six samples to ensure accurate retention time and elution order of the HPLC system. The product ion scan was obtained using first- and second-level MS data acquisition mode. The data obtained from LC–MS were analyzed using Compound Discoverer 3.3 (Thermo Fisher Scientific, USA).

2.4. Tandem MS Data Processing and Statistical Analysis

Tandem mass spectrometry (MS/MS) parameters were set to improve the mass fragmentation pattern, with an *m*/*z* scan range of 200–2000, isolation offset of *m*/*z* 0.5, collision energy (CE) of 25–45 eV, maximum IT of 50 ms, loop count of 5, MSX count of 1, resolution of 35,000, one microscan, and isolation window of *m*/*z* 2.0. The maximum AGC target for data-dependent acquisition (DDA) was set at 8.00 e^3 with unassigned charge exclusion having a dynamic exclusion rate of 10 s. An automated selection of precursor ions from an accurate mass inclusion list was prioritized for obtaining MS2 spectra as pre-annotated features after injection of two blanks and three quality checks (QCs) during system conditioning [21].

The raw data obtained from LC–MS/MS was preprocessed and analyzed using Compound Discoverer 3.3 (Thermo Fisher Scientific, USA). The analysis primarily included peak extraction, intensity detection, retention time (r.t.) correlation, filling of missing values, adjoint ion combinations, peak alignment, and normalization. The processing further facilitated metabolite identification and detection of molecular weight, retention time, and peak area. Metabolite features (m/z) were annotated using four chemical databases, namely, mzCloud (www.mzcloud.org), predicted compositions (www.thermofisher.com), metabolicatlas (metabolicatlas.org), and chemspider (www.chemspider.com) (accessed on 14 February 2023). In the absence of authenticated chemical compounds, m/z features were further reconfirmed and assigned KEEG identifier IDs after their annotation with the KEGG compound database (www.genome.jp/kegg/compound) and PLANTCYC (http://plantcyc.org) following minimum reporting standards (accessed on 14 February, 2023) [19,20].

The data were preprocessed before performing multidimensional statistical analysis, and the missing values were excluded from the original dataset. The processed data were then tabulated in MS Excel and then uploaded to the "Statistical Analysis" module of MetaboAnalyst 5.0 (www.metaboanalyst.ca, accessed on 14 February 2023). Univariate analysis included feature analysis by two-sample t-test and Wilcoxon rank-sum tests, fold change (FC) analysis to identify up- and downregulated metabolite features, volcano plot, and a correlation heatmap that aided in the verification of significantly different features. Before performing PCA model, $\log_2$ transformation was followed by Pareto scaling for scaling the obtained data. Chemometric analysis, including principal component analysis (PCA), partial least squares discriminant analysis (PLS-DA), and orthogonal PLS-DA (OPLS-DA), was performed. The VIP score obtained in the OPLS-DA model and fold-change maxima facilitated the assessment of the influence of intensity and explanatory ability of each of the metabolites in classifying and discriminating groups of samples on the basis of biologically significant metabolite features. Enrichment analysis was performed using MetaboAnalyst 5.0 by annotating KEGG IDs with "main-class" and "sub-class" metabolite chemical sets. Pathway and network analyses were performed.

A larger VIP score ($\geq$1) with fold change (FC) $\geq$2 showed a high contribution of metabolite features in the sample differentiation in OPLS-DA analysis. The features with VIP ≥ 1 and FC ≥ 2 at $p \leq 0.05$ were considered "significantly different metabolites" [21] characterized as "biomarkers". KEGG MAPPER was used to perform functional pathway annotation using "Compound Search" with major metabolic pathways to which the "annotated metabolites" or "differential metabolite biomarkers" belonged to.

2.5. Statistical Analysis

The "statistical module" of MetaboAnalyst 5.0 was used for univariate and multivariate data analysis. For other purposes, data were analyzed using SPSS 16.0 for one-way analysis of variance (ANOVA) using Student's t-test.

3. Results

A major goal of this study was to employ untargeted metabolomics to decipher metabolic responses of *S. cheesmaniae* plants against *A. solani* pathogen. The experimental setup, disease incidence, analysis procedure, and the outcomes in brief are summarized in Figure 1. *A. solani* led to a disease incidence of 17.6% based on leaf spot count after 5 days of pathogen inoculation, after which the disease did not spread intensively. LC–MS/MS-based untargeted metabolite profiling of pathogen (*A. solanii*)-challenged (MCTR) and unchallenged (MCNTR) *S. cheesmaniae* plant leaves led to the detection of 10,943 metabolite features. The analysis by "pathway hits" resulted in 92 metabolic pathways (20 at $p \leq 0.05$) mostly involved in the biosynthesis of ubiquinones, terpenoid-quinones, glycosphingolipids, flavonoids, sesqui-di- and tri-terpenoids, carotenoids, steroids, phenylpropanoids, zeatin, folate, anthocyanins, cutin and suberin, sphingolipid, porphyrins, linolenic acid, glutathione, vitamin B6, thiamine, inositol phosphate, cysteine, methionine, and ascorbate, as well as carbon fixation. Furthermore, functional analysis using "compound hits" matched 3371 compounds with the "metabolite features" having "KEGG identifiers", along with their matched forms and mass differences (Supplementary Table S1).

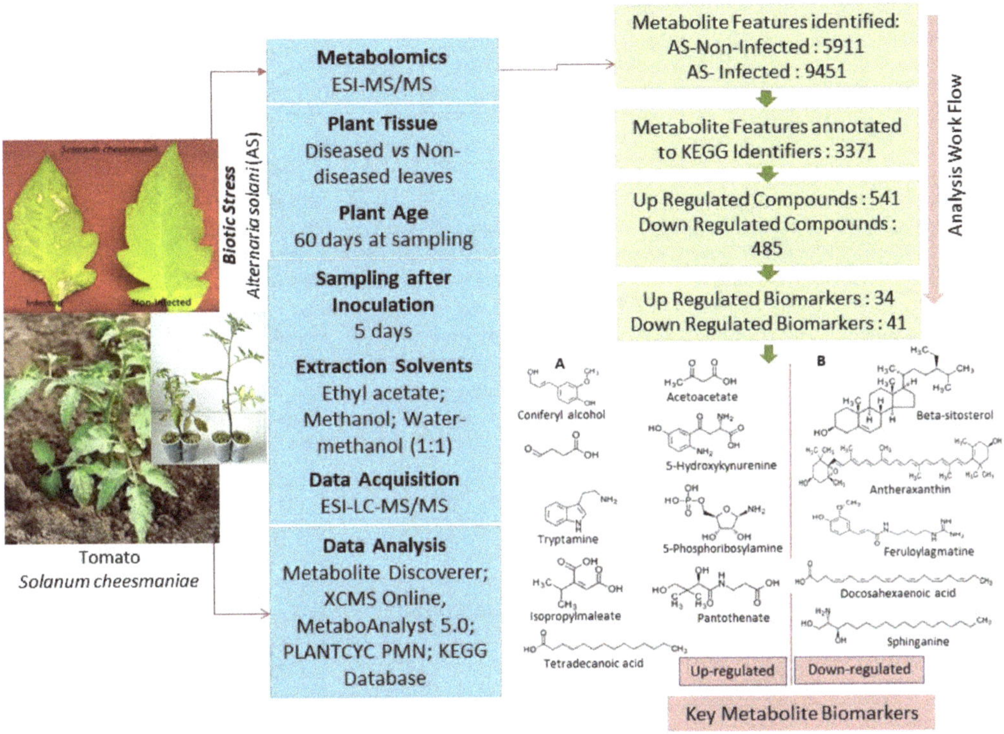

Figure 1. Representative scheme of the experimental setup.

3.1. Enrichment Analysis

All 3371 metabolite features having KEGG identifiers were annotated with the *Arabidopsis thaliana* pathway library using the "Functional Analysis" module of MetaboAnalyst 5.0. This resulted in the identification of the top 25 enriched metabolic pathways in which metabolite features were significantly linked (p-values: 4×10^{-2} to 4×10^{-13}) (Figure 2A). The pathway analysis module further identified 20 significantly enriched metabolic pathways (Figure 2B) belonging to steroid/terpenoid biosynthesis, amino-acid biosynthesis and degradation, and metabolism of arachidonic acid, tryptophan, and sphingolipids. The integrated pathway activity profile of the metabolite hits linked to different pathways after annotation against the *Arabidopsis thaliana* library in KEGG also verified the appropriateness of the functional analysis approach of metabolic features and their pathways (Figure 2C).

3.2. Univariate and Multivariate Data Analyses

Metabolite feature analysis by t-test at $p \leq 0.05$ showed 201 significant features (Figure 3A). Fold change (FC) analysis (threshold FC ≥ 2.0) identified 541 significantly up- and 485 significantly downregulated metabolite features (Figure 3B). Clear differentiation of up- and downregulated significant metabolite features on both axes was visualized using a volcano plot (Figure 3C). A correlation heatmap correlated the top 1000 metabolite features based on the interquartile range (IQR) (Figure 3D). Multivariate data analysis of samples using an unsupervised PCA score plot revealed 48.7% variance along PC1 versus 32.3% variance along PC2, indicating a clear differentiation of metabolite features in

MCTR and MCNTR (Figure 4A). The PCA loading plot (Figure 4B) identified the 10 most prominent metabolite features, with m/z 549.3029, 213.1599, 371.7534, 570.9134, 393.2858, 437.2661, 481.9997, 225.1099, 279.1377, and 205.0708 negatively contributing to PC1. In contrast, the top 10 features with m/z 374.1479, 281.1897, 391.263, 359.11, 283.1875, 377.0567, 314.1896, 284.1707, 492.3683, and 390.0878 positively contributed to PC1 at the 95% confidence level. Likewise, for PC2, the 10 prominent positively contributing features were m/z 423.32561, 532.43925, 193.14975, 610.54041, 354.17554, 198.03752, 178.05313, 625.26542, 532.4209, and 533.10217. Negatively contributing metabolite features in PC2 included m/z 360.17386, 340.28453, 415.194, 349.0899, 565.4075, 384.33342, 362.51172, 371.20361, 351.21671, and 382.20709. The metabolite features screened for supervised PLS-DA also showed clear differentiation of the two samples with component 1 (x-axis) covariance at 47.1% and component 2 covariance (y-axis) at 37.2% (Figure 5A). The variable importance in projection (VIP) scores resulted in the top 50 features having VIP scores $\geq$1 in PLS-DA model, which indicated high contribution of the feature metabolites in sample segregation (Figure 5B). Hierarchical clustering showed distinct patterns associated with metabolic changes in MCTR and MCNTR (Figure 5C). Dendrograms showed hierarchical relationships between pathogenic and nonpathogenic conditions and identified key metabolite features. One group of clusters was associated with the plant responses in control, while the other cluster showed high distinction from the nonpathogenic plant groups, demonstrating differential metabolic reprogramming in the pathogen-challenged *S. cheesmaniae* leaf samples.

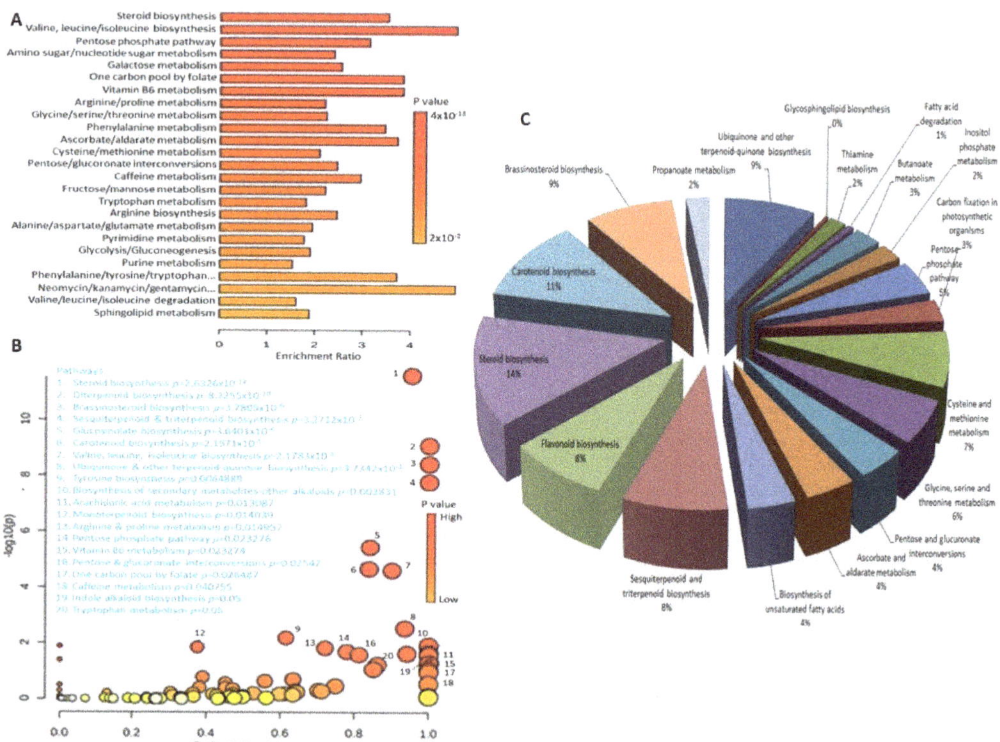

Figure 2. (**A**) Enrichment analysis of metabolite features with KEGG IDs. (**B**) Integrated pathway activity profile of significant ($p \leq 0.05$) metabolite features linked to biosynthetic pathways as analyzed by MetaboAnalyst 5.0 (GSEA algorithm) using default parameters and the *Arabidopsis thaliana* pathway as the library. (**C**) Significantly enriched major pathways. Enrichment ratio = observed hits/expected hits.

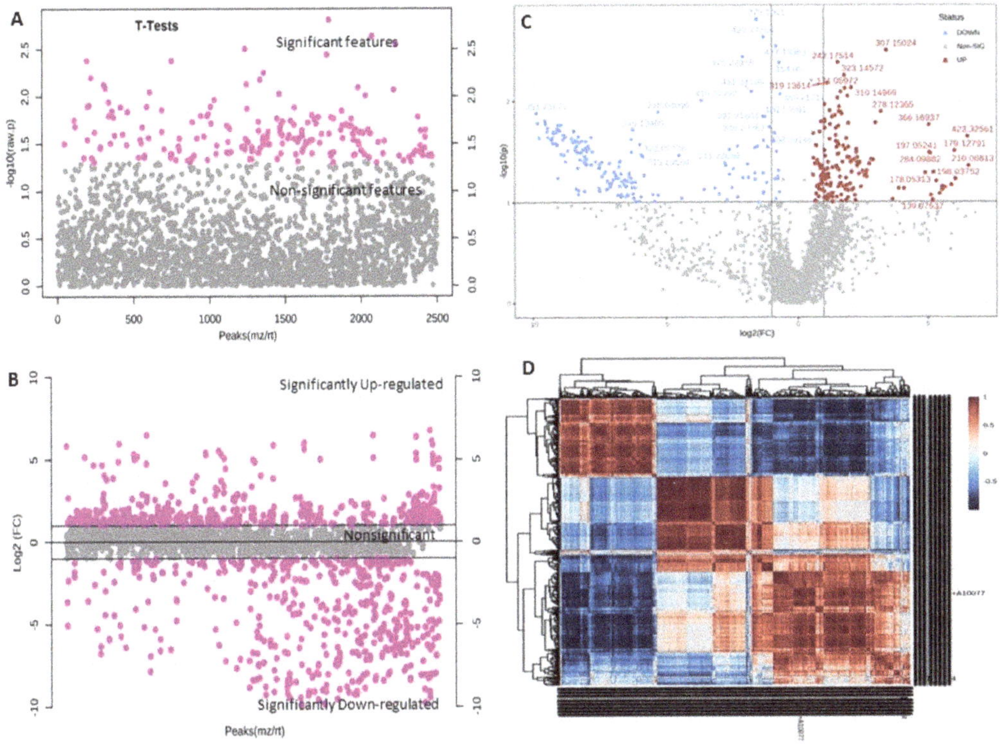

Figure 3. (**A**) Feature analysis by two-sample *t*-test and Wilcoxon rank-sum tests at $p \leq 0.05$ revealed 201 significant versus 2298 nonsignificant features. (**B**) Fold change (FC) analysis of metabolite features after the plants were exposed to pathogen. At the FC threshold of $\geq$2.0, 541 significantly upregulated, 485 downregulated, and 1472 nonsignificant features were identified. (**C**) Volcano plot indicating clear differentiation of metabolite features in the two samples. (**D**) Correlation heatmap showing correlation of the top 1000 features based on their interquartile range (IQR) in a large dataset. The distance measured is based on Pearson r values with default MetaboAnalyst parameters.

The discrimination of metabolite level between control and challenged plants was further revealed using an OPLS-DA loading S-plot, which highlighted relevant feature ions with high covariance to act as discriminatory biomarkers (Figure 6A). The "outlier" feature ions in the top right quadrant were positively correlated, while those in the lower left quadrant were negatively correlated with the response of the plants under pathogenic conditions. The features were further validated using a VIP plot with VIP scores (Figure 6B) at values $\geq$1.

3.3. Pathway Classification of Annotated Features by KEGG Mapper

All chemical query identifiers (3371) obtained from the functional annotation of the metabolite peak list when mapped against the *S. lycopersicum* database in KEGG Mapper revealed 107 pathways and 173 modules. As many as 530 compounds were mapped with metabolic pathways (sly01100): 449 related to biosynthesis of secondary metabolites, along with 25 to phenylpropanoids, 100 to cofactors, 71 to 2-oxocarboxylic acid metabolism, 31 to carbon metabolism, 65 to amino acids, 46 to ABC transporters, 42 to steroids, 36 to carotenoids, 28 to diterpenoids, 27 to ubiquinone and terpenoids/quinones, and 24 to sesquiterpenoid/triterpenoid and flavonoid biosynthesis (Figure 7). In addition to major metabolic pathways routinely involved in plant growth, a large set of metabolite features were mapped with those compounds which were directly linked to secondary metabolite biosynthesis, which have been shown to exert a direct role in the defense of plants challenged with the pathogen.

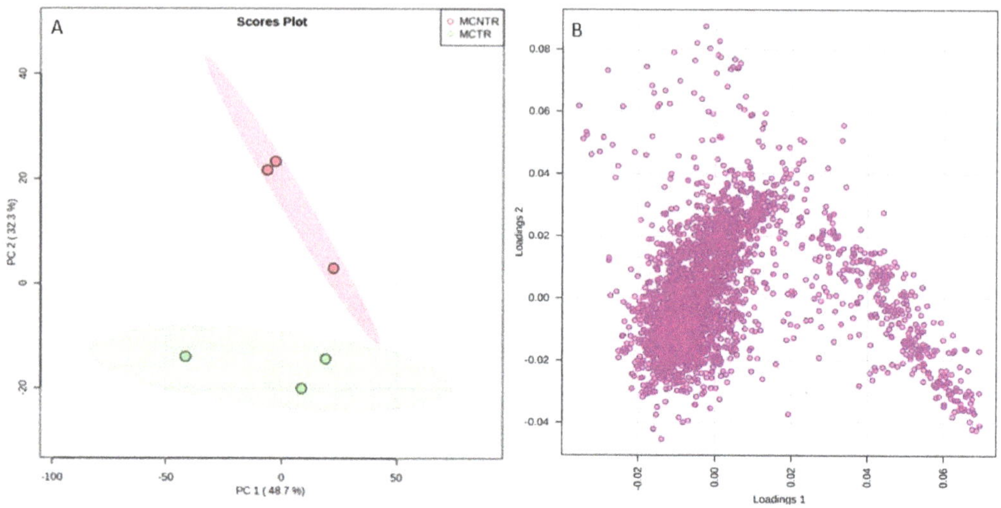

Figure 4. (**A**) Principal component analysis score plot reflecting visualization of the relationship among the samples in terms of groupings, trends, or outliers and showing differences between MCTR and MCNTR along x-axis (PC1) and y-axis (PC2). Principal components 1 and 2 explained 48.7% and 32.3% of the variance, respectively. (**B**) Loading plot describing the influence of variables in sample segregation.

3.4. Enrichment Analysis of Up- and Downregulated Compounds

FC analysis resulted in 541 significantly up- and 485 significantly downregulated features at FC $\geq$ 2.0. These features, when matched with the *S. lycopersicum* (and *S. pinnelli*) compound database in PLANT CYC PMN, revealed 179 features having KEGG IDs for upregulated and 175 features with KEGG IDs for downregulated compounds. In KEGG Mapper, the upregulated compounds were mapped with 77 pathways and 61 modules, while downregulated compounds were mapped with 88 pathways and 80 modules. The main classes of upregulated classified compound groups to which the metabolites belonged were fatty-acid conjugates, aldehydes, cholines, tryptamines, indolyl carboxylic acid, indoles, and phenols in the range of p-values 3×10^{-2} to 1×10^{-29} (Figure 8A). The subclass compound groups included chalcones, straight-chain fatty acids, vitamins K and E, biotic, pyridoxals, phosphoethanolamines, and serotonins (p-value range: 2×10^{-3} to 1×10^{-29}) (Figure 8B). Similarly, the most prominent downregulated main class compound groups were derivatives of cyclic alcohols and beta-keto acids ($p = 9\times10^{-1}$ to 2×10^{-29}) (Figure 8C). The subclass of downregulated compound groups belonged to sphinganines, ergostane steroids, peptides, prenylated hydoquinones, sphingosines, catecholamines, and phenethylamines (Figure 8D).

The results clearly reflected differentiation among the up- and down-regulated compound groups. It was confirmed that different metabolites differentially contributed to the metabolic pathways to influence biosynthesis of compounds in favor of plant responses toward pathogen challenges. Up- and down-regulated metabolites having KEGG identifiers were mapped and visualized in KEGG global metabolic network (Figure 9A) that showed compounds in the pathways. The gene–metabolite interaction network explored interactions between the functionally related metabolites and genes (Figure 9B). The input metabolites and genes (seeds) were mapped to the selected interaction network that created subnetworks 1 and 2 with 1220 and five nodes, 1879 and four edges, and 51 and two seeds (genes) for upregulated compounds. The metabolite–metabolite interaction network highlighted potential functional relationships between annotated metabolites in up- and downregulated compounds (Figure 9C). Upregulated compounds were networked with 434 nodes, 1150 edges, and 53 seeds (genes), while downregulated compounds showed 419 nodes, 908 edges, and 46 seeds (genes).

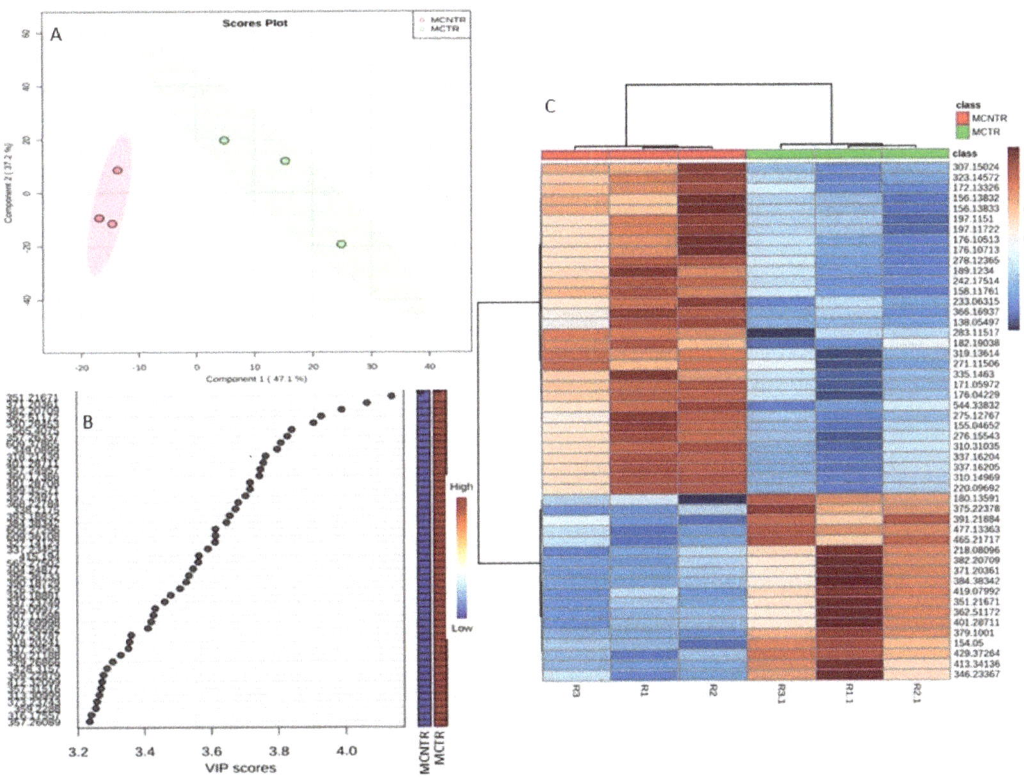

Figure 5. (**A**) PLS-DA score plot with scores (component 1 (34%) versus component 2 (42.3%) plotted from the metabolite profiles of MCTR/MCNTR samples. Both the disease-challenged (MCTR) and the unchallenged (control) (MCNTR) plants were grown under control conditions (n = 6) before metabolite extraction and analysis using ESI-MS/MS. The separation of both samples indicated consistent patterns in the metabolites which could explain differences between the diseased and non-diseased groups. (**B**) Top important features with variable importance plot (VIP). A higher VIP score for metabolites represents their importance in influencing the score most. On the Y-axis, variables are ranked in terms of their importance (variables of highest importance at the top). (**C**) Hierarchical clustering of feature abundance normalized by Pareto scaling and t-test from the normalized dataset, with Pearson as the distance measure and Ward as the clustering method. The top 50 features are shown as representative clusters. The relative abundance of the features in samples is represented by color (lower (blue) to higher (red) abundance) presented in the key. The brightness in color reflects the differential magnitude when compared with the average value. Dendrograms to the top and left show hierarchical relationships between diseased (MCTR) and non-diseased (MCNTR) samples (**top**) and identified metabolite features (**left**).

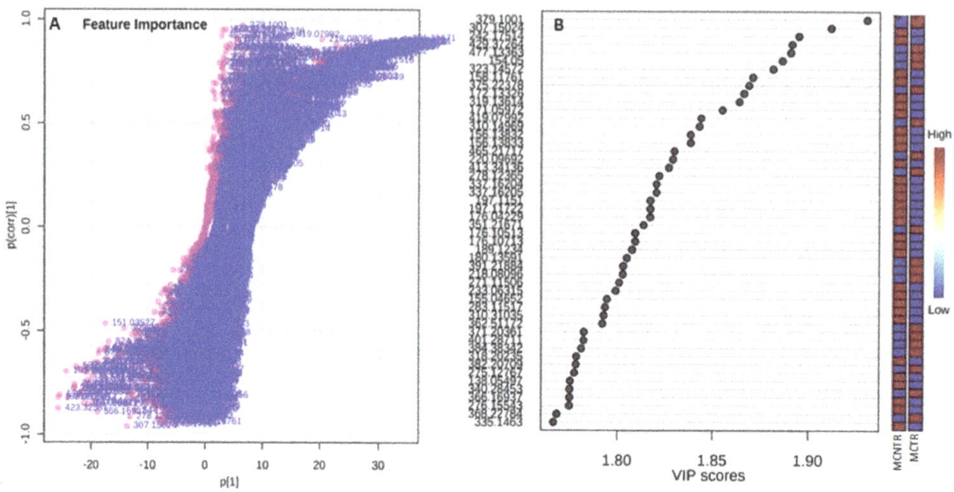

Figure 6. (**A**) OPLS-DA S-Plot (BoxPlot) facilitating visualization of the variable influence in an orthogonal PLS-DA model combining the covariance and correlation loading profile of the metabolite features in the MCTR/MCNTR samples. The plot identifies putative biomarkers (**bottom left** and **top right**) responsible for the separation of groups. (**B**) Variable importance in projection (VIP) scores identifying mass-to-charge ratio (m/z) with discrimination between sample groups. A higher VIP score denotes a better influence of the metabolite feature in discriminating between the two groups. The mini heatmap on the right indicates the concentration variations within the two groups.

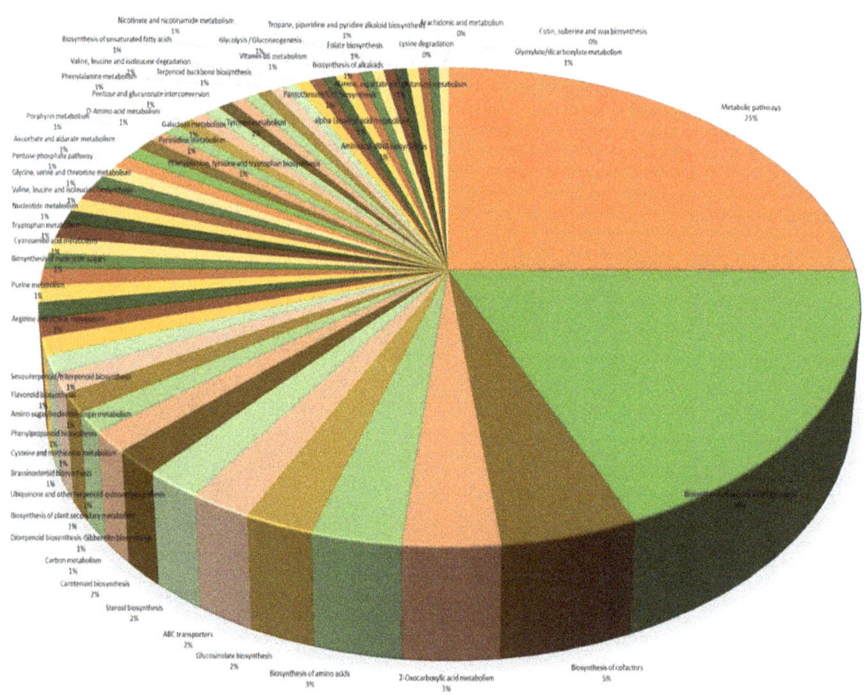

Figure 7. Enrichment pathway of the up- and down-regulated compounds in pathogen-challenged and unchallenged (control) tomato samples.

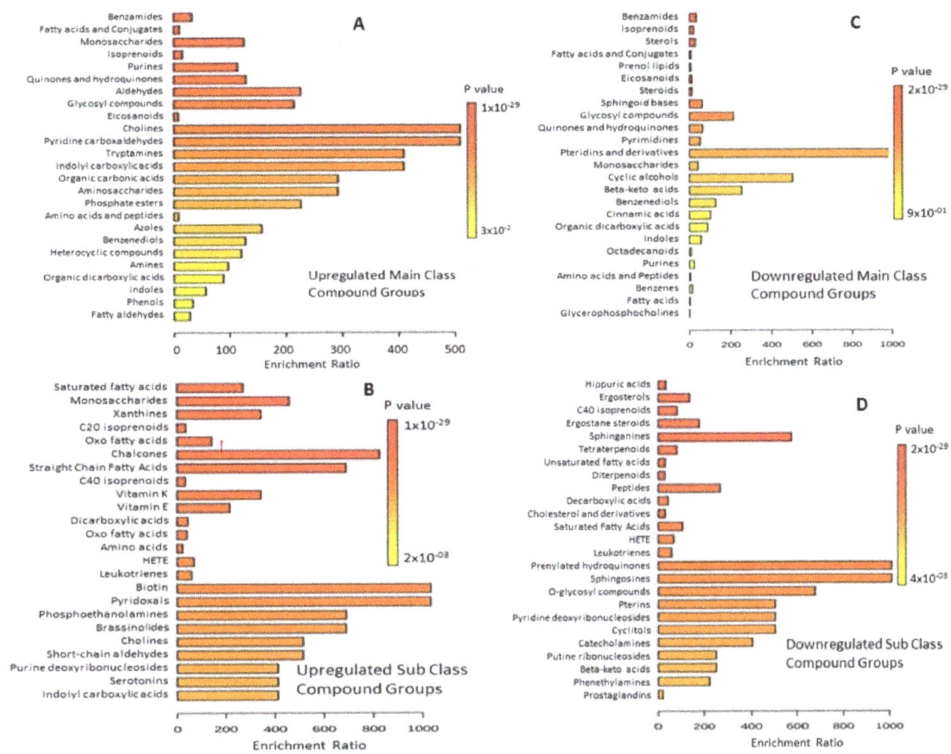

Figure 8. Metabolite sets: overview of up- and down-regulated compounds in the metabolic pathways of tomato. (**A**) up-regulated main class compound groups; (**B**) up-regulated sub-class compound groups; (**C**) down-regulated main class compound groups and (**D**) down-regulated sub-class compound groups. Annotation was performed with 464 main chemical class metabolite sets and 1072 sub chemical class metabolite sets.

3.5. Potential Upregulated Metabolites

Bioactive compounds including trans-cinnamate (C00423; *m*/*z* 146.03781), a tetrahydrofuran lignin, pinoresinol (C05366, *m*/*z* 359.13143), and 5-hydroxyconiferyl alcohol (C12205, *m*/*z* 241.07068) were identified as the metabolites involved in phenylpropanoid biosynthesis. Increased abundance of flavonoids, the prominent defense-indicator antioxidant compounds to impart improved plant resistance, was found in pathogen-challenged plant leaves, A range of upregulated metabolites involved in flavonoid biosynthesis including phloretin (C00774, *m*/*z* 275.07353), pinocembrin (5,7-dihydroxyflavanone) (C09827, *m*/*z* 257.06297), pinocembrin chalcone (C16404, *m*/*z* 257.06299), apiforol (C12124, *m*/*z* 275.07355), isoliquiritigenin (C08650, *m*/*z* 257.06298), and deoxyleucopelargonidin (C16415, *m*/*z* 275.07357) were identified. Carotenoids such as canthaxanthin (C08583, *m*/*z* 579.38672), zeaxanthin (C06098, *m*/*z* 283.20569), phoenicoxanthin (C15967, m/z 579.38674), 9′-cis-neoxanthin (C13431, *m*/*z* 615.40351), violaxanthin (C08614, *m*/*z* 615.40353), and xanthoxin (C13453, *m*/*z* 290.17496) were significantly upregulated in *S. cheesmaniae* following pathogen challenge.

Coniferyl alcohol (C00590, *m*/*z* 220.09692) involved in lignin biosynthesis was upregulated along with oxlipids, i.e., leukotriene A4 (C00909, *m*/*z* 317.21106) and icosapentaenoic acid (C06428, *m*/*z* 283.20569). Furthermore, short-chain alcohols and aldehydes, e.g., 3-hexenol (C08492, *m*/*z* 145.08606) and 3,6-nonadienal (C16323, *m*/*z* 197.11722), a post-infection hypersensitive response inducer in plants, were upregulated. Phytohormone derivative abscisic aldehyde (C13455, m/z 247.13288) and variants and intermediates of GA including GA A9 (C11863; m/z 353.13428), A20 (C02035, m/z 353.13582), A4 (C11864), and A51 (C11865, m/z 353.13583) were identified in pathogen-challenged plant leaves.

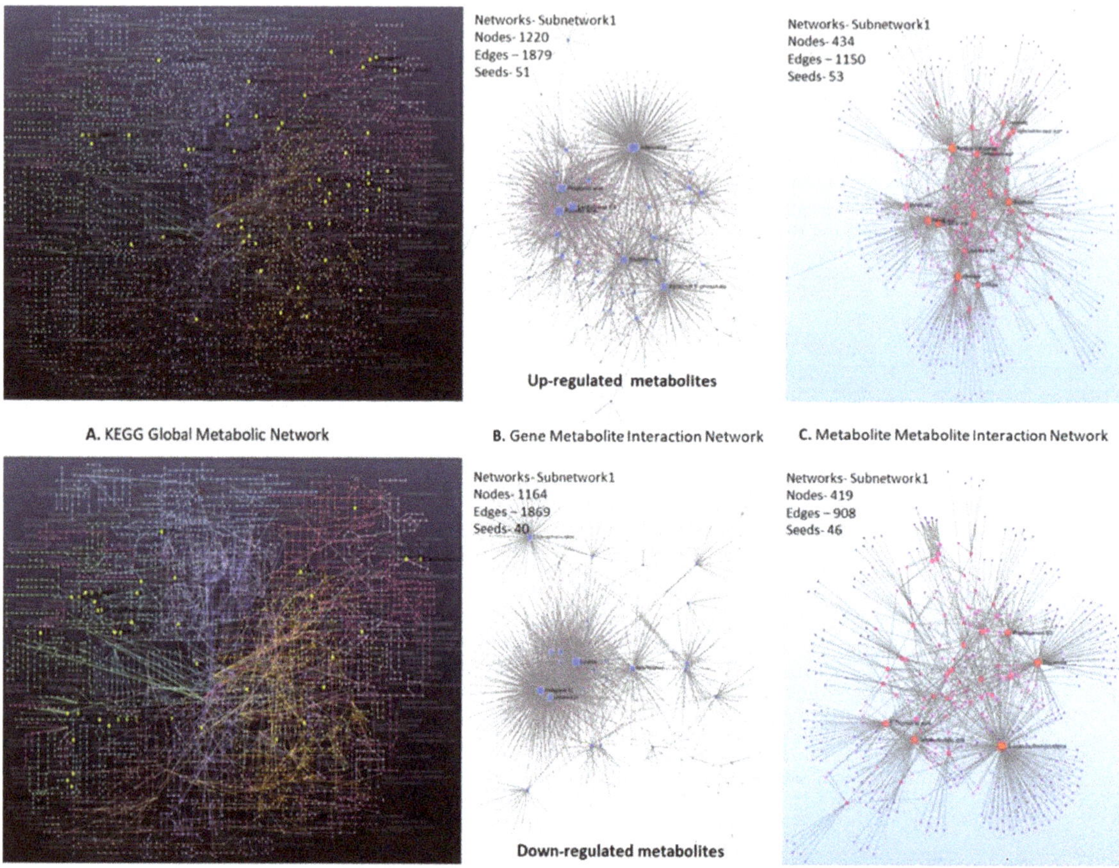

Figure 9. Network analysis of up- and down-regulated compounds in diseased and non-diseased tomato samples based on KEGG Global metabolic network: gene–metabolite interaction and metabolite–metabolite interaction.

N-acetylornithine (C00437; m/z 219.09755) is an acetylated biosynthetic intermediate of amino acids. Upregulated *O*-acetyl-L-homoserine (C01077; m/z 161.06476) is an acetylated L-homoserine involved in cysteine and methionine metabolism. Branched-chain organic acid 3-methyl-2-oxobutanoic acid (C00141; m/z 161.04424) was characteristically upregulated in leaves. Pantothenate (C00864; m/z 218.10237), biotin (C00120; m/z 259.07559), phylloquinone (C02059; m/z 529.27026), and delta-tocopherol (C14151, m/z 423.32561) were upregulated vitamin compounds in *S. cheesmaniae* leaves. Porphyrin metabolites, as exemplified by porphobilinogen (PBG) (C00931; m/z 271.09418), protoporphyrin (C02191; m/z 621.27057), pyropheophorbide *a* (C18064; m/z 579.26227), pheophorbide *a* (C18021, m/z 591.25991), and red chlorophyll catabolite (C18022, m/z 625.26542) were found upregulated in infected *S. cheesmaniae* leaves. Porphyrins as photosynthesizers are reported to exert substantial efficiency to kill plant pathogens. Likewise, compounds 3-beta-hydroxyergosta-7,24(24(1))-dien-4alpha-carboxylate (C22119, m/z 423.32561), 26-hydroxycastasterone (C19873, m/z 525.34338), typhasterol (C15793, m/z

529.27026), teasterone (C15791, *m/z* 529.27026), and brassinolide (C08814, *m/z* 525.34338) involved in brassinosteroid biosynthesis were identified as upregulated potential metabolites.

3.6. Metabolite Biomarkers/Pathways for Resistance Response in S. cheesmaniae

According to FC ≥ 2, VIP score ≥ 1, and $p < 0.05$ values, 34 up- and 41 downregulated "significantly different metabolite features" were identified as "biomarkers". Upregulated biomarker compounds included 3-hexenol, 5-phosphoribosylamine, 5-hydroxykynurenine, kaur-16-en-18-oic acid, 5-phosphoribosylamine, 2-isopropylmaleate, indolepyruvate, tetradecanoic acid, pantothenate, and O-acetyl-L-homoserine (Table 1). Downregulated biomarker metabolites were represented by pheophytin *a*, zymosterol, episterol, sphinganine, obtusifoliol, lariciresinol, etc. (Table 2). Upregulated metabolite biomarkers were critically mapped with vitamin B6 metabolism, indole alkaloid biosynthesis, and other metabolic pathways (Figure 10A). Interestingly, the downregulated metabolite biomarkers were mapped upon pathways such as biosynthesis of cutin, suberin and wax biosynthesis, secondary metabolites, phenylalanine, carotenoids, brassinosteroids, and terpenoids (Figure 10B), which are categorically known to be modulated to favor plant defense, suggesting their prominent role in plant survival.

Table 1. Up-regulated putatively identified metabolite biomarkers * on the basis of FC ≥ 2, VIP ≥ 1, and $p < 0.05$.

S. No.	Mass (*m/z*)	Matched Compound	Putatively Identified Biomarker Metabolites *	Fold Change	VIP Score (OPLS DA)
1.	229.03	C03090	5-Phosphoribosylamine	11.99	1.22
2.	278.12	C00864	Pantothenate	8.98	1.02
3.	203.06	C02631	2-Isopropylmaleate	7.51	1.34
4.	203.06	C00331	Indolepyruvate	7.50	1.34
5.	161.04	C00164	Acetoacetate	6.27	1.59
6.	161.04	C00232	Succinate semialdehyde	5.63	1.59
7.	229.20	C06424	Tetradecanoic acid	4.41	1.22
8.	162.03	C15606	1,2-Dihydroxy-5-(methylthio)pent-1-en-3-one	4.32	1.58
9.	228.08	C00881	Deoxycytidine	4.21	1.23
10	220.10	C00590	Coniferyl alcohol	3.8	1.26
11.	243.09	C00588	Choline phosphate	3.39	1.17
12.	257.10	C00670	sn-Glycero-3-phosphocholine	3.29	1.12
13.	203.05	C00331	Indolepyruvate	3.27	1.35
14.	187.02	C00121	D-Ribose	3.21	1.43
15.	219.11	C00398	Tryptamine	2.99	1.27
16.	219.10	C00437	N-Acetylornithine	2.99	1.27
17.	247.02	C00018	Pyridoxal phosphate	2.92	1.15
18.	191.03	C00438	N-Carbamoyl-L-aspartate	2.79	1.40
19.	261.04	C05817	(1R,6R)-6-Hydroxy-2-succinylcyclohexa-2,4-diene-1-carboxylate	2.65	1.09
20.	145.09	C08492	3-Hexenol	2.62	1.19
21.	271.09	C00931	Porphobilinogen	2.59	1.06
22.	169.05	C13482	Phosphodimethylethanolamine	2.47	1.54
23.	259.04	C00352	D-Glucosamine 6-phosphate	2.36	1.11
24.	257.06	C09762	Liquiritigenin	2.32	1.12
25.	277.12	C00449	N6-(L-1,3-Dicarboxypropyl)-L-lysine	2.29	1.02
26.	119.03	C00033	Acetate	2.29	1.12
27.	163.06	C06001	(S)-3-Hydroxyisobutyrate	2.28	1.57
28.	146.04	C00423	trans-Cinnamate	2.24	1.16
29.	168.05	C00436	N-Carbamoylputrescine	2.24	1.55
30.	229.03	C03090	5-Phosphoribosylamine	2.18	1.23
31.	261.11	C00140	N-Acetyl-D-glucosamine	2.15	1.09
32.	247.13	C13455	Abscisic aldehyde	2.14	1.14
33.	259.08	C00120	Biotin	2.14	1.11
34.	231.05	C16361	1,3,7-Trimethyluric acid	2.13	1.22

* Compounds putatively annotated with Plant CYC PMN database of *Solanum lycopersicum* are reported.

Table 2. Down-regulated putatively identified metabolite biomarkers * on the basis of FC ≥ 2, VIP ≥ 1, and $p < 0.05$.

S. No.	Mass (*m/z*)	Matched Compound	Putatively Identified Biomarker Metabolites *	Fold Change	VIP Score (OPLS DA)
1.	429.37	C01753	beta-Sitosterol	0.403	1.89
2.	419.08	C00448	trans,trans-Farnesyldiphosphate	0.079	1.84
3.	413.34	C15777	Episterol	0.297	1.87
4.	399.33	C05437	Zymosterol	0.001	1.76
5.	609.27	C18098	Primary fluorescent chlorophyll catabolite	0.002	1.75
6.	565.41	C08579	Antheraxanthin	0.002	1.75
7.	318.21	C00909	Leukotriene A4	0.009	1.75
8.	337.24	C06427	(9Z,12Z,15Z)-Octadecatrienoic acid	0.019	1.75
9.	357.26	C19616	18-Hydroxyoleate	0.002	1.73
10	565.39	C08583	Canthaxanthin	0.003	1.67
11.	293.21	C18218	16-Hydroxypalmitate	0.005	1.67
12.	417.34	C15882	2-Methyl-6-phytylquinol	0.007	1.66
13.	489.39	C02477	alpha-Tocopherol	0.004	1.65
14.	417.34	C15882	2-Methyl-6-phytylquinol	0.008	1.63
15.	434.27	C05797	Pheophytin a	0.009	1.59
16.	439.36	C22121	Cycloeucalenone	0.011	1.58
17.	329.27	C01530	Octadecanoic acid	0.007	1.57
18.	311.22	C04717	(9Z,11E)-(13S)-13-Hydroperoxyoctadeca-9,11-dienoic acid	0.43	1.57
19.	339.16	C20693	Carlactone	0.321	1.56
20.	380.35	C08281	Docosanoic acid	0.172	1.54
21.	549.41	C06098	Zeaxanthin	0.011	1.54
22.	423.36	C01943	Obtusifoliol	0.011	1.47
23.	337.04	C00364	dTMP	0.441	1.46
24.	324.10	C01762	Xanthosine	0.374	1.44
25.	180.10	C05332	Phenethylamine	0.032	1.44
26.	491.21	C05427	Phytyldiphosphate	0.012	1.42
27.	411.38	C00751	Squalene	0.213	1.37
28.	344.28	C02934	3-Dehydrosphinganine	0.131	1.36
29.	282.28	C00836	Sphinganine	0.432	1.26
30.	198.08	C02631	2-Isopropylmaleate	0.441	1.23
31.	337.20	C01226	12-OPDA	0.431	1.22
32.	324.20	C19691	Farnesylcysteine	0.206	1.21
33.	309.19	C05819	Menaquinol	0.314	1.18
34.	383.10	C20693	Carlactone	0.031	1.17
35.	344.28	C02934	3-Dehydrosphinganine	0.043	1.12
36.	427.14	C01674	N,N-diacetylchitobiose	0.434	1.11
37.	217.05	C00794	D-Sorbitol	0.251	1.11
38.	453.33	C15800	3-Dehydro-6-deoxoteasterone	0.011	1.09
39.	493.28	C18217	16-Feruloyloxypalmitate	0.031	1.09
40.	423.36	C01943	Obtusifoliol	0.032	1.06
41.	441.07	C10646	Lariciresinol	0.491	1.02

* Compounds putatively annotated with Plant CYC PMN database of *Solanum lycopersicum* are reported.

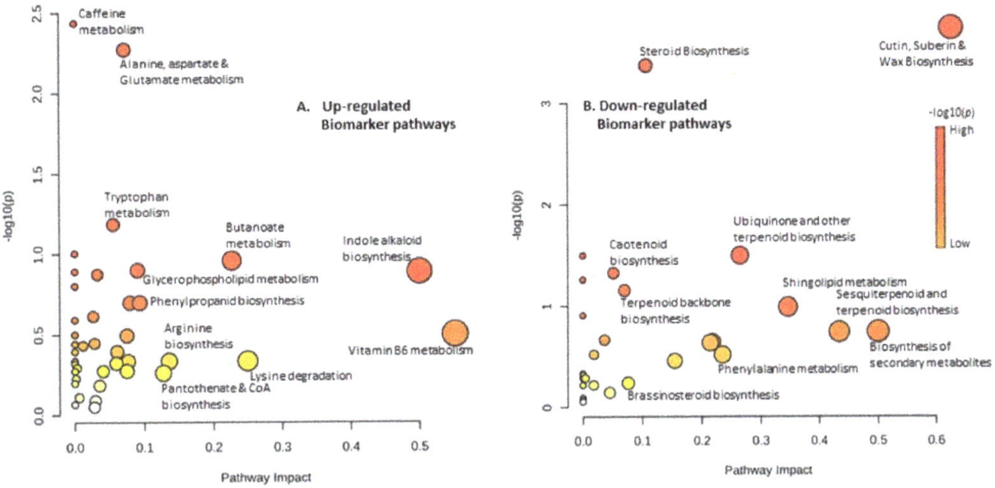

Figure 10. Pathway impact analysis of up- (**A**) and down-regulated (**B**) biomarker pathways in *S. cheesmaniae*.

4. Discussion

Tomato is a model horticultural crop plant for studying cellular, biochemical, and molecular responses linked to plant growth and development under biotic stress conditions [22]. However, mechanisms underlying plant responses in leaves of moderately disease-resistant wild tomato *S. cheesmaniae* after *A. solani* interaction in terms of metabolite profile and metabolic pathways have not yet been investigated. Untargeted metabolomics indicate differential metabolite profiles and biosynthetic pathways as plant responses against stress and aid to identify key functional biomarker metabolites [23]. Such a comprehensive approach enables metabolite profiling to analyze complex plant metabolic responses, products, and processes to identify crucial defense-related biosynthetic pathways and elucidate secondary metabolic networks [24]. Metabolite profiling of *S. cheesmaniae* plants reflected clear differences and revealed high metabolite variability in a number of metabolite features (9451 in MCTR vs. 5911 in MCNTR) and their abundance. Comparative functional analysis identified 3371 metabolite features with KEGG identifiers involved in various biosynthetic pathways. The FC ($\geq$2.0) analysis revealed 541 significantly upregulated and 485 downregulated metabolite features annotated with the *S. lycopersicum* compound library in the PLANT CYC PMN database. A total of 34 upregulated and 41 downregulated metabolite biomarkers were identified by OPLS-DA analysis in the leaves of *S. cheesmaniae* under pathogen-challenged conditions.

The enrichment analysis classified upregulated metabolites to fatty-acid conjugates, aldehydes, pyridine carboxaldehydes, tryptamines, indolyl carboxylic acid, indoles, phenols, chalcones, straight-chain fatty acids, vitamins K and E, biotin, pyridoxals, phosphoethanolamines, and brassinosteroids. These metabolite groups play a prominent role in plant defense against diseases. Fatty-acid derived compounds (C-16 and C-18) actively act as signals and modulate effector-triggered systemic immunity [25]. Aldehydes, especially volatile compounds, activate defense genes and resistance responses against pathogenic fungi [26], while pyridine carboxaldehydes are typically involved in vitamin B6 biosynthesis [27], which induces resistance against pathogens while promoting plant growth [28]. Significantly upregulated tryptamine (C00398, m/z 219.11339) minimizes fungal infection-induced damage and regulates the biosynthesis of serotonin (5-hydroxytryptamine) [29], which is a signaling molecule for stress response mechanisms in plants. The indolyl carboxylic acid group that includes indole-3-carboxylic acid derivatives are secondary metabolites which accumulate inside the cell wall in response to pathogen interaction [30], reflecting their defensive role against biotic conditions [31]. Indoles [32], phenols [33], and flavonoids including chalcones [34] have immense direct applications in plant defense, with an indirect involvement in plant growth. Straight-chain fatty acids [25], pyridoxal, vitamins K and E, and biotin [35] are antioxidants with a potential role in minimizing stress-induced damage due to ROS and improving plant development under stress [36]. Phosphatidylethanolamines (LPE) are natural phospholipids that coordinate defense responses, interfere with oxidative bursts, and improve basal immunity against pathogens [37]. Brassinosteroids play a specific role in signaling in plant–microbe interaction [38]. Phenylethylamines (PEA), tyramine, and serotonin are plant-derived monoamines identified as upregulated metabolites. All significantly upregulated compound groups directly or indirectly improve plant fitness against stress conditions and are essentially involved in growth–defense tradeoffs [39]. We expect that their abundant accumulation might have aided plants to adapt to disease stress.

Among downregulated compound groups, we reported cyclic alcohols, prenylated hydroquinones, sphingosines and sphinganines, ergostane steroids, catecholamines, beta keto acids, and phenethylamines. Cyclic secondary alcohols are often oxidized to ketones having prominent biological properties [40]. Prenylated hydroquinones exert a strong antioxidant effect [41]. Structural membrane components such as sphingolipids (sphingosines and sphinganines) act as signal molecules in cell functions against infectious bacterial and fungal pathogens [42,43]. Downregulation of these compound groups in *S. cheesmaniae* may have significance toward lowering plant resistance.

Porphyrins and their intermediates are crucial for ROS-mediated stress responses and for regulating complex networks that control stress-responsive genes [44]. In pathogen-challenged *S. cheesmaniae* plants, porphobilinogen, protoporphyrin, pheophorbide *a*, red chlorophyll catabolite, and pyropheophorbide *a* were identified as prominently upregulated metabolites of porphyrin metabolism. Porphobilinogen is a distant precursor of vitamin biosynthesis, protoporphyrin is involved in the biosynthesis of chlorophyll a, and pheophorbide *a* synthesizes red chlorophyll catabolite and pyropheophorbide *a*. Although porphobilinogen, chlorophyll catabolite pheophorbide *a*, and red chlorophyll catabolite pyropheophorbide *a* are reported to confer defense in organisms against different abiotic stressed conditions [45,46], their multifold overaccumulation in *S. cheesmaniae* leaves in response to the pathogen is supposed to play a prominent defensive role. Accumulation of pheophorbide *a* was reported to induce programmed cell death under darkness and illuminated conditions in transgenic *Arabidopsis* plants [47].

Hormones influence plant performance under biotic stresses. Plant hormones such as salicylic acid (SA), abscisic acid (ABA), jasmonic acid (JA), indole-acetic acid (IAA), and gibberellic acid (GA), alongside several other natural molecules such as brassinosteroids and strigolactones, regulate biotic stress signaling and mediate transcription factors (TFs) to regulate multiple signaling transduction pathways for plant defense against pathogenic challenges [48]. Prominent upregulation of IAA, ABA, JA, GA, GA derivatives and intermediates, brassinolide, and zeatin and its derivative as phytohormones was observed in the pathogen-challenged leaves at various levels. Diversified abundance of these compounds is believed to play a coordinated role in the plant hormone signaling cascade leading to protection against pathogens alongside the regulation of plant growth and development in *S. cheesmaniae* [48,49].

Plant–microbe interactions notably lead to multifold changes in polyamine metabolism of the host, making the interaction dynamic and complex, with profound changes in free and conjugated polyamines inside the tissues [50]. Upregulated *N*-acetylornithine, a compound of acetyl amino acids, biosynthesizes polyamines that mediate plant interactions with the pathogens. Adjio et al. [51] reported enhanced accumulation of *N*-acetylornithine in wildtype *Arabidopsis* plants under pathogen *P. syringae* infection. Overaccumulation of *O*-acetyl-L-homoserine, as reported in MCTR, may play a role in enhancing plant immunity against pathogens [52]. We presumed that all of these annotated over-accumulated metabolites in MCTR may have acted in a coordinative way to offer moderate protection in *S. cheesmaniae* against *A. solani*.

Cell-wall strengthening upon pathogen invasion is the strategic mechanism of protection in plants. Upregulation of compounds such as coniferyl alcohol and 5-hydroxyconiferyl alcohol in *S. cheesmaniae* leaves in diseased conditions may signify their role in lignin synthesis, through which plants create a physical barrier against pathogen invasion [53]. Polyunsaturated fatty acids (PUFAs) are precursors of key metabolites which mediate crosstalk between organisms and further serve as substrates to oxylipins, i.e., JA, which plays crucial role in defense against diseases [53]. Icosapentaenoic acid, a long-chain fatty acid, is a powerful elicitor that showed upregulation. The short-chain C6-alcohol 3-hexenol is an indispensable carbonyl volatile compound that offers aroma to tomato due to its unique grassy green characteristics, as well as induces defense responses in plants against diseases [54].

Vitamins as natural inducers of disease resistance [55] and carotenoids as photoprotective and antioxidant agents [56,57] are widely reported. Canthaxanthin, a red-hot keto-carotenoid pigment exerts potential antioxidant and free-radical-scavenging properties [58]. Oxygen-containing lutein and violaxanthin are pigments with stress-suppressing properties [59]. Xanthoxin, an apo-carotenoid sesquiterpenoid, is an intermediate in the biosynthesis of the plant hormone abscisic acid [59]. Phoenicoxanthin, a natural xanthophyll of carotenoid biosynthesis, plays a role in protection against oxidative stress [60]. Likewise, tocopherol, pyridoxine, and biotin induce resistance in plants against diseases, likely mediated via their antioxidant effect against generated ROS [61]. All these compounds showed a multifold increase in MCTR, offering defensive functions in *S. cheesmaniae* plants.

Flavonoids and brassinosteroids also showed multifold overaccumulation in pathogen-challenged plant leaves. Flavonoids, having the widest distribution as secondary metabolites, primarily respond to biotic stresses in plants [62]. Upregulated flavonoids included pinocembrin (5,7-dihydroxyflavanone), pinocembrin chalcone, and apiforol, a tetrahydroxyflavan that influences plant defense by mediating pathogen signaling responses, membrane permeability, quorum sensing, and pathogen virulence [63]. Chalcones phloretin and isoliquiritigenin are strong antioxidants that showed an increase [64]. Multifold upregulated brassinosteroid compounds 3-beta-hydroxyergosta-7,24(24(1))-dien-4alpha-carboxylate, 26-hydroxycastasterone, typhasterol, teasterone, and brassinolide in pathogen-challenged leaves play a prominent role as structural components of cell membranes and aid against plant stress [65]. Overaccumulation of sulfur-containing natural compounds such as 1,2-dihydroxy-5-(methylthio)pent-1-en-3-one, an aliphatic sulfide, may indicate a versatile role in plant resistance against pathogen infection [66].

Our study revealed evidence of pathogen-induced metabolic changes in *S. cheesmaniae* moderately resistant to *A. solani* and provided specific clues of metabolic pathways that play a crucial role in conferring moderate resistance toward pathogens. Pathogenic infection often shifts the secondary metabolite profile in plant tissues due to induced defense programs to confer various levels of changes in primary and secondary metabolism conferring resistance against diseases [67]. In addition to elucidating the impact of the pathogen on secondary metabolites and their pathways, which have been widely studied [68], we also critically analyzed the impact of infection on crucial metabolites and pathways linked to porphyrins, hormones, vitamins, carotenoids, and amino-acid and fatty-acid metabolism. Key upregulated biomarker metabolites revealed from this study included 5-phosphoribosylamine, 5-hydroxykynurenine, kaur-16-en-18-oic acid, pantothenate, and *O*-acetyl-L-homoserine. The predictive identification of these biomarkers in *S. cheesmaniae* independent of genetic or other climatic variation can be used to distinguish diseased and non-diseased conditions inside plants, on the basis of which diagnostic tools could be developed. This approach may also aid in identifying mQTLs to be utilized in resistance-breeding programs under biotic stress conditions.

5. Conclusions

To support the hypothesis that the intrinsic metabolite composition and chemical diversity help wild tomato plants withstand stress challenges, untargeted metabolomics of pathogen challenged and unchallenged plants was performed. We reported metabolite changes associated with moderate resistance against *A. solani* in the leaves of wild species *S. cheesmaniae* grown under normal and pathogen-challenged conditions. Plant leaf metabolite profiles were significantly differentiated, and both normal and diseased plants were discriminated not only by the presence/absence of specific metabolites as concluding factors, but also on the basis of the relative abundance of metabolites as important distinguishing criteria. Annotation of the metabolite features using the KEGG *S. lycopersicum* compound database led to the annotation of 3371 metabolite features with KEGG identifiers, which were enriched in a number of metabolic/biosynthetic pathways, mainly including secondary metabolites, cofactors, steroids, terpernoids, fatty acids, and brassinosteroids. Significantly upregulated (541) and downregulated (485) compounds distributed in different metabolite classes play a crucial role in defense, infection prevention, signaling, plant growth and development, and survival maintenance under challenged conditions. It is hypothesized that these metabolites individually and/or cumulatively influence plant responses against pathogenic interactions and provide protection against infection on a wider scale. To our knowledge, this study is the first holistic comprehensive metabolite profiling of *S. cheesmaniae* plants underlying *A. solani* infection, which led to the identification of metabolite biomarkers and their metabolic pathways. The results hold promise for developing disease-diagnostic tools based on key biomarker metabolites. The biomarker-based disease-resistance metabolic traits/biosynthetic routes could contribute to the development of mQTLs for supporting future biotic stress-breeding programs in tomato.

Supplementary Materials: The following supporting information can be downloaded at: https://www.mdpi.com/article/10.3390/metabo13050585/s1, Table S1: Supplementary Materials.

Author Contributions: Conceptualization, D.P.S.; Methodology, D.P.S., B.K.S. and M.A.F.; Software, R.P., K.K.C., M.S.F., S.S. and A.R.; Formal analysis, D.P.S., R.P. and R.S.; Investigation, D.P.S., J.K.T. and N.R.; Resources, M.S.B., S.M., S.R.Y., J.K.T. and T.K.B.; Data curation, M.S.B. and M.A.F.; Writing—original draft, D.P.S.; Writing—review and editing, R.S., A.R., B.K.S., N.R., P.M.S., T.K.B. and M.A.F.; Supervision, S.R.Y.; Funding acquisition, T.K.B. All authors have read and agreed to the published version of the manuscript.

Funding: This research received fund from Indian Council of Agricultural Research (ICAR), India.

Institutional Review Board Statement: The manuscript is approved by the PME Cell, ICAR-IIVR, Varanasi, India.

Informed Consent Statement: Not applicable.

Data Availability Statement: The data presented in this study are available on request from the corresponding author. The data are not publicly available due to privacy.

Acknowledgments: This work was carried out as part of the research project "Proteomics and metabolomics of stress-challenged tomato for functional metabolic clues of plant responses, crop quality, and yield", supported by the Indian Council of Agricultural Research (ICAR), Ministry of Agriculture and Farmers Welfare, Government of India. D.P.S. is thankful to ICAR for funding support under the "Network project on computational biology and bioinformatics", CABin, ICAR-IASRI, New Delhi.

Conflicts of Interest: The authors declare no conflict of interest.

References

1. Iqbal, Z.; Iqbal, M.S.; Hashem, A.; Abd_Allah, E.F.; Ansari, M.I. Plant defense responses to biotic stress and its interplay with fluctuating dark/light conditions. *Front. Plant Sci.* **2021**, *12*, 631810. [CrossRef] [PubMed]
2. Lamers, J.; Van Der Meer, T.; Testerink, C. How plants sense and respond to stressful environments. *Plant Physiol.* **2020**, *182*, 1624–1635. [CrossRef] [PubMed]
3. Velásquez, A.C.; Castroverde, C.D.M.; He, S.Y. Plant-pathogen warfare under changing climate conditions. *Curr. Biol.* **2018**, *28*, R619–R634. [CrossRef] [PubMed]
4. Saijo, Y.; Loo, E.P.I. Plant immunity in signal integration between biotic and abiotic stress responses. *New Phytol.* **2020**, *225*, 87–104. [CrossRef] [PubMed]
5. Abdullah, A.S.; Moffat, C.S.; Lopez-Ruiz, F.J.; Gibberd, M.R.; Hamblin, J.; Zerihun, A. Host–multi-pathogen warfare: Pathogen interactions in co-infected plants. *Front. Plant Sci.* **2017**, *8*, 1806. [CrossRef] [PubMed]
6. Schenk, P.M.; Carvalhais, L.C.; Kazan, K. Unraveling plant–microbe interactions: Can multi-species transcriptomics help? *Trends Biotechnol.* **2012**, *30*, 177–184. [CrossRef]
7. Akamatsu, A.; Shimamoto, K.; Kawano, Y. Crosstalk of signalling mechanisms involved in host defense and symbiosis against microorganisms in rice. *Curr. Genom.* **2016**, *17*, 297–307. [CrossRef]
8. Olive, A.J.; Sassetti, C.M. Metabolic crosstalk between host and pathogen: Sensing, adapting and competing. *Nat. Rev. Microbiol.* **2016**, *14*, 221–234. [CrossRef]
9. Jan, R.; Asaf, S.; Numan, M.; Lubna; Kim, K.-M. Plant secondary metabolite biosynthesis and transcriptional regulation in response to biotic and abiotic stress conditions. *Agronomy* **2021**, *11*, 968. [CrossRef]
10. Meena, K.K.; Sorty, A.M.; Bitla, U.M.; Choudhary, K.; Gupta, P.; Pareek, A.; Singh, D.P.; Prabha, R.; Sahu, P.K.; Gupta, V.K. Abiotic stress responses and microbe-mediated mitigation in plants: The Omics Strategies. *Front. Plant Sci.* **2017**, *8*, 172. [CrossRef]
11. Seymour, G.B.; Rose, J.K.C. Tomato molecular biology—Special collection of papers for molecular horticulture. *Mol. Hortic.* **2022**, *2*, 21. [CrossRef]
12. Lebeda, A.; Mieslerová, B.; Petřivalský, M.; Luhová, L.; Špundová, M.; Sedlářová, M.; Nožková-Hlaváčková, V.; Pink, D.A.C. Resistance mechanisms of wild tomato germplasm to infection of *Oidiumneo lycopersici*. *Eur. J. Plant Pathol.* **2014**, *138*, 569–596. [CrossRef]
13. Pailles, Y.; Awlia, M.; Julkowska, M.; Passone, L.; Zemmouri, K.; Negrão, S.; Schmöckel, S.M.; Tester, M. Diverse traits contribute to salinity tolerance of wild tomato seedlings from the Galapagos Islands. *Plant Physiol.* **2020**, *182*, 534–546. [CrossRef] [PubMed]
14. Yerasu, Y.R.; Murugan, L.; Halder, H.; Prasanna, H.C.; Singh, A.; Singh, B. Screening tomato genotypes for resistance to early blight and American Serpentine Leafminer. *Hortic. Environ. Biotechnol.* **2019**, *60*, 427–433. [CrossRef]
15. Adhikari, P.; Oh, Y.; Panthee, D.R. Current status of early blight resistance in tomato: An update. *Int. J. Mol. Sci.* **2017**, *18*, 2019. [CrossRef] [PubMed]
16. Allwood, J.W.; Williams, A.; Uthe, H.; van Dam, N.M.; Mur, L.A.J.; Grant, M.R.; Pétriacq, P. Unravelling plant responses to stress—The importance of targeted and untargeted metabolomics. *Metabolites* **2021**, *11*, 558. [CrossRef]
17. Mushtaq, M.Y.; Choi, Y.H.; Verpoorte, R.; Wilson, E.G. Extraction for metabolomics: Access to the metabolome. *Phytochem. Anal.* **2014**, *25*, 291–306. [CrossRef]
18. Jo, Y.-H.; Kim, S.; Kwon, S.-A.; Lee, H.J. Metabolomic analysis of ethyl acetate and methanol extracts of blueberry. *J. Korean Soc. Food Sci. Nutr.* **2014**, *43*, 419–424. [CrossRef]
19. Sumner, L.W.; Amberg, A.; Barrett, D.; Beale, M.H.; Beger, R.; Daykin, C.A.; Fan, T.W.-M.; Fiehn, O.; Goodacre, R.; Griffin, J.L.; et al. Proposed minimum reporting standards for chemical analysis Chemical Analysis Working Group (CAWG) Metabolomics Standards Initiative (MSI). *Metabolomics* **2007**, *3*, 211–221. [CrossRef]
20. Sansone, S.A.; Fan, T.; Goodacre, R.; Griffin, J.L.; Hardy, N.W.; Kaddurah-Daouk, R.; Kristal, B.S.; Lindon, J.; Mendes, P.; Morrison, N.; et al. The metabolomics standards initiative. *Nat. Biotechnol.* **2007**, *25*, 846–848. [CrossRef]
21. Rivera-Pérez, A.; Romero-González, R.; Garrido, F.A. Application of an innovative metabolomics approach to discriminate geographical origin and processing of black pepper by untargeted UHPLC-Q-Orbitrap-HRMS analysis and mid-level data fusion. *Food Res. Int.* **2021**, *150A*, 110722. [CrossRef] [PubMed]
22. Quinet, M.; Angosto, T.; Yuste-Lisbona, F.J.; Blanchard-Gros, R.; Bigot, S.; Martinez, J.-P.; Lutts, S. Tomato fruit development and metabolism. *Front. Plant Sci.* **2019**, *10*, 1554. [CrossRef]
23. Singh, D.P.; Bisen, M.S.; Shukla, R.; Prabha, R.; Maurya, S.; Reddy, Y.S.; Singh, P.M.; Rai, N.; Chaubey, T.; Chaturvedi, K.K.; et al. Metabolomics-driven mining of metabolite resources: Applications and prospects for improving vegetable crops. *Int. J. Mol. Sci.* **2022**, *23*, 12062. [CrossRef]
24. Isah, T. Stress and defense responses in plant secondary metabolites production. *Biol. Res.* **2019**, *52*, 39. [CrossRef] [PubMed]
25. Kachroo, A.; Kachroo, P. Fatty acid-derived signals in plant defense. *Annu. Rev. Phytopathol.* **2009**, *47*, 153–176. [CrossRef] [PubMed]
26. Kishimoto, K.; Matsui, K.; Ozawa, R.; Takabayashi, J. Volatile C6-aldehydes and allo-ocimene activate defense genes and induce resistance against Botrytis cinerea in Arabidopsis thaliana. *Plant Cell Physiol.* **2005**, *46*, 1093–1102. [CrossRef] [PubMed]
27. Chandrasekaran, M.; Paramasivan, M.; Chun, S.C. *Bacillus subtilis* CBR05 induces vitamin B6 biosynthesis in tomato through the de novo pathway in contributing disease resistance against *Xanthomonas campestris* pv. Vesicatoria. *Sci. Rep.* **2019**, *9*, 6495. [CrossRef]

28. Pushpalatha, H.B.; Mythrashree, S.R.; Shetty, R.; Geetha, N.; Sharathchandra, R.G.; Amruthesh, K.N.; Shetty, H.S. Ability of vitamins to induce downy mildew disease resistance and growth promotion in pearl millet. *Crop Prot.* **2007**, *26*, 1674–1681. [CrossRef]
29. Ishihara, A.; Nakao, T.; Mashimo, Y.; Murai, M.; Ichimaru, N.; Tanaka, C.; Nakajima, H.; Wakasa, K.; Miyagawa, H. Probing the role of tryptophan-derived secondary metabolism in defense responses against *Bipolaris oryzae* infection in rice leaves by a suicide substrate of tryptophan decarboxylase. *Phytochemistry* **2011**, *72*, 7–13. [CrossRef]
30. Hagemeier, J.; Schneider, B.; Oldham, N.J.; Hahlbrock, K. Accumulation of soluble and wall-bound indolic metabolites in *Arabidopsis thaliana* leaves infected with virulent or avirulent *Pseudomonas syringae* pathovar tomato strains. *Proc. Natl. Acad. Sci. USA* **2001**, *98*, 753–758. [CrossRef]
31. Böttcher, C.; Chapman, A.; Fellermeier, F.; Choudhary, M.; Scheel, D.; Glawischnig, E. The biosynthetic pathway of indole-3-carboxaaldehyde and indole-3-carboxylic acid derivatives in Arabidopsis. *Plant Physiol.* **2014**, *165*, 841–853. [CrossRef] [PubMed]
32. Shen, Q.; Liu, L.; Wang, L.; Wang, Q. Indole primes plant defense against necrotrophic fungal pathogen infection. *PLoS ONE* **2018**, *13*, e0207607. [CrossRef] [PubMed]
33. Bhattacharya, A.; Sood, P.; Citovsky, V. The roles of plant phenolics in defence and communication during *Agrobacterium* and *Rhizobium* infection. *Mol. Plant Pathol.* **2010**, *11*, 705–719. [CrossRef] [PubMed]
34. Diaz-Tielase, C.; Grana, E.; Reigosa, M.J.; Sanchez-Moreiras, A.M. Biological activities and novel applications of chalcones. *Plantadaninha* **2016**, *34*, 607–616. [CrossRef]
35. León, R.-H.; Hernández-Equihua, M.G.; Boone-Villa, D.; Jacobo, G.C.M.; Aguilera-Méndez, A. Biotin supplementation alters root system architecture and development in *Arabidopsis thaliana*. *Plant Root* **2019**, *13*, 29–40.
36. Olorunnisola, O.S.; Ajayi, A.F.; Okeleji, L.O.; Oladipo, A.A.; Emorioloye, J.T. Vitamins as antioxidants. *J. Food Sci. Nutr. Res.* **2019**, *2*, 214–235.
37. Völz, R.; Park, J.Y.; Harris, W.; Hwang, S.; Lee, Y.-H. Lyso-phosphatidylethanolamine primes the plant immune system and promotes basal resistance against hemibiotrophic pathogens. *BMC Biotechnol.* **2021**, *21*, 12. [CrossRef]
38. Yu, M.H.; Zhao, Z.Z.; He, J.X. Brassinosteroid signaling in plant-microbe interactions. *Int. J. Mol. Sci.* **2018**, *19*, 4091. [CrossRef]
39. Hout, B.; Yao, J.; Montgomery, B.L.; He, S.Y. Growth–Defense Tradeoffs in Plants: A Balancing Act to Optimize Fitness. *Mol. Plant* **2014**, *7*, 1267–1287. [CrossRef]
40. Patocka, J.; Kuca, K. Biologically active alcohols: Cyclic alcohols. *Mil. Med. Sci. Lett. Vojenske Zdr. Listy* **2013**, *82*, 162–171. [CrossRef]
41. Giner, R.M.; Ríos, J.L.; Máñez, S. Antioxidant activity of natural hydroquinones. *Antioxidants* **2022**, *11*, 343. [CrossRef]
42. Wu, Y.; Liu, Y.; Gulbins, E.; Grassmé, H. The anti-infectious role of sphingosine in microbial diseases. *Cells* **2021**, *10*, 1105. [CrossRef] [PubMed]
43. Wang, J.; Chen, Y.L.; Li, Y.K.; Chen, D.K.; He, J.F.; Yao, N. Functions of sphingolipids in pathogenesis during host–pathogen interactions. *Front. Microbiol.* **2021**, *12*, 701041. [CrossRef] [PubMed]
44. Phung, T.H.; Jung, H.I.; Park, J.H.; Kim, J.G.; Back, K.; Jung, S. Porphyrin biosynthesis control under water stress: Sustained porphyrin status correlates with drought tolerance in transgenic rice. *Plant Physiol.* **2011**, *157*, 1746–1764. [CrossRef] [PubMed]
45. Quesada, V.; Sarmiento-Mañús, R.; González-Bayón, R.; Hricová, A.; Ponce, M.R.; Micol, J.L. Porphobilinogen deaminase deficiency alters vegetative and reproductive development and causes lesions in *Arabidopsis*. *PLoS ONE* **2013**, *8*, e53378. [CrossRef] [PubMed]
46. Aubry, S.; Fankhauser, N.; Ovinnikov, S.; Pružinská, A.; Stirnemann, M.; Zienkiewicz, K.; Herrfurth, C.; Feussner, I.; Hörtensteiner, S. Pheophorbide *a* may regulate jasmonate signaling during dark-induced senescence. *Plant Physiol.* **2020**, *182*, 776–791. [CrossRef]
47. Hirashima, M.; Tanaka, R.; Tanaka, A. Light-independent cell death induced by accumulation of pheophorbide *a* in *Arabidopsis thaliana*. *Plant Cell Physiol.* **2009**, *50*, 719–729. [CrossRef]
48. Chen, K.; Li, G.-J.; Bressan, R.A.; Song, C.-P.; Zhu, J.-K.; Zhao, Y. Abscisic acid dynamics, signaling, and functions in plants. *J. Integr. Plant Biol.* **2021**, *62*, 25–54. [CrossRef] [PubMed]
49. Tal, L.; Gil, M.X.A.; Guercio, A.M.; Shabek, N. Structural aspects of plant hormone signal perception and regulation by ubiquitin ligases. *Plant Physiol.* **2020**, *182*, 1537–1544. [CrossRef]
50. Jiménez-Bremont, J.F.; Marina, M.; de la Luz Guerrero-González, L.; Rossi, F.R.; Sánchez-Rangel, D.; Rodríguez-Kessler, M.; Ruiz, O.A.; Gárriz, A. Physiological and molecular implications of plant polyamine metabolism during biotic interactions. *Front. Plant Sci.* **2014**, *18*, 95. [CrossRef]
51. Adio, A.M.; Casteel, C.L.; De Vos, M.; Kim, J.H.; Joshi, V.; Li, B.; Juéry, C.; Daron, J.; Kliebenstein, D.J.; Jander, G. Biosynthesis and defensive function of *N*δ-acetylornithine, a jasmonate-induced Arabidopsis metabolite. *Plant Cell* **2011**, *23*, 3303–3318. [CrossRef]
52. Zeier, J. New insights into the regulation of plant immunity by amino acid metabolic pathways. *Plant Cell Environ.* **2013**, *36*, 2085–2301. [CrossRef] [PubMed]
53. Dye, S.M.; Yang, J.; Bostock, R.M. Eicosapolyenoic fatty acids alter oxylipin gene expression and fatty acid hydroperoxide profiles in tomato and pepper roots. *Physiol. Mol. Plant Pathol.* **2020**, *109*, 101444. [CrossRef]
54. Yang, F.; Zhang, Q.; Yao, Q.; Chen, G.; Tong, H.; Zhang, J.; Su, Q.; Zhang, Y. Direct and indirect plant defenses induced by (Z)-3-hexenol in tomato against whitefly attack. *J. Pest Sci.* **2020**, *93*, 1243–1254. [CrossRef]

55. Hatem, B.; Gargouri, M.; Mliki, A.; Brini, F.; Chong, J.; Moez, J. Vitamins for enhancing plant resistance. *Planta* **2016**, *244*, 529–543. [CrossRef]
56. Swapnil, P.; Meena, M.; Singh, S.K.; Dhuldhaj, U.P.; Marwal, H.A. Vital roles of carotenoids in plants and humans to deteriorate stress with its structure, biosynthesis, metabolic engineering and functional aspects. *Curr. Plant Biol.* **2021**, *26*, 100203. [CrossRef]
57. Sun, T.; Rao, S.; Zhou, X.; Li, L. Plant carotenoids: Recent advances and future perspectives. *Mol. Hortic.* **2022**, *2*, 3. [CrossRef]
58. Rebelo, B.A.; Farrona, S.; Ventura, M.R.; Abranches, R. Canthaxanthin, a red-hot carotenoid: Applications, synthesis, and biosynthetic evolution. *Plants* **2020**, *9*, 1039. [CrossRef]
59. Simkin, A.J. Carotenoids and apocarotenoids in planta: Their role in plant development, contribution to the flavour and aroma of fruits and flowers, and their nutraceutical benefits. *Plants* **2021**, *10*, 2321. [CrossRef]
60. Latowski, D.; Kuczyńska, P.; Strzałka, K. Xanthophyll cycle—A mechanism protecting plants against oxidative stress. *Redox Rep.* **2011**, *16*, 78–90. [CrossRef]
61. Liu, W.; Feng, Y.; Yu, S.; Fan, Z.; Li, X.; Li, J.; Yin, H. The flavonoid biosynthesis network in plants. *Int. J. Mol. Sci.* **2021**, *22*, 12824. [CrossRef] [PubMed]
62. Geng, D.; Shen, X.; Xie, Y.; Yang, Y.; Bian, R.; Gao, Y.; Li, P.; Sun, L.; Feng, H.; Ma, F.; et al. Regulation of phenylpropanoid biosynthesis by MdMYB88 and MdMYB124 contributes to pathogen and drought resistance in apple. *Hortic. Res.* **2020**, *7*, 102. [CrossRef]
63. Pun, M.; Khazanov, N.; Galsurker, O.; Weitman, M.; Kerem, Z.; Senderowitz, H.; Yedidia, I. Phloretin, an apple phytoalexin, affects the virulence and fitness of *Pectobacterium brasiliense* by interfering with quorum-sensing. *Front. Plant Sci.* **2021**, *12*, 671807. [CrossRef] [PubMed]
64. Rogowska, A.; Szakiel, A. The role of sterols in plant response to abiotic stress. *Phytochem. Rev.* **2020**, *19*, 1525–1538. [CrossRef]
65. Künstler, A.; Gullner, G.; Ádám, A.L.; Nagy, J.K.; Király, L. The versatile roles of sulfur-containing biomolecules in plant defense—A road to disease resistance. *Plants* **2020**, *9*, 1705. [CrossRef] [PubMed]
66. Kaur, S.; Samota, M.K.; Choudhary, M.; Choudhary, M.; Pandey, A.K.; Sharma, A.; Thakur, J. How do plants defend themselves against pathogens-Biochemical mechanisms and genetic interventions. *Physiol. Mol. Biol. Plant.* **2022**, *28*, 485–504. [CrossRef] [PubMed]
67. Castro-Moretti, F.R.; Gentzel, I.N.; Mackey, D.; Alonso, A.P. Metabolomics as an emerging tool for the study of plant–pathogen interactions. *Metabolites* **2020**, *10*, 52. [CrossRef] [PubMed]
68. Pretorius, C.J.; Tugizimana, F.; Steenkamp, P.A.; Piater, L.A.; Dubery, I.A. Metabolomics for biomarker discovery: Key signatory metabolic profiles for the identification and discrimination of oat cultivars. *Metabolites* **2021**, *11*, 165. [CrossRef]

metabolites

MDPI

Article

Metabolomic Reconfiguration in Primed Barley (*Hordeum vulgare*) Plants in Response to *Pyrenophora teres* f. *teres* Infection

Claude Y. Hamany Djande, Fidele Tugizimana, Paul A. Steenkamp, Lizelle A. Piater and Ian A. Dubery *

Research Centre for Plant Metabolomics, Department of Biochemistry, University of Johannesburg, P.O. Box 524, Auckland Park, Johannesburg 2006, South Africa; claudeh@uj.ac.za (C.Y.H.D.); ftugizimana@uj.ac.za (F.T.); psteenkamp@uj.ac.za (P.A.S.); lpiater@uj.ac.za (L.A.P.)
* Correspondence: idubery@uj.ac.za; Tel.: +27-11-5592401

Abstract: Necrotrophic fungi affect a wide range of plants and cause significant crop losses. For the activation of multi-layered innate immune defences, plants can be primed or pre-conditioned to rapidly and more efficiently counteract this pathogen. Untargeted and targeted metabolomics analyses were applied to elucidate the biochemical processes involved in the response of 3,5-dichloroanthranilic acid (3,5-DCAA) primed barley plants to *Pyrenophora teres* f. *teres* (*Ptt*). A susceptible barley cultivar ('Hessekwa') at the third leaf growth stage was treated with 3,5-DCAA 24 h prior to infection using a *Ptt* conidia suspension. The infection was monitored over 2, 4, and 6 days post-inoculation. For untargeted studies, ultra-high performance liquid chromatography coupled with high-resolution mass spectrometry (UHPLC–MS) was used to analyse methanolic plant extracts. Acquired data were processed to generate the data matrices utilised in chemometric modelling and multi-dimensional data mining. For targeted studies, selected metabolites from the amino acids, phenolic acids, and alkaloids classes were quantified using multiple reaction monitoring (MRM) mass spectrometry. 3,5-DCAA was effective as a priming agent in delaying the onset and intensity of symptoms but could not prevent the progression of the disease. Unsupervised learning methods revealed clear differences between the sample extracts from the control plants and the infected plants. Both orthogonal projection to latent structure-discriminant analysis (OPLS-DA) and 'shared and unique structures' (SUS) plots allowed for the extraction of potential markers of the primed and naïve plant responses to *Ptt*. These include classes of organic acids, fatty acids, amino acids, phenolic acids, and derivatives and flavonoids. Among these, 5-oxo-proline and citric acid were notable as priming response-related metabolites. Metabolites from the tricarboxylic acid pathway were only discriminant in the primed plant infected with *Ptt*. Furthermore, the quantification of targeted metabolites revealed that hydroxycinnamic acids were significantly more prominent in the primed infected plants, especially at 2 d.p.i. Our research advances efforts to better understand regulated and reprogrammed metabolic responses that constitute defence priming in barley against *Ptt*.

Keywords: barley; *Pyrenophora teres*; dichloroanthranilic acid; hydroxycinnamic acids; metabolomics; net-blotch disease; priming; defence-related metabolites

1. Introduction

Understanding plant defence mechanisms is critical in plant breeding practices to control plant disease. The rising global food demand and the indiscriminate application of chemical fertilisers highlight the necessity for sustainable crop production systems. Novel approaches for improved resistance are required to address these concerns, as phytopathogenic microorganisms represent a severe threat to global food production and ecosystem stability. Plants use an intricate and multi-layered innate immune system to ward off attacks by potential pathogens present in the environment. This involves the activation and regulation of signalling cascades and actions at the gene expression-level and protein biosynthesis-level, as well as the structural reinforcement of the cell wall [1–3].

Barley (*Hordeum vulgare* L.) is one of the most versatile cereals cultivated in different regions. However, the crop has shown susceptibility to fungal pathogens, including the ascomycete *Pyrenophora teres* f. *teres* (*Ptt*), the causal agent of the 'net blotch net form' (NBNF) disease. Worldwide, net blotch is among the most prevalent destructive disease of barley, leading to enormous yield losses of up to 70% in some cases, depending on the susceptibility of the cultivars, the trends in the environmental conditions, and the virulence of the pathogen [4–8]. This interdependence between the plant's susceptibility, the pathogen's virulence (both having strong genetic links), and environmental factors is illustrated in the disease triangle (Figure S1). The triangle is complete when all of the following three conditions are met: a susceptible host, a virulent pathogen, and a favourable environment. An alteration or manipulation of one triangle component may affect the disease's progression [9,10].

Plant–pathogen interactions commence immediately after the host makes contact with the microbe-associated molecular pattern (MAMP) molecules derived from the pathogen. The latter uses a wide spectrum of physical and/or enzymatic tools to access plant nutrients. In turn, the plant recognises the pathogen signals and activates constitutive pre-formed and specific inducible defence systems. Fungal infections significantly impact the physiology of the host, including its gene expression profiles, primary metabolism, oxidative stress, and regulation of signalling pathways. Through the phenomenon known as priming, plants can be prepared or induced to easily adjust to such environmental changes, resulting in a faster and more efficient response to the pathogen and helping the plant to resist the attempted attack, thus enabling it to sustain less damage.

Despite lacking an adaptive immune system, it is now well known that plants can 'remember' or' learn' from previous attacks (e.g., rewiring the stress memory), resulting in stronger and faster responses to low-stimulus concentrations. This phenomenon is known as defence priming, which is regarded as a component of induced resistance (IR) [11]. Through 'priming' or 'preconditioning' mechanisms, information from the complex immune system may be stored as a 'memory', enabling the plant to quickly and more efficiently activate/stimulate defences to restrict pathogen growth and multiplication [11,12]. The primed condition can be established by preconditioning or pretreatment with particular substances, biological agents, or physical factors to increase survivability and the effectiveness of defence mechanisms under potential stress [13,14]. While the mechanism behind defence priming is still unclear, the different phases involved remain, and they are as follows: (i) the priming phase corresponding to the perception of the stimulus; (ii) the post-challenged primed state involving exposure to a secondary stimulus; and (iii) the transgenerational primed state where primed parents pass down the primed state to future generations [12,15–17].

Metabolites not only define and determine the fundamental physiological and agronomic features of plants but also shape the biochemical dynamics of defence responses to stress conditions, which carries imprints of environmental and genetic factors [18]. In an earlier study [19], we found that treatment with 3,5-dichloroanthranilic acid (3,5-DCAA) as a priming agent orchestrated metabolic perturbations in the investigated barley cultivars. Changes were characterised by the up- and down-regulation of defence-related metabolites and precursors, which define a 'state of alertness' in the plants. Mass spectrometry-based metabolomics approaches, targeted or untargeted, are essential components of system biology investigations. Over the last decade, a lot of attention has been paid to applying metabolomics tools and approaches to shed more light on the mechanisms involved in plant–pathogen interactions [20–22]. For the present study, the 'Hessekwa' cultivar, which has shown susceptibility to *Ptt* in trial studies (Figure S2), was selected to evaluate the effect of the pathogen on naïve and 3,5-DCAA-primed plants. Both untargeted and targeted metabolomics approaches with computational tools were used to monitor and explore the metabolic reconfiguration that occurs in barley plants (primed or naïve) and the characteristics of their response to pathogenic infection.

2. Materials and Methods

2.1. Barley Plant Material and Growth Conditions

'Hessekwa' is an experimental barley cultivar that was developed for the Western Cape region of South Africa (a winter rainfall area). The South African Barley Breeding Institute (SABBI, Bredasdorp, Western Cape, South Africa) provided the cultivar seeds. The seeds were grown as previously described [23]. Briefly, the cultivar was grown in a plant growth room under well-controlled conditions, which were as follows: 12 h fluorescent light (≈85 µmol m^{-2} s^{-2}) and 12 h dark cycle at 22–27 °C. The seeds were surfaced-sterilised (70% ethanol) and soaked in sterile water for 2 h before they were planted in pasteurised (at 70 °C) soil (Germination mix, Culterra, Muldersdrift, South Africa). About 40 seeds were planted per pot (12 cm in diameter and 8 cm depth). The plants were watered twice a week with distilled water containing a water-soluble chemical fertiliser (Multisol 'N', Culterra, Muldersdrift, South Africa). This cultivar was selected based on a preliminary study of five barley cultivars: 'Erica', 'Agulhas', 'Elim', 'S16', and 'Hessekwa'. The results of the preliminary study identified 'Hessekwa' as the most susceptible to *Ptt*. In addition to being an important cultivar under development in South Africa, the susceptibility of the cultivar to the pathogen of interest was a part of the criteria for selection.

2.2. Fungal Isolate, Culture, and Sporulation

The *P. teres* f. *teres* isolate W1-1, the causative pathogen of net blotch net form disease (NBNF) on barley, was provided by Dr. Jordi Muria-Gonzalez, Curtin University, WA, Australia. Growth of the fungus first commenced on V8-PDA (V8 vegetable extract-potato dextrose agar) medium and sub-cultured into a barley–oat–agar (BOA, pH 7) solid media for improved sporulation (adapted from [7,24]). The plates were incubated for 10 d under a 12 h/12 h photoperiod at 22 °C. *Ptt* sporulation was induced by placing plates hydrated with 500 µL sterile water under black light (365 nm UV light) for 20 h. To induce conidia formation, fungal plates were further incubated for 24 h at 15 °C in the dark (JM Gonzalez, personal communication). Figure S3 shows a plate before and after enhancing sporulation. For spore collection, 5 mL of sterile water containing 0.05% of Tween-20 was added to the plates, and a soft brush was used to dislocate them. This process was performed twice to maximise the number of conidia collected. The resulting liquid was vortexed and filtered through muslin cloth before determining the conidia concentration using a haemocytometer and light microscope at 40X magnification.

2.3. Plant Treatment/Inoculation and Experimental Design

Barley shoot tissues were treated with 3,5-DCAA 24 h prior to infection (when the plants were 20 d post-planting). The choice to use 3,5-DCAA as a priming agent was based on metabolomic data [19]. The inducer was dissolved in dimethylsulphoxide (DMSO; 1 µL mL^{-1}) and subsequently diluted into a buffered wetting agent dissolved in dH_2O, pH 6.0 ('Insure', Efekto, Pretoria, South Africa), to a concentration of 200 µM and sprayed on shoot tissue for systemic uptake.

The harvested *Ptt* conidia were diluted to a concentration of 1×10^5 mL^{-1}, and the suspension was sprayed onto the shoot tissues of the primed and naïve barley plants (5 mL per pot). The plants were incubated for 24 h at 18 °C and in humid conditions to favour conidia solution uptake before returning to the normal above-mentioned growing conditions. The plants were inoculated when they turned 21 d old and were harvested at 2, 4, and 6 d post-infection (d.p.i.).

Overall, the experimental design included the following: (i) the naïve plants (control; no treatment); (ii) the Tween-20 control, which served as the vehicle control for the *Ptt* conidia; (iii) the DMSO control, which served as the vehicle control for the 3,5-DCAA agent; (iv) 3,5-DCAA, which served as a control for the primed and infected conditions (PI); (v) infected naïve plants (*Ptt*) and (vi) infected primed plants (DCAA + *Ptt*). All of the above were involved in three independent biological replicates. Upon harvesting, the leaves were cut from the roots and snap-frozen in liquid nitrogen to quench metabolic activity.

2.4. Metabolite Extraction and Sample Preparation

Harvested leaf tissues were pulverised in a mortar and pestle with liquid nitrogen. To 1 g of leaf powder, 10 mL of 80% methanol was added, and the mixture was homogenised using an Ultra-Turrax homogeniser (CAT, Ballrechten-Dottingen, Germany) and a probe sonicator (Bandelin, Sonopolus, Berlin, Germany) set at 55% power for 10 s. The homogenates were centrifuged for 20 min at 5100× *g* and 4 °C, and the resulting hydro-methanolic supernatants were concentrated to 1 mL on a rotary evaporator, transferred to 2 mL Eppendorf tubes, and evaporated further to complete dryness in a dry bath at 45 °C within a fume hood. The extracts were dissolved in 50% methanol and then filtered through 0.22 µm nylon filters into chromatography vials before ultra-high performance liquid chromatography–quadrupole time-of-flight mass spectrometry (UHPLC–qTOF–MS).

2.5. Ultra-High Performance Liquid Chromatography–High Definition Mass Spectrometry (UHPLC-HDMS) for Untargeted Metabolomics

The UHPLC–qTOF–MS analyses of the extracts were performed on a Waters Acquity UHPLC hyphenated with a Waters SYNAPT G1 high resolution using accurate mass spectrometer system in V-optics (Waters Corporation, Milford, MA, USA) as a detector. The aqueous methanol extracts were separated using the Waters HSS T3 C18 column (150 mm × 2.1 mm × 1.8 µm) at 60 °C. The T3 column has the advantage of separating compounds ranging from polar to non-polar despite its status as a C18-based reverse phase column. A binary solvent system made up of water (eluent A) and acetonitrile (Romil Pure Chemistry, Cambridge, UK; eluent B) (both containing 0.1% formic acid) was used for the concave gradient elution at a flow rate of 0.4 mL min^{-1}. The elution was initiated with 5% B for 1 min and gradually increased to 95% B after over 24 min. The concentration of B was kept constant at 95% for 2 min and finally changed back to the initial conditions after 27 min. Before the next injection, the conditions were set in a manner that allowed the analytical column to calibrate for 3 min. The total run time was 30 min, and each sample's injection volume was 2 µL. Each sample was examined in three technical replicates to account for analytical variability. Quality control samples (QCs) were run at the beginning and end of the batch and in between (every 15 injections) to monitor the condition of the LC-MS system and evaluate the dependability and reproducibility of the analysis. The sample order was randomised, and blanks (50% methanol) were injected to monitor background noise, the possible carry-over of samples, and any solvent contamination.

High-resolution, accurate mass MS analyses were operated in negative and positive electrospray ionisation (ESI) modes. The capillary voltage was set at 2.5 kV; the sampling and extraction cone voltages were 40 V and 4.0 V, respectively. The source temperature was fixed at 120 °C, and the desolvation temperature was set at 450 °C. The cone and desolvation gas flows were 50 L h^{-1} and 550 L h^{-1}, respectively. Nitrogen was used as the nebulisation gas at a flow rate of 700 L h^{-1}. A mass range of 50 to 1200 m/z was selected with a scan time of 0.1 s. The reference mass calibrant, leucine enkephalin (50 pg mL^{-1}, $[M–H]^{-}$ = 554.2615 and $[M + H]^{+}$ = 556.2766) was sampled every 15 s and produced an average intensity of 350 counts per scan. The reference allowed the processing software (MassLynx XS^{TM} 4.1, Waters Corporation, Milford, MA, USA) to automatically correct slight deviations in the centroid masses observed in the samples, thereby providing exact mass measurements. This resulted in a typical mass accuracy of 1 to 3 mDa. In addition, a data-independent acquisition (DIA) method (MS^{E}) involving fragmentation at different collision energies (0–40 eV) was applied to generate fragmentation data to comprehensively extract the structural information of detected analytes and assist in metabolite annotation.

2.6. LC-ESI-QqQ-MS System for Targeted Metabolite Analyses

The triple quadrupole (QqQ) mass spectrometry platform used was an LC–MS 8050 instrument (Shimadzu, Kyoto, Japan), and it was equipped with an electrospray ionisation (ESI) source and an ultra-fast liquid chromatograph (UFLC) as a front-end. The multiple reaction monitoring (MRM) method was used for the absolute quantification of the targeted metabolites. The MRM-MS conditions were developed and optimised via direct infusion (using ESI source of MS), and the collision energy (CE) was optimised for each transition using the 'MRM optimisation method tool', an integral component of LabSolutions LC–MS software (Shimadzu Corporation, Kyoto, Japan). The tool automates the process by collecting product ion scan data and finding the optimum CE for each transition.

The targeted defence-related compounds were the amino acids phenylalanine (Phe), tyrosine (Tyr), and tryptophan (Trp); phenolic acids such as cinnamic acid, caffeic acid, ferulic acid, and sinapic acid; and, finally, two alkaloids, namely, hordenine and gramine. All the standards were obtained from Sigma-Aldrich (St. Louis, MO, USA) and BDH (Poole, UK). All working solutions of the mixed standards (concentrations ranging from 5×10^{-2} µg mL^{-1} to 4.5 µg mL^{-1}) were dissolved in 50% MeOH. Both the samples and the working solutions were analysed on the UFLC system using a C18 reverse phase chromatography column (Pinnacle DB Aqueous C18, 100 mm × 2.1 mm, 3 µm particle size) (Restek, Bellefonte, PA, USA). The injection volume was 1 µL, and the constant flow rate was 0.4 mL min^{-1}. A binary gradient made of MilliQ water with 0.1% formic acid (eluent A) and pure-grade MeOH (Romil Chemistry, Cambridge, UK) with 0.1% formic acid (eluent B) was used. The elution was initiated with 5% B for 1 min and gradually increased to 95% B after over 25 min. The concentration of B was kept constant at 95% for 2 min and finally changed back to the initial conditions after 27 min.

The MRM-MS method was developed and used for quantitative MS analyses. The MS conditions were as follows: the heating gas flow was set at 10 L min^{-1}, the interface temperature was set at 300 °C, the interface voltage was set at 4 kV, the DL and heat block temperatures were set at 250 °C and 400 °C, respectively. Nitrogen gas was used as a drying gas at a flow rate of 10 L min^{-1} and as a nebulising gas at a flow rate of 3 L min^{-1}. The optimal parameters for each metabolite and the equation generated from the standards are reported in Table S1. The concentrations are reported in Table S2.

2.7. Data Processing and Data Mining

Data pre-processing was carried out using the MarkerLynx-XS™ application manager of the MassLynx-XS™ 4.1 software (Waters, Manchester, UK). The software creates a matrix with individual mass spectra information, variable retention time (Rt) and m/z pairs, and integrated and normalised peak areas. The following parameters were used: Rt = 0.7–24 min (ESI$^-$) and Rt = 0.7–20 min (ESI$^+$) range and 100–1200 Da for the mass range. The mass tolerance was set to vary by up to 0.05 Da, and the Rts was set to vary by up to 0.2 min. The intensity threshold was set at 50 counts. Only data matrices with noise levels under the MarkerLynx cut-off of 50% were considered for further chemometric and statistical studies. The generated data matrices were imported into SIMCA (Soft Independent Modelling of Class Analogy) software, version 14 (Sartorius, Umeå, Sweden), for chemometric analyses. The principal components analysis (PCA) unsupervised approach and a supervised learning model, orthogonal projection to latent structures-discriminant analysis (OPLS-DA), were used within the SIMCA package. The models were consistently validated using permutation (n = 100) tests analyses and a seven-fold cross-validation (CV) method computed within the software [21,25]. OPLS-DA-generated loadings 'S-plots' were assessed for the selection of variables. Features with both high correlation and covariation were considered, [p(corr) $\geq$ 0.5, ≤ -0.5 and (p1) $\geq$ 0.1, ≤ -0.1], and variable importance in projection (VIP) plots were used to evaluate the statistical significance of each feature (cut-off > 1) and avoid possible bias from occurring during feature selection [16]. In addition, 'Shared-and-Unique-Structures' (SUS) plots were generated to reveal the shared and distinct metabolic features of the primed and naïve infected plants. The SUS plots were generated using the p(corr) value obtained in the OPLS-DA models, and they show metabolites that co-ordinately contribute to the primed and naïve infected conditions. Only components contributing to the strength of the prediction capacity of the model (R1 significant components) were considered.

2.8. Metabolite Annotation and Metabolic Pathway and Network Analyses

Metabolite annotation was based on MS data (accurate mass and mass fragmentation patterns) as described [23,26], focusing mostly on discriminant metabolites. The annotations were made according to the Metabolomics Standard Initiative (MSI) level 2 [27]. The estimated molecular formula of a chosen ion (feature corresponding to a metabolite) was manually searched using bioinformatics tools and databases such as PubChem [28], ChemSpider [29], Lipidmaps [30], and the Dictionary of Natural Products [31]. The mass spectral fragmentation patterns were examined and confirmed via comparison with information on barley in the literature and metabolome data from related plants. The MetPA (Metabolomics Pathway Analysis) [32]) module of the MetaboAnalyst bioinformatics platform (Version 5.0) [33] was used to perform pathway analysis on the successfully annotated metabolites that were selected based on the OPLS-DA data. Metabolites with known KEGG (Kyoto Encyclopedia of Genes and Genomics, [34]) identities (KEGG identifiers) were entered into the MetPA tool, and enrichment analysis was used to evaluate the potential biological roles thereof. The pathway analysis calculated the significance of the impacted pathways based on the pathway impact values (x-axis) obtained from our pathway topology analysis and the p values acquired from our pathway enrichment analysis (y-axis). Furthermore, using MetaMapp [35], correlation network analysis was performed for a global visualisation of the metabolic alterations. This was performed utilising the fold-changes and p values determined from OPLS-DA-derived descriptive statistics as well as the MetaMapp-encoded chemical structures of the annotated metabolites from the PubChem and KEGG databases (Table S3). The subsequent metabolic networks were displayed using the network visualisation tool Cytoscape (version 3.9.1; [36]), and the similarity cut-off between metabolites was determined using a Tanimoto score threshold of 0.7.

3. Results and Discussion

The ability of a plant to overcome a pathogen attack determines the level of resistance thereof. The molecular communication between both organisms depends on the genetic make-up of both participants and may result in resistance by the host or invasion by the pathogen. A resistant (R) plant will successfully overcome the attack, while the pathogen will defeat a susceptible (S) one, causing infectious disease. Partially resistant or tolerant cultivars lie within an R–S continuum, with the outcome frequently determined by abiotic environmental factors. As previously mentioned, resistance can be enhanced, and an IR phenotype can involve both primed and induced defence responses and can be brought into play at a local or systemic level against microbial pathogens that can exhibit varied lifestyles. With xenobiotic agents/inducers like 3,5-DCAA, the relative significance of primed vs. induced defences may be dependent on a range of factors, such as the concentration and chemical nature of the stimulus and the bio-availability and perception thereof, as well as the kinetics of the response and timing of analysis. Moreover, the effectiveness of IR responses can depend on the plant species, the developmental age of the plant, the tissue type, and the virulence of the pathogen [14].

3.1. Evaluation of the Net Blotch Net Form Disease on Barley

As indicated in the experimental section, the 'Hessekwa' cultivar of barley (naïve and pre-treated with 3,5-DCAA) was infected with a 1×10^5 mL^{-1} conidia suspension of the fungal pathogen *Ptt*. This was followed by assessing the plants' phenotypic characteristics and evaluating the disease severity index. Symptom development and the corresponding disease severity score ranging from 0 to 10 (as reported by Tekauz) [37] are illustrated in Figure 1A,B, respectively. In Figure 1A, an early manifestation of fungal infection (at 2 d.p.i.) was phenotypically characterised by fusiform and circular dot-like lesions, progressing into dark brown longitudinal and transverse striations constituting the net-like pattern on the shoot tissues of the naïve infected plants. From 4 d.p.i., the symptoms on primed and naïve plants are slightly comparable; however, there is a shrinking observed on naïve-treated plants but not on plants pre-treated with 3,5-DCAA. In addition, chlorotic areas could also be highlighted around the necrotic lesions observed on both challenged leaves from 4 d.p.i for the 3,5-DCAA-treated plants and at 6 d.p.i for naïve plants. The zoomed-in photograph highlights the observed net-type lesions (characteristic of the NBNF disease).

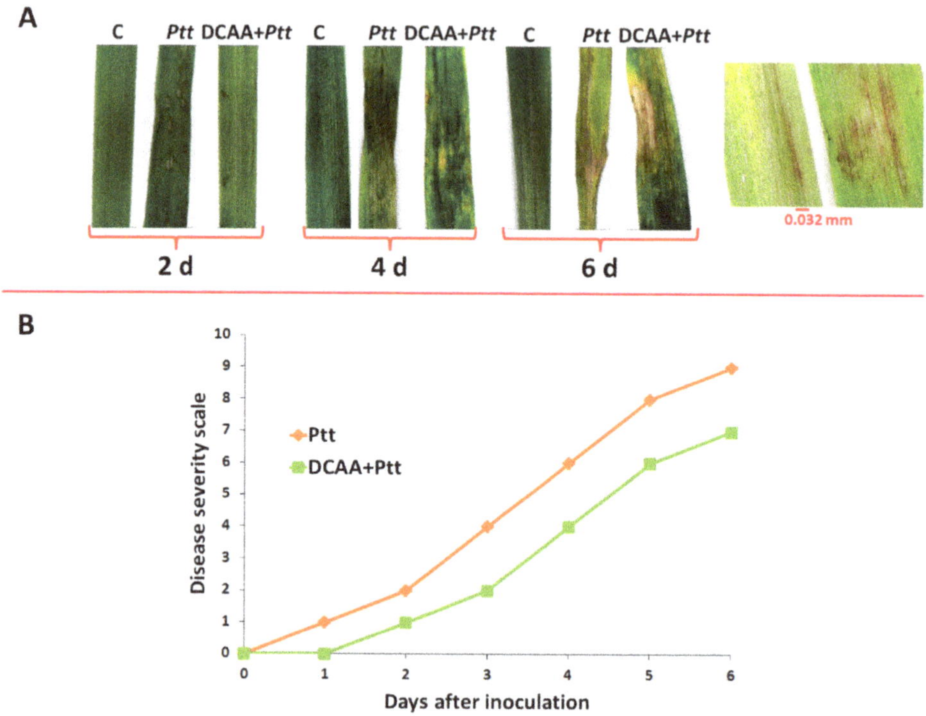

Figure 1. Fungal infection symptoms and disease severity assessment. (**A**): Barley shoot tissue at 2, 4, and 6 d post-inoculation with *P. teres* f. *teres* conidia. (**B**): Net blotch net form (NBNF) disease symptom evaluation on a numerical scale ranging from 0 to 10. 0 = No symptoms; 1 = Resistant; 2 = Resistant to moderately resistant; 3 = Moderately resistant; 4 = Moderately resistant to moderately susceptible; 5 = Moderately resistant to moderately susceptible; 6 = Moderately resistant to moderately susceptible; 7 = Moderately susceptible; 8 = Moderately susceptible to susceptible; 9 = Susceptible; 10 = Very susceptible.

The disease severity index was as described in the legend of Figure 1A. Looking at the disease severity scale shown in Figure 1B, symptoms could be observed as early as 24 h post-inoculation in the naïve infected plants showing a disease index of 1. On the 3,5-DCAA treated shoot tissues, the first symptoms were observed at 2 d.p.i. In both cases, the steady development of phenotypic symptoms was observed to reach a score of 9 (for the naïve plants) and 7 (for the primed plants) at over 6 d.p.i, corresponding to the plant's susceptibility state. As mentioned in the introduction, disease progression depends on the virulence of the pathogen, the environment, and the susceptibility of the cultivar. Here, environmental variables were controlled, and preliminary studies revealed 'Hessekwa' cultivar as susceptible to the *Ptt* W1-1 isolate (Figure S2).

The NBNF disease is a destructive disease of barley that causes important yield losses in susceptible cultivars. *Ptt* can colonise the host by producing necrotic and chlorotic symptoms due to net blotch in plant tissue [5,38,39]. In susceptible barley cultivars, germinated conidia produce the appressorium through the plant membrane at 1 d.p.i. [40,41]. This is followed by fungal growth inside the leaf tissue, resulting in necrosis and chlorosis from

4 d.p.i. The interaction between the susceptible barley plant and the *Ptt* pathogen results in the colonisation of the plant tissue, the release of nutrients and energy production for fungal growth, the suppression of plant defence responses, and, eventually, induced cell death. The delayed development of symptoms in the 3,5-DCAA-treated plants inversely correlates to the time deployed to respond to the fungal attack. This behaviour supports the post-challenged primed state characterised by the quicker and stronger defence response of the primed plants compared to the naïve infected plants [17,42]. Although the disease severity index of the primed infected plants at 6 d.p.i. is still reduced compared to that of the naïve infected plants, the pathogen can seemingly overcome the host defences at that time point. Subsequently, untargeted and targeted metabolomic approaches were used to pinpoint discriminatory metabolites and identify the biological pathways characterising the responses of the primed and naïve plants to the fungal infections.

3.2. Chromatographic and Chemometric Profiling of Primed and Naïve Barley Plants Following Infection by the Fungus P. teres f. teres

To elucidate the metabolic profiles of the primed and naïve barely plants responding to *P. teres* f. *teres* infection, the methanolic extracts were analysed via high-definition, accurate mass UHPLC-MS. Although the base peak intensity (BPI) chromatograms show only the most intense peaks at a specific retention time, they revealed subtle quantitative and qualitative differences across extracts (Figure S4). The data obtained from our LC-MS analyses were further analysed using multivariate statistical tools to simplify the visualisation of the complex multidimensional datasets, detect outliers, uncover the underlying trends, and extract relevant information for biological interpretation. From the unsupervised method, the computed PCA scores plots (Figures 2A–C and S5), distinct clustering separating all of the control and primed samples from the infected ones was observed at all time-points. Although this result was not so evident on day 2 and day 4, with the ESI (−) data, at day 6, a clear separation of the naïve infected and primed infected clusters was observed. In addition, all control samples (no vehicle control, DMSO control, and Tween-20/H_2O control), as well as the 3,5-DCAA-treated samples, grouped together on the PCA score plots, showing no major variations between these conditions. As a result, the naïve plant condition (no vehicle control, C) was then used as a comparative reference for primed infected and naïve infected conditions in downstream data mining processes.

As mentioned in the Materials and Methods section, OPLS-DA was applied as a supervised learning method to characterise and distinguish the classes and groups observed in the profile generated using the above-mentioned explorative models. The evaluation of the OPLS-DA models was performed by conducting a cross-validation (CV)-ANOVA (with a p value < 0.05 being considered a significant model) [25]. In addition, out of the hundred randomly permutated models, none outperformed the original ones. Figure 3 (based on data from the 2 d.p.i timepoint) serves as a representative example of the steps followed, with Figure 3A showing the correlation between the original and the permutated Y-vectors, as well as the R^2 and Q^2 of the models. The represented permutation analysis plot displays the R^2 and Q^2 values of the permutated models (left-bottom corner), which are lower than those of the original (right-top corner). Statistically significant features were selected using the loading S-plots generated from the OPLS-DA data and models (Figure 3B). The S-plots identified features originating from metabolites with high covariation and high correlation that discriminates between the conditions being compared. Furthermore, taking into consideration the aim of this study, which was to investigate the changes that occur after both the naïve and primed 'Hessekwa' barley cultivars become infected, another variable selection method, known as the 'Shared-and-Unique Structures' (SUS) loadings plot, was used (Figure 3C), and underlying response markers associated with the post-challenged trends observed after employing the unsupervised models could thus be extracted. Only metabolites with high covariance and high correlation and those with a VIP score > 1 were considered potential biomarkers (Figure 3D). All annotated discriminant metabolites or potential biomarkers are summarised in Table S4.

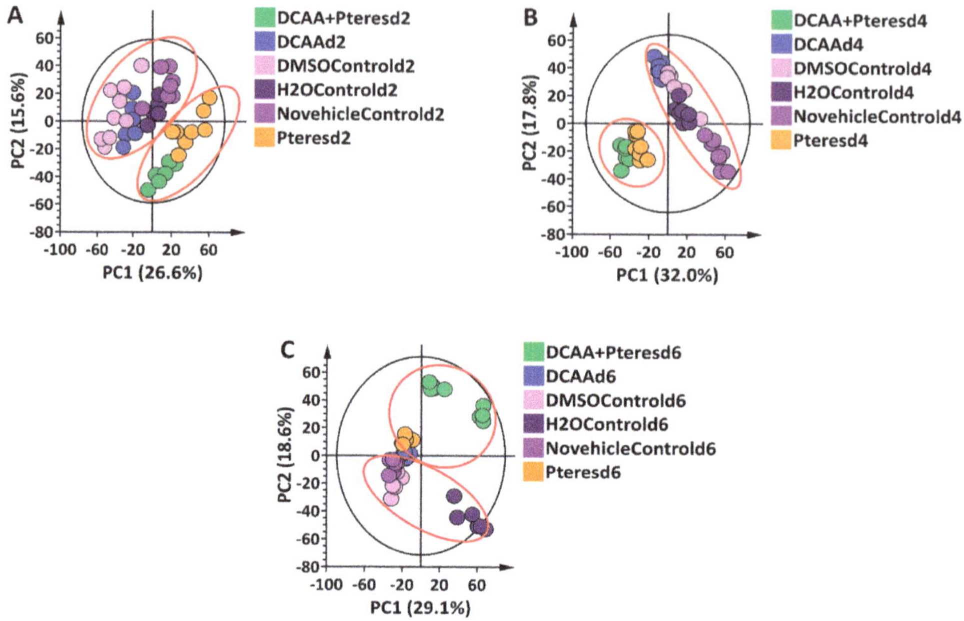

Figure 2. Principal component analysis (PCA) score plots of the ESI (–) data from shoot extracts of the 'Hessekwa' cultivar of *Hordeum vulgare* treated with 3,5-DCAA and infected with *P. teres* f. *teres*. All data were *Pareto*-scaled, and the calculated Hotelling's T^2 with a 95% confidence interval is represented by the ellipses present in each PCA score plot. (**A**): Six-component model of all conditions, d2, explaining 69.2% variation and predicting 45.1% variation; (**B**): Six-component model of all conditions, d4, explaining 74.8% variation and predicting 54.7% variation; (**C**): Seven-component model of all conditions, d6, explaining 72.5% variation and predicting 55.2% variation. The groupings of the controls and the treatments are highlighted in red.

3.3. Defence-Related Metabolic Reprogramming in 3,5-DCAA-Primed and Naïve Barley Plants Following Infection by the Fungus P. teres f. teres

The metabolic reconfiguration of primed and naïve 'Hessekwa' barley plants following infection with *Ptt* at different time points involved primary metabolites such as amino acids, organic acids, fatty acids, and derivatives and secondary metabolites such as phenolic acids and derivatives, flavonoids, and alkaloids (Table S4). Phenolic acids and derivatives were the largest class of annotated discriminant metabolites, followed by flavonoids and fatty acids (Figure 4A). In addition, differential distribution patterns were observed on the generated heatmap dendrogram, highlighting differences in the relative concentrations of the metabolites between the infected and the control samples and between the naïve and primed infected samples (Figure 4B). Annotated metabolites included oxo-proline, citric acid, linolenic acid derivative isomer I, linolenoylglycerol, hordatine A hexose isomer II, hydroxytryptamine, isoscoparin 7-O-[6′′-sinapoyl]-glucoside, and lutonarin (an isoorientin glucoside/a glucoside of luteolin), all discriminant metabolites in the primed infected samples. Discriminant metabolites in the naïve infected condition included hydroxylinolenic acid, hydroxyjasmonate sulphate, caffeoyl shikimate derivative isomer I and II, gallic acid monohydrate, and 6-prenylnaringenin (Table S4). The up- or down-regulation of these metabolites were characteristics of the infection status of the primed and naïve plants at different time points (2, 4, and 6 d.p.i.). Annotated metabolites were investigated for pathway analysis to reveal their implication in protection or defence mechanisms.

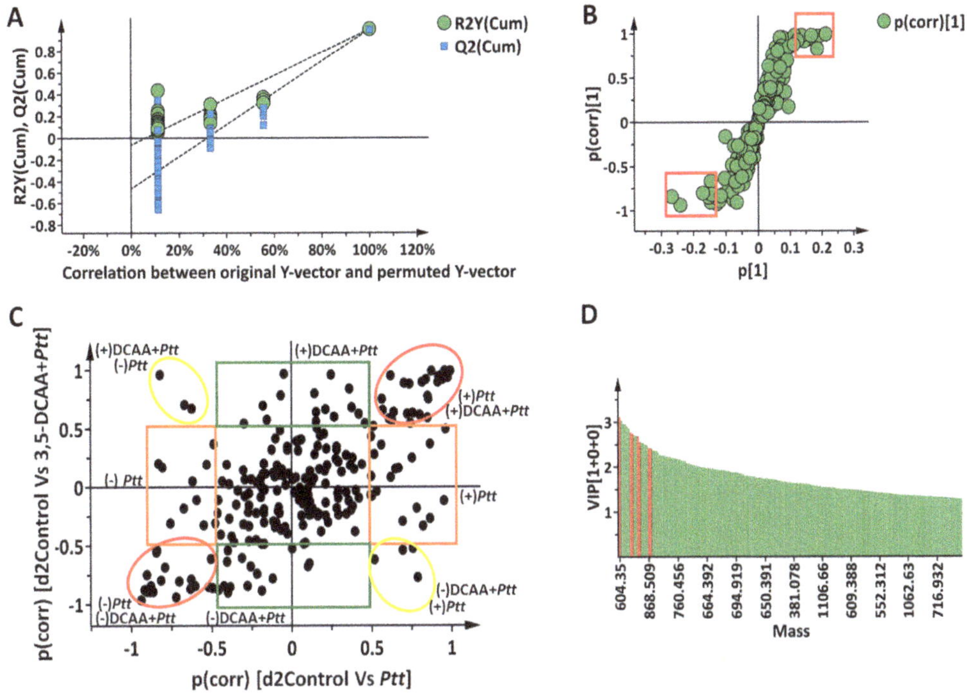

Figure 3. Orthogonal projection to latent structures discriminant analysis: selection of discriminant metabolites at 2 d.p.i. (**A**): Permutation test (n = 100) for the OPLS-DA model of 'Control vs. *Ptt*' (x-axis, component 1 + 2 + 0; $R^2X = 0.694$, $R^2Y = 0.999$, $Q^2 = 0.985$. CV-ANOVA = 2.15×10^{-9}); (**B**): OPLS-DA loading S-plot. The variables highlighted in the red squares are statistically significant in discriminating the two compared groups. The covariance (magnitude) and correlation (reliability) of the samples in the model are represented on the axes as p[1] and p(corr), respectively. (**C**): Shared-and-unique structures (SUS) plot analysis for correlating the predictive component of the OPLS-DA models 'Control vs. *Ptt*' and 'control vs. DCAA + *Ptt*' (y-axis, Component 1 + 2 + 0; $R^2X = 0.789$, $R^2Y = 0.994$, $Q^2 = 0.975$; CV-ANOVA = 3.44×10^{-8}) at 2 d.p.i. The most discriminant features in both models are highlighted with circles. Yellow circles correspond to the discriminant features inversely correlated between the two models, and red circles indicate those similarly correlating. The unique discriminant features for each model have been highlighted using orange (for *Ptt*) and green (for DCAA + *Ptt*) rectangles. As mentioned, only features with a high correlation were selected (p(corr) ≥ 0.5, ≤ -0.5). Up-regulation and down-regulation are shown as positive and negative values, respectively. (**D**): Variable importance in projection (VIP) scores for each selected metabolite >1.

The MetPA module of the MetaboAnalyst bioinformatics tool package (http://www.metaboanalyst.ca/ (accessed on 30 November 2022) was used to determine the key metabolic pathways characterising the primed and naïve defence responses. The MetPA analyses of both states (naïve infected and primed infected) with the KEGG identifiers revealed the involvement of similar pathways but with different levels of significance and impact (Figure 3B,C; Tables S5 and S6). These included phenylpropanoid biosynthesis; phenylalanine (Phe), tyrosine (Tyr), and tryptophan (Trp) metabolism (both individually and as a group); and isoquinoline alkaloid biosynthesis. The primed infected plants displayed metabolites from the TCA cycle with moderate impact and significance. The mentioned pathways are among the most prevalent in the plant kingdom. In addition to having implications on defence against biotic and abiotic stresses, these pathways are also connected to the modulation of physiological and biochemical processes [43,44]. The flagged metabolic pathways and temporal dynamics of selected defence-related discriminant barley metabolites are discussed below. In general, these patterns indicate differential reprogramming over time (either high or low accumulation at specific time points that reflect early, late, or oscillatory responses) [45]. These fluctuations may mirror the availability of precursor metabolites, the flux through induced metabolic pathways, and the extent of degradation or conversion of specific metabolites.

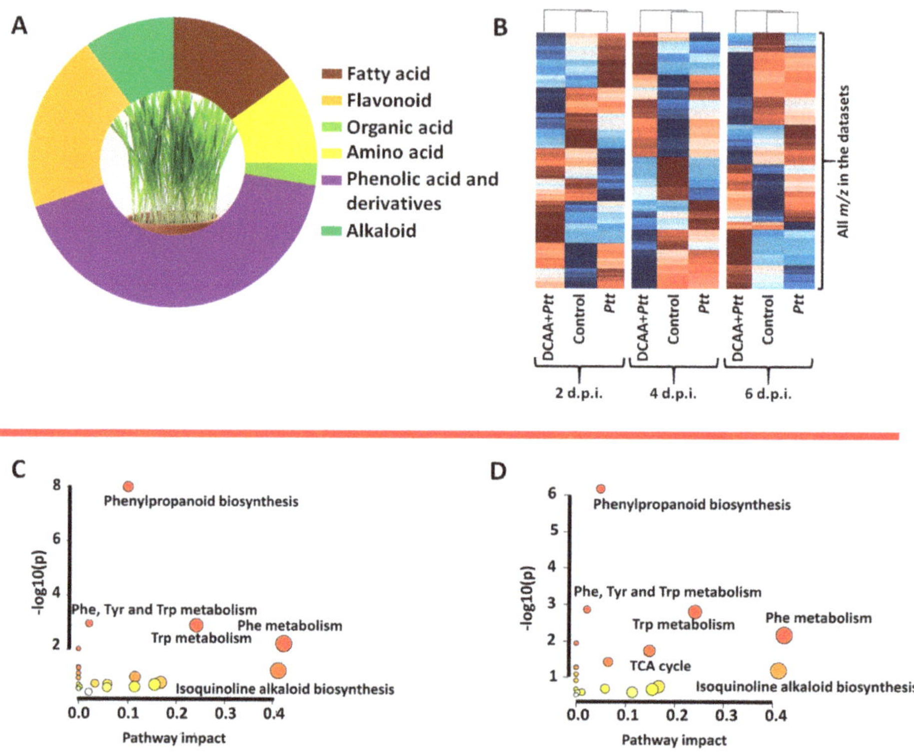

Figure 4. Discriminant metabolites in the 3,5-DCAA-primed and naïve barley plants after infection with *Ptt* and summary of the MetPA-generated metabolic pathways in the naïve and primed infected plants. (**A**): Metabolite class distribution. (**B**): Dendrogram heatmap showing the extent of metabolomic reconfiguration as reflected by the relative concentration of all metabolites across all conditions (Control, DCAA + *Ptt*, and *Ptt*) and all timepoints (2, 4, and 6 d.p.i.). Brown = high concentration; blue = low concentration. (**C**): Pathway mapping in naïve infected plants. (**D**): Pathway mapping in primed infected plants revealing some of the main metabolic pathways involved in the plants' (naïve and primed) interaction with *Ptt*. The colour and sizes of the circles are based on *p* values and pathway impact values, respectively. The circles with large pathway impact values and small *p* values are significantly perturbed.

3.3.1. Amino Acids

Significant pathways in both barley states also included the biosynthesis of Phe, Tyr, and Trp (Figure 5) and subsequent Phe and Trp metabolism. The MRM quantification profiles of the main components of the Phe, Tyr, and Trp biosynthesis pathways are shown via the red arrows in Figure 5 and Table S2. The concentrations of the three aromatic amino acids were significantly higher in the primed state at an early stage of infection (2 d.p.i. and 4 d.p.i). The concentration of these amino acids dropped at 6 d.p.i. in the primed plants and increased in the naïve infected plants. Alterations in the level of amino acids have previously been reported as a characteristic of the post-challenged phase [46,47]. Phe, Tyr, and Trp are mainly precursors in the biosynthesis of secondary metabolites, including all the ones mentioned above. In addition to their role as building blocks in protein synthesis, these amino acids are known for their actions in scavenging free radicals and the modulation of stomatal exposure and as osmolytes [48].

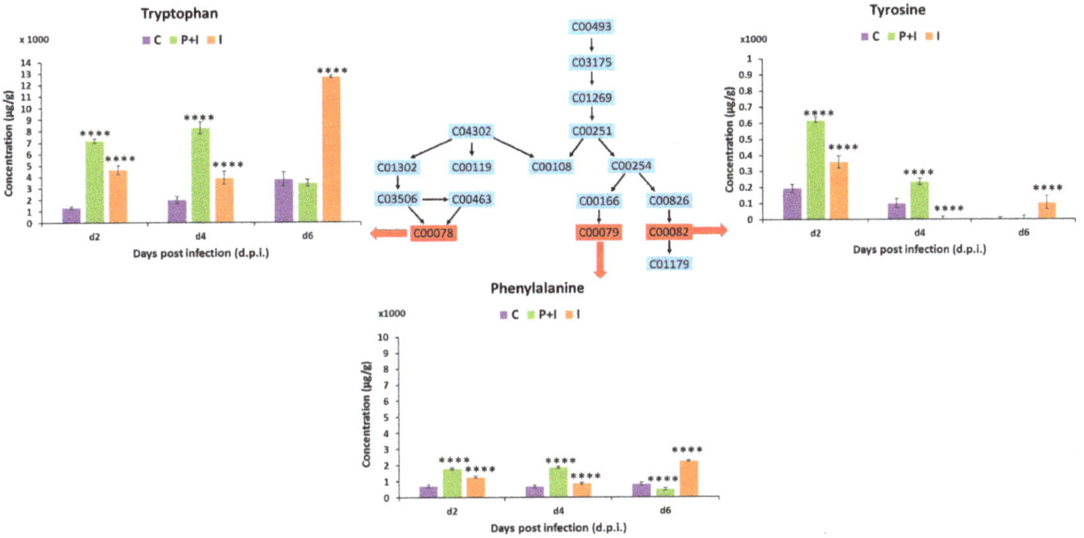

Figure 5. Illustration of the Phe, Tyr, and Trp biosynthesis pathways and MRM quantification of the main participants in the primed and naïve shoot tissues of barley infected with the fungal pathogen *P. teres* f. *teres*. The bar graphs were generated using the average integrated peak area ($n = 9$). The error bars represent the standard deviation. Concentrations are expressed in $\mu g\ g^{-1}$ (Mean +/− standard deviation; Table S2). The statistical significance of the differences observed was evaluated using Student's t-test, and all p values < 0.0001 (****) were considered as significant.

Other changes in amino acids included the up-regulation of oxo-proline (oxo-Pro) in the primed infected plants (Table S4). Plants accumulate Pro in response to several environmental stresses. Pro is a multi-functional amino acid, and its abundance may result from de novo synthesis or reduced degradation and utilisation. In the context of stress adaptation, Pro imparts stress tolerance by preserving the osmotic balance and cell turgidity and indirectly modulating the metabolism of ROS. In addition, the crosstalk of Pro with other osmoprotectants and signalling molecules like abscisic acid, glycine betaine, nitric oxide, hydrogen sulphide, etc., helps in strengthening the protective mechanism under stressful environments [49].

3.3.2. Tricarboxylates and Fatty Acids

Specifically significant in primed infected plants, the TCA cycle is one of the key metabolic pathways. Changes affecting organic acids (e.g., through reversible transaminase reactions) and amino acids might affect nitrogen and carbon metabolism and link the associated metabolic cycles in support of anaplerotic reactions to replenish the citric acid cycle if it becomes depleted of intermediates in response to biosynthetic demands. Furthermore, alterations in carboxylic acid levels were reported in plants during stress responses, and it was suggested that the tricarboxylates could modulate signal transduction cascades linked to plant defence responses [50,51]. In this context, citrate and fumarate were reported as inducers of defence priming in *Arabidopsis thaliana* through complex signalling pathways [51].

Fatty acids (FAs) are ubiquitous, and their oxygenated derivatives (oxilipins) play crucial roles in the plant system [52]. Among the FAs annotated here, hydroxylinolenic and alpha-linolenic acids were found to be up-regulated in naïve infected barley at 2 and 4 d.p.i., respectively. In the same plants and at the early infection stage, 12-hydroxyjasmonate sulphate was also up-regulated. These compounds are precursors and/or active participants in signalling pathways and regulatory activities [53]. It has been suggested that FAs act by inserting themselves into the lipid bilayers of the fungal membrane. This results in the loss of fungal membrane integrity which uncontrollably releases proteins and electrolytes and eventually leads to the disintegration of the cell cytoplasm [54,55]. The antifungal activity of linolenic acid and other unsaturated FAs was assessed in a study conducted by Walters et al. [56], and the compound inhibits more than 50% of mycelial growth of a range of fungi, including *Pyrenophora avenae* and *Rhizoctonia solani*.

3.3.3. Hydroxycinnamic Acids

The most significant pathway in both the naïve and primed plants upon infection (as indicated by the topological characteristics) was the phenylpropanoid biosynthesis pathway, which results in the formation of cinnamic acid and hydroxylated derivatives such as caffeic acid, ferulic acid, and sinapic acid. The MRM quantification profiles of each phenolic acid are shown in Figure 6 and Table S2. Among all the phenolic acids, the concentration of cinnamic and caffeic acids significantly ($p < 0.05$) increased at the early infection stage (2 d.p.i.) in the primed infected plants compared to the control and the naïve infected plants. At 4 d.p.i., a significant decrease in ferulic acid was observed in the primed infected plants. These compounds might be associated with delayed infection or the faster response observed at 2 d.p.i. (Section 3.1). Interestingly, concentrations of sinapic acid were significantly higher in the naïve infected plant at 6 d.p.i. This suggests that sinapic acid (3,5-dimethoxy-4-hydroxycinnamic acid) is directly implicated in the barley–*P. teres* interaction and during the later stages of infection. Similarly to this study, in another study, rust infection in wheat crops was associated with an increase in ferulic and sinapic acid levels [57].

Phenolic acids are important specialised secondary metabolites of plants and are required for several physiological and mechanical processes [44,58]. Plants produce these phenolics for growth and development and, importantly, for protection (e.g., as phytoanticipin/phytoalexin antimicrobials and anti-oxidants, as well as for the strengthening of the cell wall through lignin synthesis and deposition). Through the activation of the phenylpropanoid pathway and associated sub-pathways, plants are shielded against upcoming attacks by phytopathogens and the spread of infection due to these induced defence mechanisms that are expressed both at the location of the attack and distant to it. As well as providing the chemical building blocks of the defence response, a network of interconnected signal transduction pathways regulates induced resistance, with phenolic acids serving as crucial signalling molecules [59,60].

3.3.4. Hydroxycinnamic Acid Amides and Hordatines

As phenolic acid derivatives, except for *p*-coumaroylhydroxyagmatine and *p*-coumaroylputrescine, all hydroxycinnamic acid amides (HCAAs) were down-regulated in both infected states and at all time points (Table S4). Backes et al. [41] reported the accumulation of coumaroylhydroxyagmatine in barley after infection with *Ptt*. Adding a hydroxyl group on coumaroylagmatine may increase the compound's activity. Down-regulated HCAAs included the barley-specific metabolites hordatines A and B, as well as their precursor, *p*-coumaroylagmatine (Figure 7). The relative concentrations of these compounds were higher in the untreated control compared to infected plants (both naïve and primed), suggesting that their utilisation, conversion, or degradation is part of the defence against pathogen ingress.

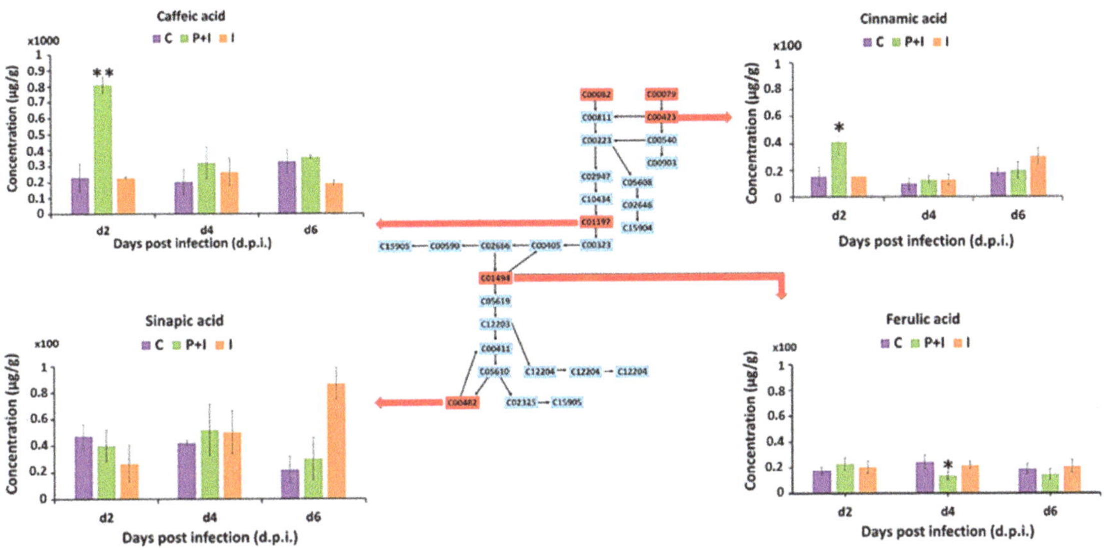

Figure 6. Illustration of the phenylpropanoid pathway and MRM quantification of main participants (cinnamic acid, caffeic acid, ferulic acid, and sinapic acid) in the primed and naïve shoot tissue of barley infected with the fungal pathogen *P. teres* f. *teres*. The bar graphs were generated using the average integrated peak area ($n = 9$). The error bars represent the standard deviation. Concentrations are expressed in µg g^{-1} (Mean +/− standard deviation; Table S2). The statistical significance of the observed differences was evaluated using Student's *t*-test, and *p* values < 0.05 (*) and 0.01 (**) were considered as significant.

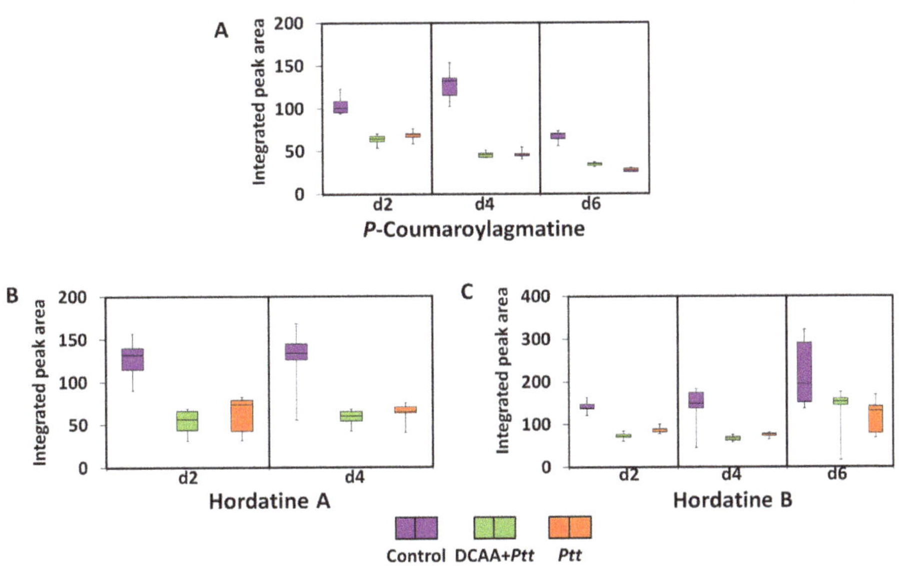

Figure 7. Box-and-whisker plots illustrating the relative quantitative profiles of discriminant barley-specific metabolites and a precursor resulting from priming treatment with 3,5-DCAA and infection with *Ptt*. (**A**): *p*-coumaroylagmatine, (**B**): hordatine A, and (**C**): hordatine B.

Several studies have revealed the pivotal role of HCAAs in plant–pathogen interactions [61–64]. HCAA biosynthesis emanates from the phenylpropanoid pathway, and the occurrence in plants is associated with plant disease resistance. They confer cellular protection against pathogen attack by reinforcing cell walls or by acting directly as antimicrobial agents [61,65]. Hordatine A and B have been recognised as strong antifungal compounds of barley as early as in the 1960s [66–69]. The presence of these compounds as signatory metabolites reiterate their function in plant defence; however, the down-regulation observed might be associated with the level of susceptibility of the plant to *Ptt* and the inability to produce sustainable levels that allow for the plant to overcome the pathogen attack. It is also possible that the pathogen might target this important component of inducible plant defence as part of effector-triggered susceptibility (ETS). Phenolic compounds may protect against fungal attack through the inhibition of chitin and glycan biosynthesis, which results in the fungal plasma membrane and cell wall disruption. Such action causes the inhibition of metabolic enzymes and mitochondrial processes [44,59].

3.3.5. Flavonoids

Another metabolite class derived from the phenylpropanoid pathway is the flavonoid pathway. Among these, a prenylated naringenin was down-regulated in the naïve infected plants at a late stage (6 d.p.i.) (Table S4). Naringenin is a central precursor in the biosynthesis of flavonoids. Derivatives of apigenin and luteolin were also found to be discriminative across time points (Figure 8). The relative concentrations of isovitexin decrease at 2 and 4 d.p.i, while that of its diglycosylated derivative, isovitexin 7,6″-di-O-glucoside, increase. The relative concentrations of lutonarin were down-regulated at the early infection stage but higher in primed plants compared to the naïve ones. Similarly, at 2 d.p.i., isoscoparin 7-O-[6″-sinapoyl]-glucoside was down-regulated, and its concentration was higher in the primed infected plants than in the naïve infected plants. At 4 d.p.i., there was an up-regulation of the metabolite, and again, the level seemed higher in the primed infected plants than the naïve infected plants. Together with the phenolic acids, flavonoids are the most conserved class of compounds and the largest group of phytochemicals. They are involved in various processes such as auxin transport, signalling, plant growth, and pigmentation. The main function ascribed to flavonoids is their antioxidant abilities, which help to quench reactive oxygen species (ROS) in the host and the pathogen during plant response [70–73]. Lutonarin isolated from barley was demonstrated to have inhibitory activity on free radicals [74]. The capacity of flavonoids to inhibit the growth of fungal spores and limit the elongation of mycelium hyphae is well recognised [75,76]. It is believed that flavonoids contribute to strengthening plant structures and function as a physical barrier against fungus invasion [76]. In wheat and maize, flavonoids are involved in resistance against *F. graminearum* and *F. verticillioides*, respectively [77,78]; this study marks the first time in the literature that their role in plant defence against the fungal pathogen *P. teres* f. *teres* has been highlighted.

3.3.6. Alkaloids

Furthermore, two important alkaloids in barley were also found and quantified using the MRM method as described in the Materials and Methods section: hordenine and gramine (Figure 9 and Table S2). The concentration of these metabolites fluctuated over time. However, hordenine concentration was significantly higher in the primed infected plants at 2 d.p.i. and in the naïve infected plants at 2 and 6 d.p.i. With gramine, higher concentrations were observed at 4 and 6 d.p.i. for all infected plant conditions. Hordenine, an isoquinoline alkaloid (second most impactful pathway; Figure 4A,B; Tables S5 and S6) and gramine, an indole alkaloid, were both reported as antifungal metabolites. Lovett and Hoult [79] demonstrated the inhibition capacity of both alkaloids against *Drechslera teres*, the anamorph stage of *Ptt*.

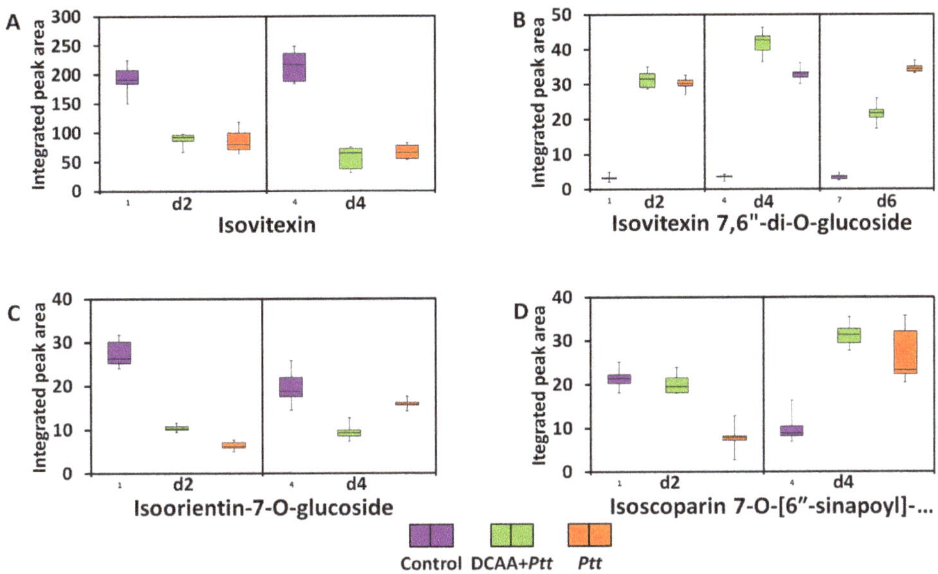

Figure 8. Relative quantitative profile of discriminant selected barley flavonoids (apigenin and luteolin derivatives) resulting from priming treatment with 3,5-DCAA and infection with *Ptt*. (**A**): isovitexin, (**B**): isovitexin 7,6″-di-O-glucoside, (**C**): isoorientin-7-O-glucoside (lutonarin), and (**D**): isoscoparin 7-O-[6″-sinapoyl]-glucoside.

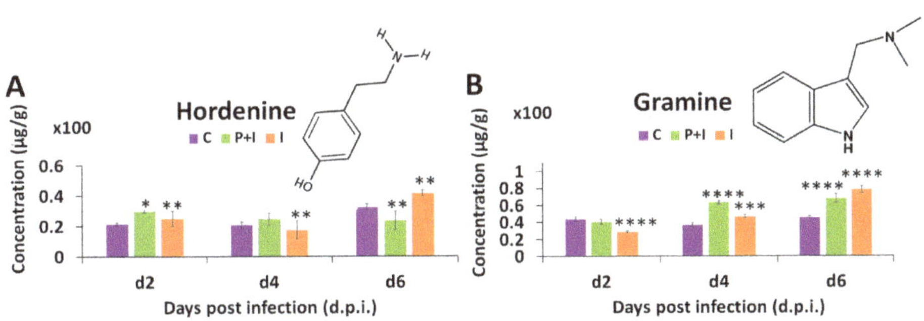

Figure 9. MRM quantification of hordenine and gramine from primed and naïve shoot tissues of barley infected with the fungal pathogen *P. teres* f. *teres*. The bar graphs were generated using the average integrated peak area (n = 9). The error bars represent the standard deviation. Concentrations are expressed in µg g^{-1} (Mean +/− standard deviation; Table S2). The statistical significance of the differences observed was evaluated using Student's t-test, and p values < 0.05 (*), 0.01 (**), 0.001 (***), and 0.0001 (****) were considered as significant.

N-isoferuloylhydroxytryptamine and *p*-coumaroyltryptamine were also annotated as discriminant metabolites as they were up-regulated in both primed and naïve infected plants at 2 and 6 d.p.i. These are neutral HCAAs that are generally classified as alkaloids. Together with hydroxytryptamine (up-regulated at 2 d.p.i. in the primed infected plants), they were not previously identified as constituent metabolites in barley cultivar profiling [23,26]. However, in a recent study where DCAA and other dichlorinated compounds were used to induce systemic acquired resistance in barley, they were signatory biomarkers that accumulated in shoot tissues in response to treatment [22]. This points to their antifungal function in plant defence and their role as phytoalexins in barley. The antimicrobial activity of tryptamine derivatives was also described in barley under attack by *Fusarium culmorum*, responsible for Fusarium root rot disease [80], and the build-up of *p*-coumaroyltryptamine was noted in *A. thaliana* leaf tissue after 24 h of infection with *Pseudomonas syringae* [81].

3.4. Mapped Metabolic Networks

Metabolic (correlation) network analysis enables the characterisation of the complex relationship(s) in measured metabolites. Thus, in addition to the pathway enrichment analyses (Figure 4), a network demonstrating metabolic interaction patterns between metabolites identified by OPLS-DA (comparing the naïve and primed infected states) was constructed to understand the possible regulatory elements of the metabolic responses of primed and naïve plants to *Ptt* at the early infection stage (2 d.p.i.) (Figure 10). MetaMapp analysis makes use of statistical data such as fold changes and *p* values to construct metabolic networks showing significantly different metabolites in primed infected samples (green) and naïve infected plants (orange) [47]. Relational patterns in the experimental data are depicted by metabolic networks that identify altered graph neighbourhoods. Since these relationships are not dependent on predefined pathways, they allow for the description of the altered metabolomes established by new pathway connections as a result of the experimental conditions. For example, the agmatines and the hordatines are found in one cluster, indicating the structural relatedness, further affirming that the agmatines serve as precursors to the biosynthesis of the hordatines.

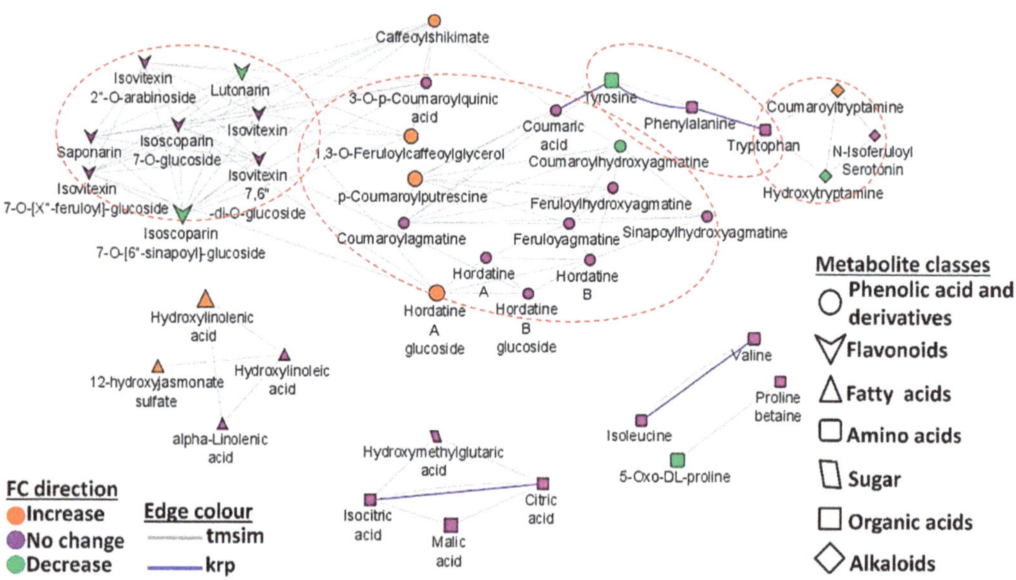

Figure 10. Mapped metabolic networks of all annotated metabolites identified via OPLS-DA analysis based on 2 d.p.i. data. The green nodes represent the metabolites that are up-regulated in the primed infected plants, and the orange ones represent those that are up-regulated in the naïve infected plants. There is an overlap between the class of phenolic acids and derivatives and that of flavonoids. The node size reflects the magnitude of the fold change (FC). The blue lines represent the connections of compounds via KEGG reaction pair (krp), and the grey lines connects the nodes (metabolites) based on their chemical similarity.

In addition, in the network topology, phenolic acids, flavonoids, and alkaloids, derived from caffeoylshikimate and amino acids such as Tyr, Phe, and Trp, appear to form the core hub. Isolated interaction could also be responsible for forming the fatty acid, organic acid, and other amino acid (proline betaine, oxo-proline, valine, and isoleucine) clusters. In addition, the biological interaction observed between Tyr, Phe, and Trp highlights the strong connection between these metabolites and reiterates their combined role as precursors of a broad spectrum of secondary metabolites. It can also be proposed that there is alignment between the production and utilisation of the three aromatic amino acids in regulating the production of these defensive secondary metabolites.

4. Conclusions

The mechanism(s) underlying plant priming is still very unclear at a biochemical level; however, two of the main characteristics are the early and long-lasting responses. This study focused on evaluating the dynamic intricacies of the metabolic changes in barley plants primarily preconditioned with 3,5-DCAA in response to a secondary stress—infection by *P. teres* f. *teres*. The primed plants were first characterised by a delay in the development of the distinctive visual symptoms associated with *P. teres* f. *teres* infection. A strong and rapid response was observed, portraying a state of alertness. However, this response was not maintained over time as the pathogen apparently managed to overcome the induced resistance.

These observations were further investigated using untargeted metabolomics (in combination with chemometric tools) and targeted metabolomic approaches which provided insights into the reconfigured primed and natural response of barley to *Ptt*. Metabolites and the measurement of dynamic changes in their profiles informatively reflected the differential and functional features of the naïve and 3,5-DCAA-treated plants' metabolism in response to *Ptt* inoculation and subsequent infection. Both primary and secondary metabolism were affected, and one of the most significant pathways affected was the phenylpropanoid pathway. Higher concentrations of the aromatic amino acids Phe, Tyr, and Trp were found to be associated with the early responses in the barley-infected shoot tissues. Because the hordatine metabolites are well-known antifungal agents in barley plants, their down-regulation here may be linked to plant susceptibility due to the inability to synthesise sufficient amounts to repel the attack.

The variations observed at the beginning of infection (2 d.p.i.) in the primed plants were linked to the early and stronger response typical of the post-primed phase. The metabolites responsible for such delays and related to the priming response were highlighted; e.g., isovitexin-7,6″-diglucoside accumulated and was more prevalent in the primed plants at 2 and 4 d.p.i. Others included cinnamic and caffeic acids, citric acid, and oxo-proline. These findings support the speculation on the ability of 3,5-DCAA as a priming agent to enhance the defence response. Despite the aforementioned metabolic reconfiguration of the primed barley, the defence-related metabolite production therein was not sustainable to ward-off the attack. The secondary effects brought on by cell death may also impact and obfuscate the observable patterns in the data. The data obtained under the experimental conditions did not allow for the depiction of a possible mechanism underlying priming by 3,5-DCAA. However, the obtained descriptive insights serve as a departure point for further research on the use of dichlorinated inducers as priming agents in barley and other plants to manipulate/enhance the defence response to fungal attack and subsequent infection.

Supplementary Materials: The following are available online at https://www.mdpi.com/article/10.3390/metabo13090997/s1: Figure S1. Disease triangle illustrating factors contributing to the progression of disease. Figure S2. Preliminary screening of cultivars from the Western Cape region of South Africa. Figure S3. Fungal growth before (A) and after (B) conidia induction under near-UV light (long wave). Figure S4. Ultra-high performance liquid chromatography—mass spectrometry (UHPLC–MS) base peak intensity (BPI) chromatograms (negative and positive ionisation) of barley treated with 3,5-DCAA and infected with *P. teres* f. *teres* and evaluated over 2, 4, and 6 d.p.i. Figure S5. Principal component analysis (PCA) score plots of ESI (+) data from shoot extracts of the 'Hessekwa' cultivar of *Hordeum vulgare*. Table S1. Optimum parameters for UFLC–MRM–MS quantitative analysis listing targeted standards and standard curve equation for quantification. Table S2. Absolute quantification of selected metabolites in extracts from shoot tissues of primed and naïve barley plants following infection with *P. teres* f. *teres*. Table S3. Annotated metabolites used for correlation network analyses. Discriminant metabolites were extracted from OPLS-DA, comparing the naïve infected (reference) vs. primed infected metabolites (Condition_A). Table S4. List of all annotated (putatively identified) discriminant metabolites from leaves of the barley cultivar 'Hessekwa' (treated/untreated with 3,5-DCAA and infected with *P. teres* f. *teres* and harvested at 2, 4, and 6 d.p.i). Table S5. Metabolic pathways generated from Metabolomics Pathway Analysis (MetPA) in MetaboAnalyst 5.0 and involving annotated metabolites in primed infected barley plants. Table S6. Metabolic pathways generated from Metabolomics Pathway Analysis (MetPA) in MetaboAnalyst 5.0 and involving annotated metabolites in naïve infected barley plants.

Author Contributions: Conceptualisation: I.A.D.; methodology: C.Y.H.D., F.T. and P.A.S.; formal analysis: C.Y.H.D. and P.A.S.; investigation: C.Y.H.D. and I.A.D.; data curation: C.Y.H.D.; writing—original draft preparation: C.Y.H.D.; writing—review and editing: I.A.D., F.T. and L.A.P.; supervision: I.A.D. All authors have read and agreed to the published version of the manuscript.

Funding: This research received no external funding.

Institutional Review Board Statement: Not applicable.

Informed Consent Statement: Not applicable.

Data Availability Statement: The study design information, LC-MS data, data processing, and analyses are reported on and incorporated into the main text. Raw data, analyses and data processing information, and the meta-data were deposited into the Mass Spectrometry Interactive Virtual Environment, MassiVE (https://massive.ucsd.edu) at https://massive.ucsd.edu/MSV000092480/ (19 July 2023).

Acknowledgments: We would like to thank the South African Barley Breeding Institute (SABBI) for providing the cultivar seeds used in our study.

Conflicts of Interest: The authors declare no conflict of interest.

References

1. Ngou, B.P.M.; Ding, P.; Jones, J.D.G. Thirty years of resistance: Zig-zag through the plant immune system. *Plant Cell* **2022**, *34*, 1447–1478. [CrossRef]
2. Balmer, A.; Pastor, V.; Gamir, J.; Flors, V.; Mauch-Mani, B. The 'prime-ome': Towards a holistic approach to priming. *Trends Plant Sci.* **2015**, *20*, 443–452. [CrossRef]
3. Djami-Tchatchou, A.T.; Maake, M.P.; Piater, L.A.; Dubery, I.A. Isonitrosoacetophenone drives transcriptional reprogramming in *Nicotiana tabacum* cells in support of defense. *PLoS ONE* **2015**, *10*, e0117377. [CrossRef] [PubMed]
4. Ellwood, S.R. Transposable element genomic fissuring in *Pyrenophora teres* is associated with genome expansion and dynamics of host–pathogen genetic interactions. *Front. Genet.* **2018**, *9*, 130.
5. Liu, Z.; Ellwood, S.R.; Oliver, R.P.; Friesen, T.L. *Pyrenophora teres*: Profile of an increasingly damaging barley pathogen. *Mol. Plant Pathol.* **2011**, *12*, 1–19. [CrossRef]
6. Wallwork, H.; Butt, M.; Capio, E. Pathogen diversity and screening for minor gene resistance to *Pyrenophora teres* f. *teres* in barley and its use for plant breeding. *Australas. Plant Pathol.* **2016**, *45*, 527–531. [CrossRef]
7. Backes, A.; Hausman, J.F.; Renaut, J.; Ait Barka, E.; Jacquard, C.; Guerriero, G. Expression analysis of cell wall-related genes in the plant pathogenic fungus *Drechslera teres*. *Genes* **2020**, *11*, 300. [CrossRef]
8. Clare, S.J.; Wyatt, N.A.; Brueggeman, R.S.; Friesen, T.L. Research advances in the *Pyrenophora teres*–barley interaction. *Mol. Plant Pathol.* **2020**, *21*, 272–288. [CrossRef] [PubMed]
9. González-Domínguez, E.; Fedele, G.; Salinari, F.; Rossi, V. A general model for the effect of crop management on plant disease epidemics at different scales of complexity. *Agronomy* **2020**, *10*, 462. [CrossRef]
10. Islam, W. Plant disease epidemiology: Disease triangle and forecasting mechanisms in highlights. *Hosts Viruses* **2018**, *5*, 7–11. [CrossRef]
11. Conrath, U.; Beckers, G.J.M.; Langenbach, C.J.G.; Jaskiewicz, M.R. Priming for enhanced defense. *Annu. Rev. Phytopathol.* **2015**, *53*, 97–119. [CrossRef]
12. Pastor, V.; Luna, E.; Mauch-Mani, B.; Ton, J.; Flors, V. Primed plants do not forget. *Environ. Exp. Bot.* **2013**, *94*, 46–56. [CrossRef]
13. Martinez-Medina, A.; Flors, V.; Heil, M.; Mauch-Mani, B.; Pieterse, C.M.; Pozo, M.J.; Ton, J.; van Dam, N.M.; Conrath, U. Recognizing plant defense priming. *Trends Plant Sci.* **2016**, *21*, 818–822. [CrossRef] [PubMed]
14. De Kesel, J.; Conrath, U.; Flors, V.; Luna, E.; Mageroy, M.H.; Mauch-Mani, B.; Pastor, V.; Pozo, M.J.; Pieterse, C.M.; Ton, J.; et al. The induced resistance lexicon: Do's and don'ts. *Trends Plant Sci.* **2021**, *26*, 685–691. [CrossRef] [PubMed]

15. Hilker, M.; Schwachtje, J.; Baier, M.; Balazadeh, S.; Bäurle, I.; Geiselhardt, S.; Hincha, D.K.; Kunze, R.; Mueller-Roeber, B.; Rillig, M.C.; et al. Priming and memory of stress responses in organisms lacking a nervous system. *Biol. Rev.* **2016**, *91*, 1118–1133. [CrossRef] [PubMed]
16. Schwachtje, J.; Whitcomb, S.J.; Firmino, A.A.P.; Zuther, E.; Hincha, D.K.; Kopka, J. Induced, imprinted, and primed responses to changing environments: Does metabolism store and process information? *Front. Plant Sci.* **2019**, *10*, 106. [CrossRef]
17. Tugizimana, F.; Mhlongo, M.I.; Piater, L.A.; Dubery, I.A. Metabolomics in plant priming research: The way forward? *Int. J. Mol. Sci.* **2018**, *19*, 1759.
18. Allwood, J.W.; Heald, J.; Lloyd, A.J.; Goodacre, R.; Mur, L.A. Separating the Inseparable: The Metabolomic Analysis of Plant–Pathogen Interactions. In *Plant Metabolomics. Methods in Molecular Biology*; Hardy, N., Hall, R., Eds.; Humana Press: Totowa, NJ, USA, 2012; Volume 860, pp. 31–49. [CrossRef]
19. Hamany Djande, C.Y.; Steenkamp, P.A.; Piater, L.A.; Tugizimana, F.; Dubery, I.A. Metabolic Reprogramming of Barley in Response to Foliar Application of Dichlorinated Functional Analogues of Salicylic Acid as Priming Agents and Inducers of Plant Defence. *Metabolites* **2023**, *13*, 666. [CrossRef]
20. Balmer, D.; Flors, V.; Glauser, G.; Mauch-Mani, B. Metabolomics of cereals under biotic stress: Current knowledge and techniques. *Front. Plant Sci.* **2013**, *4*, 82. [CrossRef] [PubMed]
21. Castro-Moretti, F.R.; Gentzel, I.N.; Mackey, D.; Alonso, A.P. Metabolomics as an emerging tool for the study of plant–pathogen interactions. *Metabolites* **2020**, *10*, 52. [CrossRef] [PubMed]
22. Mhlongo, M.I.; Piater, L.A.; Madala, N.E.; Labuschagne, N.; Dubery, I.A. The chemistry of plant–microbe interactions in the rhizosphere and the potential for metabolomics to reveal signaling related to defense priming and induced systemic resistance. *Front. Plant Sci.* **2018**, *9*, 112. [CrossRef] [PubMed]
23. Hamany Djande, C.Y.; Piater, L.A.; Steenkamp, P.A.; Tugizimana, F.; Dubery, I.A. A metabolomics approach and chemometric tools for differentiation of barley cultivars and biomarker discovery. *Metabolites* **2021**, *11*, 578. [CrossRef]
24. Muria-Gonzalez, M.J.; Zulak, K.G.; Allegaert, E.; Oliver, R.P.; Ellwood, S.R. Profile of the *in vitro* secretome of the barley net blotch fungus, *Pyrenophora teres* f. *teres*. *Physiol. Mol. Plant Pathol.* **2020**, *109*, 101451. [CrossRef]
25. Eriksson, L.; Trygg, J.; Wold, S. CV-ANOVA for significance testing of PLS and OPLS®models. *J. Chemom.* **2008**, *22*, 594–600. [CrossRef]
26. Hamany Djande, C.Y.; Steenkamp, P.A.; Piater, L.A.; Tugizimana, F.; Dubery, I.A. Hordatines and associated precursors dominate metabolite profiles of barley (*Hordeum vulgare* L.) seedlings: A metabolomics study of five cultivars. *Metabolites* **2022**, *12*, 310. [CrossRef]
27. Sumner, L.W.; Amberg, A.; Barrett, D.; Beale, M.H.; Beger, R.; Daykin, C.A.; Fan, T.W.M.; Fiehn, O.; Goodacre, R.; Griffin, J.L.; et al. Proposed minimum reporting standards for chemical analysis: Chemical analysis working group (CAWG) metabolomics standards initiative (MSI). *Metabolomics* **2007**, *3*, 211–221. [CrossRef]
28. PubChem. Available online: https://pubchem.ncbi.nlm.nih.gov (accessed on 28 April 2022).
29. ChemSpider. Available online: http://www.chemspider.com/ (accessed on 3 March 2022).
30. Lipidmaps. Available online: https://www.lipidmaps.org (accessed on 18 March 2022).
31. Dictionary of Natural Products. Available online: www.dnp.chemnetbase.com (accessed on 10 January 2022).
32. Xia, J.; Wishart, D.S. MetPA: A web-based metabolomics tool for pathway analysis and visualization. *Bioinformatics* **2010**, *26*, 2342–2344. [CrossRef]
33. Chong, J.; Xia, J. MetaboAnalystR: An R package for flexible and reproducible analysis of metabolomics data. *Bioinformatics* **2018**, *34*, 4313–4314. [CrossRef] [PubMed]
34. Kyoto Encyclopedia of Genes and Genomics. Available online: https://www.genome.jp (accessed on 20 February 2022).
35. Barupal, D.K.; Haldiya, P.K.; Wohlgemuth, G.; Kind, T.; Kothari, S.L.; Pinkerton, K.E.; Fiehn, O. MetaMapp: Mapping and visualizing metabolomic data by integrating information from biochemical pathways and chemical and mass spectral similarity. *BMC Bioinform.* **2012**, *13*, 1–15. [CrossRef]
36. Cytoscape. Available online: https://cytoscape.org. (accessed on 20 February 2022).
37. Tekauz, A. A numerical scale to classify reactions of barley to *Pyrenophora teres*. *Can. J. Plant Pathol.* **1985**, *7*, 181–183. [CrossRef]
38. Lightfoot, D.J.; Able, A.J. Growth of *Pyrenophora teres* in planta during barley net blotch disease. *Australas. Plant Pathol.* **2010**, *39*, 499–507. [CrossRef]
39. Masi, M.; Zorrilla, J.G.; Meyer, S. Bioactive metabolite production in the genus Pyrenophora (Pleosporaceae, Pleosporales). *Toxins* **2022**, *14*, 588. [CrossRef] [PubMed]
40. Ismail, I.A.; Able, A.J. Gene expression profiling of virulence-associated proteins in planta during net blotch disease of barley. *Physiol. Mol. Plant Pathol.* **2017**, *98*, 69–79. [CrossRef]
41. Backes, A.; Guerriero, G.; Barka, E.A.; Jacquard, C. *Pyrenophora teres*: Taxonomy, morphology, interaction with barley, and mode of control. *Front. Plant Sci.* **2021**, *12*, 614951. [CrossRef] [PubMed]
42. Balmer, A.; Pastor, V.; Glauser, G.; Mauch-Mani, B. Tricarboxylates induce defense priming against bacteria in *Arabidopsis thaliana*. *Front. Plant Sci.* **2018**, *9*, 1221. [CrossRef]
43. Wang, M.; Ding, Y.; Wang, Q.; Wang, P.; Han, Y.; Gu, Z.; Yang, R. NaCl treatment on physio-biochemical metabolism and phenolics accumulation in barley seedlings. *Food Chem.* **2020**, *331*, 127282. [CrossRef]

44. Pratyusha, S. Phenolic Compounds in the Plant Development and Defense: An Overview. In *Plant Stress Physiology—Perspectives in Agriculture*; Hasanuzzaman, M., Nahar, K., Eds.; IntechOpen: London, UK, 2022; p. 102873.
45. Mhlongo, M.I.; Piater, L.A.; Steenkamp, P.A.; Labuschagne, N.; Dubery, I.A. Metabolomic evaluation of tissue-specific defense responses in tomato plants modulated by PGPR-priming against *Phytophthora capsici* infection. *Plants* **2021**, *8*, 1530. [CrossRef]
46. Tugizimana, F.; Steenkamp, P.A.; Piater, L.A.; Labuschagne, N.; Dubery, I.A. Unravelling the metabolic reconfiguration of the post-challenge primed state in *Sorghum bicolor* responding to *Colletotrichum sublineolum* infection. *Metabolites* **2019**, *9*, 194. [CrossRef]
47. Othibeng, K.; Nephali, L.; Ramabulana, A.T.; Steenkamp, P.; Petras, D.; Kang, K.B.; Opperman, H.; Huyser, J.; Tugizimana, F. A metabolic choreography of maize plants treated with a humic substance-based biostimulant under normal and starved conditions. *Metabolites* **2021**, *11*, 403. [CrossRef]
48. Guidea, A.; Zăgrean-Tuza, C.; Moț, A.C.; Sârbu, C. Comprehensive evaluation of radical scavenging, reducing power and chelating capacity of free proteinogenic amino acids using spectroscopic assays and multivariate exploratory techniques. *Spectrochim. Acta Part A Mol. Biomol. Spectrosc.* **2020**, *233*, 118158. [CrossRef]
49. El Moukhtari, A.; Cabassa-Hourton, C.; Farissi, M.; Savouré, A. How does proline treatment promote salt stress tolerance during crop plant development? *Front. Plant Sci.* **2020**, *11*, 1127. [CrossRef] [PubMed]
50. Hildebrandt, T.M.; Nunes Nesi, A.; Araújo, W.L.; Braun, H.P. Amino acid catabolism in plants. *Mol. Plant* **2015**, *8*, 1563–1579. [CrossRef] [PubMed]
51. da Camara, N.; Dubery, I.A.; Piater, L.A. Proteome Analysis of *Nicotiana tabacum* Cells following Isonitrosoacetophenone Treatment Reveals Defence-Related Responses Associated with Priming. *Plants* **2023**, *12*, 1137. [CrossRef]
52. Pretorius, C.J.; Zeiss, D.R.; Dubery, I.A. The presence of oxygenated lipids in plant defence in response to biotic stress: A metabolomics appraisal. *Plant Signal. Behav.* **2021**, *16*, 1989215. [CrossRef]
53. Walley, J.W.; Kliebenstein, D.J.; Bostock, R.M.; Dehesh, K. Fatty acids and early detection of pathogens. *Curr. Opin. Plant Biol.* **2013**, *16*, 520–526. [CrossRef] [PubMed]
54. Avis, T.J.; Bélanger, R.R. Specificity and mode of action of the antifungal fatty acid cis-9-heptadecenoic acid produced by *Pseudozyma flocculosa*. *Appl. Environ. Microbiol.* **2001**, *67*, 956–960. [CrossRef] [PubMed]
55. Guimarães, A.; Venâncio, A. The potential of fatty acids and their derivatives as antifungal agents: A review. *Toxins* **2022**, *14*, 188. [CrossRef]
56. Walters, D.; Raynor, L.; Mitchell, A.; Walker, R.; Walker, K. Antifungal activities of four fatty acids against plant pathogenic fungi. *Mycopathologia* **2004**, *157*, 87–90.
57. Atta, B.M.; Saleem, M.; Ali, H.; Bilal, M.; Fayyaz, M. Application of Fluorescence Spectroscopy in Wheat Crop: Early Disease Detection and Associated Molecular Changes. *J. Fluoresc.* **2020**, *30*, 801–810. [CrossRef]
58. Aoun, M. Host defense mechanisms during fungal pathogenesis and how these are overcome in susceptible plants: A review. *Int. J. Bot.* **2017**, *13*, 82–102. [CrossRef]
59. Wallis, C.M.; Galarneau, E.R.A. Phenolic compound induction in plant-microbe and plant-insect interactions: A meta-analysis. *Front. Plant Sci.* **2020**, *11*, 580753. [CrossRef] [PubMed]
60. Dehghanian, Z.; Habibi, K.; Dehghanian, M.; Aliyar, S.; Lajayer, B.A.; Astatkie, T.; Minkina, T.; Keswani, C. Reinforcing the bulwark: Unravelling the efficient applications of plant phenolics and tannins against environmental stresses. *Heliyon* **2022**, *1*, e09094. [CrossRef] [PubMed]
61. Zeiss, D.R.; Piater, L.A.; Dubery, I.A. Hydroxycinnamate amides: Intriguing conjugates of plant protective metabolites. *Trends Plant Sci.* **2021**, *26*, 184–195. [CrossRef] [PubMed]
62. Hu, C.; Chen, P.; Zhou, X.; Li, Y.; Ma, K.; Li, S.; Liu, H.; Li, L. Arms Race between the Host and Pathogen Associated with Fusarium Head Blight of Wheat. *Cells* **2022**, *11*, 2275. [CrossRef] [PubMed]
63. Liu, S.; Jiang, J.; Ma, Z.; Xiao, M.; Yang, L.; Tian, B.; Yu, Y.; Bi, C.; Fang, A.; Yang, Y. The Role of Hydroxycinnamic Acid Amide Pathway in Plant Immunity. *Front. Plant Sci.* **2022**, *13*, 922119. [CrossRef] [PubMed]
64. Liu, S.; Xie, L.; Su, J.; Tian, B.; Fang, A.; Yu, Y.; Bi, C.; Yang, Y. Integrated Metabolo-transcriptomics Reveals the Defense Response of Homogentisic Acid in Wheat against *Puccinia striiformis* f. sp. *tritici*. *J. Agric. Food Chem.* **2022**, *70*, 3719–3729. [CrossRef]
65. Campos, L.; Lison, P.; López-Gresa, M.P.; Rodrigo, I.; Zacares, L.; Conejero, V.; Bellés, J.M. Transgenic tomato plants overexpressing tyramine N-hydroxycinnamoyltransferase exhibit elevated hydroxycinnamic acid amide levels and enhanced resistance to *Pseudomonas syringae*. *Mol. Plant-Microbe Interact.* **2014**, *27*, 1159–1169. [CrossRef]
66. Stoessl, A. The antifungal factors in barley. IV. Isolation, structure, and synthesis of the hordatines. *Can. J. Chem.* **1967**, *45*, 1745–1760. [CrossRef]
67. Bollina, V.; Kushalappa, A.C.; Choo, T.M.; Dion, Y.; Rioux, S. Identification of metabolites related to mechanisms of resistance in barley against *Fusarium graminearum*, based on mass spectrometry. *Plant Mol. Biol.* **2011**, *77*, 355–370. [CrossRef]
68. Stoessl, A.; Unwin, C.H. The antifungal factors in barley. V. antifungal activity of the hordatines. *Can. J. Bot.* **1970**, *48*, 465–470. [CrossRef]
69. Mikkelsen, B.L.; Olsen, C.E.; Lyngkjær, M.F. Accumulation of secondary metabolites in healthy and diseased barley, grown under future climate levels of CO2, ozone and temperature. *Phytochemistry* **2015**, *118*, 162–173. [CrossRef] [PubMed]
70. Jung, M.J.; Heo, S.I.; Wang, M.H. HPLC analysis and antioxidant activity of *Ulmus davidiana* and some flavonoids. *Food Chem.* **2010**, *120*, 313–318. [CrossRef]

71. Agati, G.; Azzarello, E.; Pollastri, S.; Tattini, M. Flavonoids as antioxidants in plants: Location and functional significance. *Plant Sci.* **2012**, *196*, 67–76. [CrossRef]
72. Bag, S.; Mondal, A.; Majumder, A.; Mondal, S.K.; Banik, A. Flavonoid mediated selective cross-talk between plants and beneficial soil microbiome. *Phytochem. Rev.* **2022**, *21*, 1739–1760. [CrossRef] [PubMed]
73. Wu, J.; Lv, S.; Zhao, L.; Gao, T.; Yu, C.; Hu, J.; Ma, F. Advances in the study of the function and mechanism of the action of flavonoids in plants under environmental stresses. *Planta* **2023**, *257*, 108. [CrossRef]
74. Koga, R.; Meng, T.; Nakamura, E.; Miura, C.; Irino, N.; Yahara, S.; Kondo, R. Model examination for the effect of treading stress on young green barley (*Hordeum vulgare*). *Am. J. Plant Sci.* **2013**, *4*, 2013. [CrossRef]
75. Khlestkina, E. The adaptive role of flavonoids: Emphasis on cereals. *Cereal Res. Commun.* **2013**, *41*, 185–198. [CrossRef]
76. Mierziak, J.; Kostyn, K.; Kulma, A. Flavonoids as important molecules of plant interactions with the environment. *Molecules* **2014**, *19*, 16240–16265. [CrossRef]
77. Buśko, M.; Góral, T.; Ostrowska, A.; Matysiak, A.; Walentyn-Góral, D.; Perkowski, J. The effect of Fusarium inoculation and fungicide application on concentrations of flavonoids (apigenin, kaempferol, luteolin, naringenin, quercetin, rutin, vitexin) in winter wheat cultivars. *Am. J. Plant Sci.* **2014**, *5*, 3727–3736. [CrossRef]
78. Venturini, G.; Toffolatti, S.L.; Assante, G.; Babazadeh, L.; Campia, P.; Fasoli, E.; Salomoni, D.; Vercesi, A. The influence of flavonoids in maize pericarp on Fusarium ear rot symptoms and fumonisin accumulation under field conditions. *Plant Pathol.* **2015**, *64*, 671–679. [CrossRef]
79. Lovett, J.V.; Hoult, A.H.C. Biological activity of barley secondary metabolites. In Proceeding of the 7th Australian Society of Agronomy Conference, Adelaide, Australia, 19–24 September 1993; pp. 158–161.
80. Ube, N.; Yabuta, Y.; Tohnooka, T.; Ueno, K.; Taketa, S.; Ishihara, A. Biosynthesis of phenylamide phytoalexins in pathogen-infected barley. *Int. J. Mol. Sci.* **2019**, *20*, 5541. [CrossRef] [PubMed]
81. Macoy, D.M.J.; Uddin, S.; Ahn, G.; Peseth, S.; Ryu, G.R.; Cha, J.Y.; Lee, J.Y.; Bae, D.; Paek, S.M.; Chung, H.J.; et al. Effect of Hydroxycinnamic Acid Amides, Coumaroyl Tyramine and Coumaroyl Tryptamine on Biotic Stress Response in Arabidopsis. *J. Plant Biol.* **2022**, *65*, 145–155. [CrossRef]

metabolites

Article

Metabolic Rewiring in Tea Plants in Response to Gray Blight Disease Unveiled by Multi-Omics Analysis

Shiqin Zheng [1,2,3], Zhenghua Du [2], Xiaxia Wang [2], Chao Zheng [2], Zonghua Wang [3,4,*] and Xiaomin Yu [2,*]

1 Tea Research Institute, Fujian Academy of Agricultural Sciences, Fuzhou 350013, China; azhyeah@163.com
2 Center for Plant Metabolomics, Haixia Institute of Science and Technology, Fujian Agriculture and Forestry University, Fuzhou 350002, China; zhenghuadu@fafu.edu.cn (Z.D.); wangxiaxia530@fafu.edu.cn (X.W.); zhengchaotea@fafu.edu.cn (C.Z.)
3 State Key Laboratory of Ecological Pest Control for Fujian and Taiwan Crops, College of Plant Protection, Fujian Agriculture and Forestry University, Fuzhou 350002, China
4 Fuzhou Institute of Oceanography, Minjiang University, Fuzhou 350108, China
* Correspondence: wangzh@fafu.edu.cn (Z.W.); xmyu0616@fafu.edu.cn (X.Y.)

Abstract: Gray blight disease, which is caused by *Pestalotiopsis*-like species, poses significant challenges to global tea production. However, the comprehensive metabolic responses of tea plants during gray blight infection remain understudied. Here, we employed a multi-omics strategy to characterize the temporal transcriptomic and metabolomic changes in tea plants during infection by *Pseudopestalotiopsis theae*, the causal agent of gray blight. Untargeted metabolomic profiling with ultra-performance liquid chromatography–quadrupole time-of-flight mass spectrometry (UPLC-QTOFMS) revealed extensive metabolic rewiring over the course of infection, particularly within 24 h post-inoculation. A total of 64 differentially accumulated metabolites were identified, including elevated levels of antimicrobial compounds such as caffeine and (−)-epigallocatechin 3-gallate, as well as oxidative catechin polymers like theaflavins, theasinensins and theacitrins. Conversely, the synthesis of (+)-catechin, (−)-epicatechin, oligomeric proanthocyanidins and flavonol glycosides decreased. Integrated omics analyses uncovered up-regulation of phenylpropanoid, flavonoid, lignin biosynthesis and down-regulation of photosynthesis in response to the pathogen stress. This study provides novel insights into the defense strategies of tea plants against gray blight disease, offering potential targets for disease control and crop improvement.

Keywords: tea plant; gray blight disease; flavonoid; lignin; defense-related metabolites; metabolomics; transcriptomics

1. Introduction

The tea plant (*Camellia sinensis*) is an important cash crop cultivated primarily for its leaves, which are used as raw material for tea production. China is the world's largest tea producer and exporter, with an annual production of tea leaves exceeding 13 million tons in 2021, supporting the livelihoods of millions in the tea industry [1]. However, tea production faces persistent challenges from fungal diseases [2]. *Pestalotiopsis-like* species, known as the causal agents of gray blight disease in tea plants, are a highly destructive group of phytopathogens [3].

Pestalotiopsis-like species, which are classified into *Pseudopestalotiopsis*, *Neopestalotiopsis* and *Pestalotiopsis*, infect both tender shoots and mature leaves [4]. The initial symptoms of gray blight appear as small brown concentric spots on wounded leaves, which gradually expand into large necrotic lesions with black and brown colors. In severe cases, these lesions spread throughout the leaf, leading to defoliation [5]. Disease severity is exacerbated under warm and humid conditions, negatively impacting tea yield and quality [2]. Various *Pestalotiopsis*-like species have caused significant yield losses in major tea-producing countries like China, India and Japan [6–8]. Current disease management relies heavily on synthetic fungicides such as methyl benzimidazole carbamates and dithiocarbamates. However, these approaches come with high costs, environmental concerns and an increased risk of fungal resistance development [2,9,10]. Therefore, developing effective control measures requires a profound understanding of the defense mechanisms employed by the tea plant against the causal agent of gray blight disease.

Plants have evolved a sophisticated innate immune system to fend off pathogens through multilayered defense responses [11]. Physical structures such as thorns, trichomes, rigid cell walls and waxy cuticles form the first line of defense by preventing pathogen entry. In addition to these physical barriers, plants can recognize invading pathogens and initiate complex signaling cascades that trigger the targeted production of pathogenesis-related proteins, enzymes and specialized metabolites [12]. Among these metabolites are antimicrobial compounds, including constitutively expressed phytoanticipins and induced phytoalexins synthesized de novo in response to infection. These metabolites can directly inhibit microbial growth or elicit additional immune responses, serving to contain pathogens within the infected plant tissues [13]. Tea plants (*Camellia sinensis*), in particular, reply on the production of specialized metabolites (e.g., phenolics, alkaloids and terpenes) as a key defense strategy against biotic and abiotic stresses [14]. Phenolic compounds, such as (−)-epigallocatechin (EGC), (−)-epicatechin (EC) and their gallate esters, play a crucial role in leaf resistance. Among these phenolics, (−)-epigallocatechin 3-gallate (EGCG), the most abundant catechin in fresh tea leaves, is particularly noteworthy. EGCG has been shown to possess potent antibacterial, antifungal and antiviral properties [15,16]. Through a multifaceted mechanism of action, EGCG effectively combats fungi via disruption of cell structures, inhibition of key enzymes and synergistic potentiation of antifungal drugs, enabling robust activity against fungal invaders [16]. When infected by pathogens such as *Colletotrichum fructicola* or *Pseudopestalotiopsis* isolates, tea plants up-regulate EGCG or other catechin levels in correlation with enhanced resistance [17,18]. Nonetheless, the full scope of the specialized metabolic adaptation of tea plants during infection remains poorly understood.

In our previous study, we identified a novel pathogenic strain, *Pseudopestalotiopsis theae* CYF27, as the cause of gray blight disease in tea plants [19]. To further explore the molecular responses of tea plants to *P. theae* infection, we conducted an integrated multi-omics analysis. Through untargeted metabolomics, we characterized the temporal dynamics of the tea plant metabolome during *P. theae* infection. Additionally, through RNA sequencing, we uncovered transcriptional reprogramming events that underlie the immune response. This integrated omics approach provides new perspectives on the intricate host–pathogen interactions over time and may help inform the development of sustainable disease control strategies.

2. Materials and Methods

2.1. Chemical and Reagents

MS-grade acetonitrile and methanol were purchased from Thermo Fisher Scientific, Inc. (Pittsburgh, PA, USA), and formic acid was obtained from Honeywell Fluka (Seelze, Germany). Ultra-pure water was prepared with a Milli-Q purification system (Millipore, Bedford, MA, USA). Reference compounds were obtained from Sigma-Aldrich (St. Louis, MO, USA), ChemFaces (Wuhan, China), Yuanye Biotechnology Inc. (Shanghai, China) and BioBioPha Co., Ltd. (Kunming, China).

2.2. Plant Materials, Pathogen Inoculation and Sampling

One-year-old tea seedlings (*C. sinensis* cv. "Tieguanyin") were obtained from a tea plantation in Anxi, Fujian, China (118°13′ E, 25°08′ N). The seedlings were cultivated hydroponically using Hoagland's nutrient solution (pH 5.6) in a greenhouse at Fujian Agriculture and Forestry University (Fuzhou, China) under the following conditions: 25 ± 3 °C, 16 h photoperiod and 65 ± 5% relative humidity. To ensure the optimal growth of tea seedlings, the nutrient solution was aerated using an oxygen pump and replaced every seven days. After growing for over 30 days, seedlings that displayed uniform growth and were free from any signs of disease or insect infestation were selected as test materials.

The pathogenic strain *P. theae* CYF27, which has been previously reported [19], was used for inoculation in this study. CYF27 was first incubated at 25 °C on potato dextrose agar (PDA) medium for 20 days. PDA discs (6 mm) containing mycelia and conidia were excised to serve as the inocula. Prior to inoculation, tea leaves were rinsed with sterile water and air-dried to remove any residues from the surface. Each tea seedling received inoculation at three fully expanded leaves with two inoculation sites per leaf. Each site was gently scratched three times (~5 mm in length) with a sterile needle. A 5 mm disc of PDA medium bearing mycelia and conidia was placed onto the upper surface of the wounded tea leaf. The inoculated leaves were covered with cotton dampened in sterile water and wrapped in plastic. The seedlings were then transferred to an inoculation chamber (26 ± 3 °C, 16 h photoperiod, 90 ± 5% relative humidity) for 24 h of dark incubation. At 48 h post-inoculation, the coverings were removed to allow the seedlings to continue growing until harvest. The control samples were wounded tea leaves inoculated with blank PDA discs of the same size (hereafter referred as "CK"). For metabolome and transcriptome analyses, the entire fungal-inoculated leaves (hereafter referred as "PT"), along with the corresponding whole leaves from the CK samples, were harvested at 0, 1, 3 and 6 days post-inoculation (dpi) in three independent biological replicates. Each replicate consisted of three individual seedlings. Samples were snap-frozen in liquid nitrogen and subsequently stored at −80 °C for further analyses.

2.3. Metabolite Extraction and UPLC-QTOFMS Analysis

To prepare samples for metabolite analysis, the freeze-dried samples were ground to fine powders with pre-chilled mortars and pestles. Approximately 30 mg of leaf powder was weighted and extracted with 1.0 mL of 70% aqueous methanol. The resulting mixture was subjected to sonication for 20 min, followed by centrifugation at 12,000× *g* for 10 min. The supernatant was then filtered using a 0.22 µm PVDF filter (Millipore).

The metabolomics data were acquired on a Waters Acquity UPLC system (Milford, MA, USA) coupled with a Waters SYNAPT G2-*Si* HDMS QTOF mass spectrometer (Manchester, UK). The instrument was operated in electrospray ionization (ESI) mode and controlled with MassLynx 4.2 software (Waters). UPLC separation was carried out on a Waters Acquity UPLC HSS T3 column (1.8 µm, 2.1 × 100 mm). The mobile phase consisted of water containing 0.1% formic acid (Solvent A) and acetonitrile containing 0.1% formic acid (Solvent B). Gradient elution was performed as per the following program: 0–2 min (1–7% B), 2–13 min (7–40% B), 13–14 min (40–99% B), 14–18 min (held at 99% B) and 5 min of re-equilibration time before the next injection. The CK and PT samples were injected randomly. The ESI parameters were as follows: source temperature, 100 °C; desolvation temperature, 350 °C; cone gas flow, 50 L/h; desolvation gas flow, 800 L/h; capillary voltage, 1.28 kV; sampling cone voltage, 40 V. Data were acquired in both negative and positive ionization modes, operating in a full-scan mode over a mass range of 50–1200 m/z. The MS^e data were collected in the continuum mode with a collision energy ramp ranging from 10 to 50 eV. Online calibration of MS data was achieved by continually infusing leucine enkephalin (Waters, Milford, MA, USA) at a rate of 2 ng/min.

2.4. Metabolite Annotation and Multivariate Analysis

Progenesis QI software (v2.4, Nonlinear Dynamics) was used to process the raw data acquired in the positive and negative ionization modes separately. We detailed the process of data processing and metabolite identification in a prior study [20]. Briefly, raw data were imported into Progenesis QI for peak alignment, picking and normalization (with normalization to all compounds) using the default settings. To achieve the fusion of precursor ions, different adduct ion forms were grouped. These included $[M-H]^-$, $[2M-H]^-$ and $[M+FA-H]^-$ in ESI^- and $[M+H]^+$, $[2M+H]^+$ and $[M+Na]^+$ in ESI^+. The generated matrix involving the information of retention time, m/z and normalized peak abundance from each mode was exported and combined for subsequent multivariate statistical analyses in Simca-P 14.1 software (Umetrics, Umea, Sweden). After *Pareto* scaling, principal component analysis (PCA) and orthogonal partial least-squares discriminant analysis (OPLS-DA) were applied to identify metabolites with significant differences between the CK and PT groups at each time point, based on a fold change (FC) threshold > 2, variable importance in projection (VIP) value > 1 and *p*-value < 0.05. The differential metabolites were annotated by referencing an in-house MS database as well as public databases such as Metlin, MassBank and HMDB [20–23].

2.5. Transcriptome Analysis

To explore the temporal changes in transcriptional response during *P. theae* infection, we performed RNA-seq analysis on pathogen-inoculated leaf samples harvested at 0, 1, 3 and 6 dpi (designated as PT0, PT1, PT3 and PT6, respectively). Briefly, total RNA was extracted using the RNAprep Pure Plant Kit (Tiangen, Beijing, China) according to the manufacturer's instructions. Sequencing libraries were prepared using the NEBNext Ultra RNA Library Prep Kit for Illumina (NEB, Ipswich, MA, USA). The resulting library was then paired-end-sequenced (150 bp reads) on an Illumina HiSeq2500 platform (Novogene Biotech, Beijing, China) following standard protocols. The raw sequence data were deposited in the Genome Sequence Archive (GSA) database under the accession no. CRA011958.

After quality trimming, the clean reads were mapped to the reference transcriptome (*Camellia sinensis* cv. "Suchazao") from the TPIA database (http://tpia.teaplants.cn, accessed on 16 June 2022) using the Salmon program (v0.9.0) [24] after the exclusion of reads that contained *P. theae* sequences. Gene expression was quantified as transcripts per million (TPM). DESeq2 was employed to identify differentially expressed genes (DEGs) with a false discovery rate (FDR) < 0.05 and $|\log_2\text{fold change}| > 1$. Gene ontology (GO) enrichment analysis were performed using the clusterProfiler package [25]. Kyoto Encyclopedia of Genes and Genomes (KEGG) enrichment analysis was performed using the KOBAS software [26].

2.6. Quantitative Real-Time PCR (qRT-PCR)

To validate the expression levels of DEGs, qRT-PCR was performed using SYBR Premix Ex Taq^{TM} II (Takara, Dalian, China) on a CFX96TM real-time PCR system (Bio-Rad, Hercules, CA, USA) as previously described [27]. Primers for 12 randomly chosen DEGs were designed using the Primer Premier software (v5.00) (Table S1). The relative expression of genes was determined using the $2^{-\Delta\Delta Ct}$ method [28] with the *CsGAPDH* gene (KA295375.1) as an endogenous control. Each biological sample was analyzed in triplicate.

2.7. Statistical Analysis

Data were analyzed using GraphPad Prism 8, TBtools (v.1.131) [29] and Microsoft Excel 2010. Results were reported as the mean $\pm$ standard deviation (SD) obtained from at least three replicates. The statistical discrepancy between treatments was evaluated using the nonparametric Mann–Whitney *U* test or one-way ANOVA with a post hoc Fisher's least significant difference test as appropriate. In all cases, differences were considered significant at $p < 0.05$.

3. Results and Discussion

3.1. Morphological Analysis of Gray Blight Symptom Development in Tea Leaves

The onset and progression of disease symptoms on tea leaves inoculated with *P. theae* were monitored over the infection process (Figure 1A). At 0 and 1 dpi, no symptoms were observed on the wounded leaves. At 3 dpi, the symptoms began to manifest, appearing as small brown lesions specifically at the sites where the leaves had been pierced. These lesions gradually enlarged as the infection developed. By 6 dpi, the individual lesions merged together, forming prominent necrotic brown spots across the injured areas. Quantitative analysis showed that the lesion size increased by 4.5-fold from 3 to 6 dpi (Figure 1B).

Interestingly, a similar study conducted on the "Suchazao" tea cultivar using a different *Pseudopestalotiopsis* isolate reported a similar lesion onset at 4 dpi, but the subsequent progression was much slower. In fact, it took 13 dpi for the lesions in "Suchazao" to coalesce into large brown spots resembling those observed at 6 dpi in our study [18]. This disparity implies that the rate at which symptoms spread may vary depending on the specific combination of the tea cultivar and the pathogenic *Pseudopestalotiopsis* isolate under investigation.

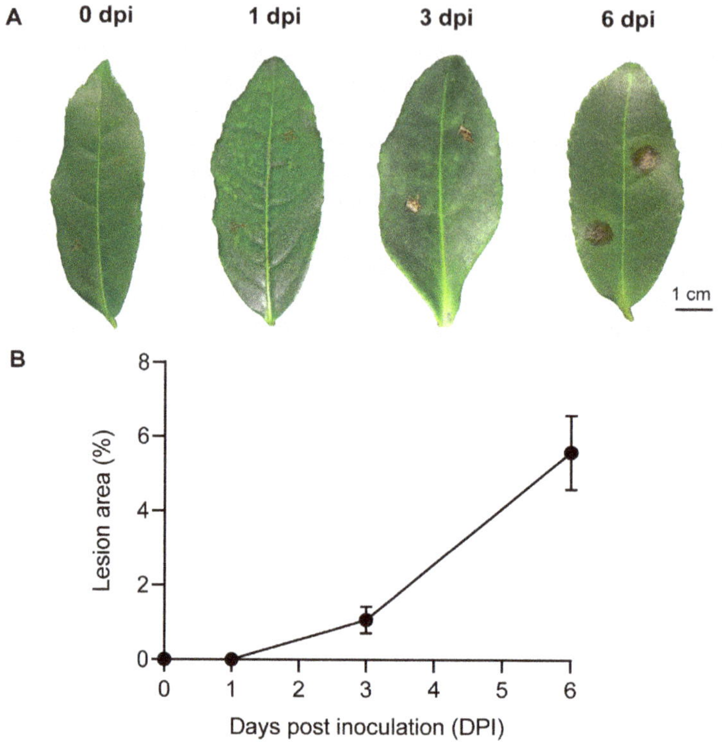

Figure 1. Development of gray blight disease on tea leaves over time. Intact leaves were inoculated with mycelia and conidia mixture of *P. theae*. (**A**) Representative symptoms on tea leaves using intact plant inoculation. Scale bar = 1 cm. (**B**) Statistics of disease development on tea leaves assessed at 0, 1, 3 and 6 dpi following fungal inoculation. The combined leaf disease area (cm^2) from two inoculations per leaf was calculated as a percentage of the total leaf area using ImageJ software.

3.2. Metabolic Analysis of Tea Plant Response to *P. theae* Infection

The changes in catechin contents in tea leaves afflicted by gray blight disease have been explored previously [18]. However, the impact of *P. theae* infection on other defense-related metabolites (e.g., phytoanticipins and phytoalexins) in tea plants remains largely unknown. To gain comprehensive insights into how tea metabolism is altered during pathogenesis, we conducted non-targeted metabolomics profiling of tea leaves subjected to *P. theae* inoculation or mock inoculation at 0, 1, 3 and 6 dpi. The representative total ion chromatograms of tea leaf metabolic fingerprints at 1 dpi in the negative mode are presented in Figure S1. After excluding fungal metabolites, we detected a total of 5362 metabolic features in the negative ionization mode and 3048 in the positive mode across all samples. Signals occurring at low abundance (maximum abundance < 1000) were removed, retaining 677 and 295 features in the negative and positive modes, respectively. Chemometrics analyses were then performed on the merged data from both modes. Our analysis unveiled dynamic metabolic shifts following fungal inoculation, as indicated by the log_2-fold differences in the relative metabolite abundance between the infected groups and the control groups (Figure 2A). As expected, the metabolite profiles of tea samples at 0 dpi were similar between the CK and PT groups. However, at 1 dpi, the infected group displayed the most intense metabolic response among all time points examined, with sharp increases in both up-regulated and down-regulated metabolic features. Studies have shown that the early plant–pathogen interactions typically take place within 24 h post-inoculation, during which pathogens can fully invade and colonize the host tissues [30]. Hence, the direct contact between the tea plant and *P. theae* likely triggered the activation of the plant's defense system at 1 dpi. When comparing the pathogen-inoculated group to the mock-inoculated group, there was an increased abundance of 222, 212 and 184 metabolic features and a decreased abundance of 231, 172 and 136 features at 1, 3 and 6 dpi, respectively (Figure 2A). From 1 to 6 dpi, the number of up-regulated features in the PT group compared to CK only slightly declined, whereas the number of down-regulated features decreased by 41%. This sustained up-regulation and lessening down-regulation of metabolic features in response to fungal infection suggest complex and dynamic interactions between the tea plant and the pathogen.

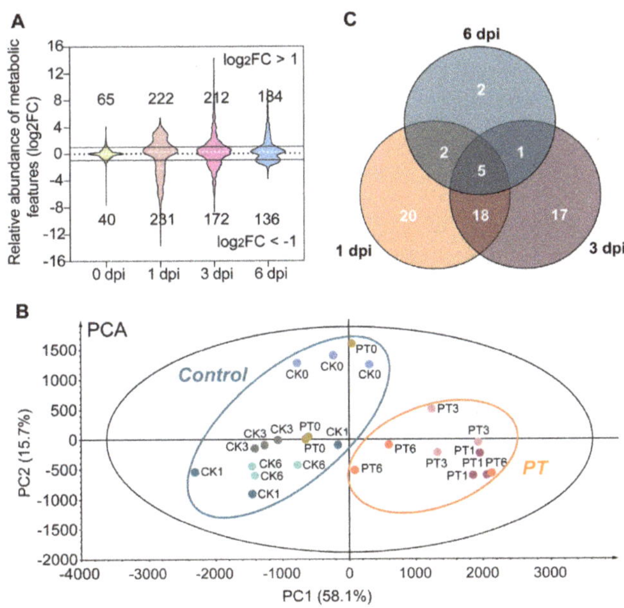

Figure 2. Time course analysis of tea metabolome reconfiguration in response to gray blight infection. (**A**) Overview of altered metabolic features. The violin plot shows the fold change values for the fungal-inoculated group relative to the mock-inoculated group at each time point. (**B**) Principal component analysis of tea leaves inoculated with *P. theae* or the mock control. (**C**) Venn diagram of differential metabolites between fungal-inoculated and mock-inoculated groups at each time point. PT, tea leaves inoculated with *P. theae*. CK, tea leaves inoculated with the mock control.

PCA analysis showed that, except for samples harvested at 0 dpi, the pathogen-inoculated samples were clearly distinguished from the mock-inoculated samples, with the first two principal components (PCs) explaining 73.8% of the total variance (Figure 2B). This again highlights the significant impact of *P. theae* infection on metabolite rewiring in tea plants. Within the PT group, one biological replicate harvested at 6 dpi deviated from the other two replicates, which made clustering by inoculation duration less distinct in the PCA plot.

To further elucidate the differences between the fungal-inoculated and mock-inoculated samples, we employed OPLS-DA analysis, a robust supervised multivariate method known for its ability to identify statistically significant variables [31]. The OPLS-DA score plots visually depict the separation between the sample groups based on their metabolite profiles at each time point (Figure S2). This analysis enabled us to observe distinct cluster patterns and confirm the significant metabolic alterations associated with the fungal infection. A total of 64 metabolites displayed statistically significant differences ($p < 0.05$, VIP > 1 and $|FC| > 2$), as shown in Table 1. These differentially accumulated metabolites (DAMs) were classified into 10 categories: polymerized catechin derivatives (23), flavonol glycosides (13), flavanols (8), flavone and glycosides (2), flavanone glycosides (1), xanthine alkaloids (1), amino acids (1), hydrolysable tannins (1), phenolic acids and derivatives (1) and unknown compounds (13). Flavonoid compounds made up the vast majority (72%) of the identified DAMs, pointing to their key roles in controlling gray blight infection.

A Venn diagram was constructed to visualize the overlap and differences in annotated DAMs between the treatment and control groups (Figure 2C). The number of DAMs showed a notable decrease as the infection progressed, with 45 identified at 1 dpi, 41 at 3 dpi and only 10 at 6 dpi. This indicates that the tea plant's metabolic response was most pronounced immediately after pathogen invasion but tapered over time. This decline could signify a transition from acute defense to a more balanced state. Alternatively, the pathogen's manipulation or suppression of certain metabolic pathways may contribute to the decrease. The metabolic alterations coincided with the phenotypic observation of progressively enlarged necrotic spots as the inoculation duration increased (Figure 1). Among the DAMs shared across all time points (Figure 2C and Table 1), four metabolites—caffeine, theaflavin 3,3′-digallate, EGCG and theacitrin C—showed a persistent induction in response to the *P. theae* challenge compared to CK. As a naturally occurring alkaloid prevalent in plants, caffeine functions as a broad-spectrum deterrent against herbivores and pathogens through its antimicrobial properties [32]. Notably, previous studies have demonstrated the ability of caffeine to inhibit the growth of the tea pathogens *P. theae* and *C. fructicola* even at low concentrations [17,33]. This underscores the probable essential role of caffeine in the innate defense response of tea plants to fungal infection. Similarly, reports from various plant–fungus interaction systems have documented a sustained increase in different flavonoid compounds as an integral component of innate immunity [34]. The consistent accumulation of these specialized metabolites in infected tea leaves throughout the infection process may suggest a strategic modulation of metabolism that imparts long-lasting resistance against pathogen challenges. Further evaluation is warranted to explore the potential of these metabolites as candidate markers of resistance and to elucidate how tea plants coordinate chemical defense.

3.3. Metabolic Reprogramming of Flavonoid Biosynthesis in Tea Plants against P. theae Infection

The heatmap visualization of the annotated DAMs reveals fascinating patterns of flavonoid metabolism in tea plants upon pathogen attack (Figure 3A–C). During early infection (1 and 3 dpi), the levels of EGCG, (−)-epigallocatechin (EGC), (−)-epicatechin gallate (ECG) and (−)-epigallocatechin 3-*O*-(3-*O*-methyl)gallate (EGCG3″Me) notably increased, while the EC and C levels decreased. Tea polyphenols, with EGCG, EGC, ECG and EC being the dominant compounds in tea leaves, demonstrate effective antifungal properties by inhibiting the growth and spore germination of various phytopathogens [16]. Additionally, studies conducted in Kenya and India have indicated an inverse correlation between the content of tea polyphenols and susceptibility to the pathogen *P. theae* [35]. These studies suggest the potential of tea polyphenols as antifungal agents for disease control, although the contribution of each polyphenol component to the antifungal activity is yet to be established. Of note, our results differ from those reported by Wang et al., who observed increased EC and GC contents in gray blight-inflicted tea leaves at 4 dpi, with minimal changes in EGCG and ECG [18]. The reasons for the disparities remain unclear but could relate to differences in *P. theae* isolates and tea cultivars used in the two studies. Therefore, further studies are needed to comprehend the influence of these factors on catechin biosynthesis during fungal infection and to clarify the antifungal effects of individual catechins against *P. theae*.

Table 1. Identification or tentative identification of significantly differential metabolites in response to *P. theae* infection.

No.	Metabolite Assignment	RT (min)	Metabolite Class	Adducts	Formula	Theoretical *m/z*	Measured *m/z*	PPM Error	MS/MS Fragments	Time Point
1	theanine *	1.40	amino acids	$[M-H]^-$	$C_7H_{14}N_2O_3$	173.0926	173.0931	2.89	155.0826, 128.0945	3 dpi
2	theogallin *	2.92	phenolic acids and derivatives	$[M-H]^-$	$C_{14}H_{16}O_{10}$	343.0665	343.0672	2.04	191.0568, 169.0143, 125.0244	3 dpi
3	prodelphinidin B	4.14	polymerized catechin derivatives	$[M-H]^-$	$C_{30}H_{26}O_{14}$	609.1244	609.1250	0.99	441.0826, 423.0716, 305.0665, 125.0242	1, 3 dpi
4	EGC-(2->7,4->8)-EGCG	4.16	polymerized catechin derivatives	$[M-H]^-$	$C_{37}H_{28}O_{18}$	759.1197	759.1191	−0.79	605.0929	1, 3 dpi
5	catechin-4-ol 3-*O*-hexoside	4.53	flavanols	$[M-H]^-$	$C_{21}H_{24}O_{12}$	467.1190	467.1188	−0.43	305.0659, 125.0246	3 dpi
6	prodelphinidin trimer GC-C-C isomer 1	4.74	polymerized catechin derivatives	$[M-H]^-$	$C_{45}H_{38}O_{20}$	897.1878	897.1879	0.11	771.1569	3 dpi
7	EC-GC dimer	4.81	polymerized catechin derivatives	$[M+H]^+$	$C_{30}H_{26}O_{13}$	595.1452	595.1440	−2.02	279.0922	3 dpi
8	theacitrin A	4.86	polymerized catechin derivatives	$[M-H]^-$	$C_{37}H_{28}O_{18}$	759.1197	759.1194	−0.40	741.1062, 571.0869	1 dpi
9	prodelphinidin trimer GC-C-C isomer 2	4.86	polymerized catechin derivatives	$[M-H]^-$	$C_{45}H_{38}O_{20}$	897.1878	897.1875	−0.33	771.158	3 dpi
10	epigallocatechin *	4.91	flavanols	$[M-H]^-$	$C_{15}H_{14}O_7$	305.0661	305.0666	1.64	179.0352, 125.0251	1, 6 dpi
11	unknown	4.93	unknown	$[M-H]^-$	$C_{45}H_{40}O_{20}$	899.2035	899.2035	0.00	ND	3 dpi
12	(E)GC+(E)C+(E)C isomer 1	5.14	polymerized catechin derivatives	$[M-H]^-$	$C_{45}H_{38}O_{19}$	881.1929	881.1926	−0.34	423.0722, 305.0671, 287.0561, 125.0248	3 dpi
13	catechin *	5.33	flavanols	$[M-H]^-$	$C_{15}H_{14}O_6$	289.0712	289.0718	2.08	179.0347, 137.0241, 123.0451	1 dpi
14	strictinin	5.34	hydrolysable tannins	$[M-2H]^{2-}$	$C_{27}H_{22}O_{18}$	633.0728	633.0730	0.32	316.0328, 300.9989, 275.0194	1 dpi
15	(E)GC+(E)C+(E)C isomer 2	5.56	polymerized catechin derivatives	$[M-H]^-$	$C_{45}H_{38}O_{19}$	881.1929	881.1927	−0.23	305.0649, 287.0550	3 dpi
16	caffeine *	5.56	alkaloids	$[M+H]^+$	$C_8H_{10}N_4O_2$	195.0882	195.0893	5.64	138.0671, 110.0720	0, 1, 3, 6 dpi
17	procyanidin dimer	5.68	polymerized catechin derivatives	$[M-H]^-$	$C_{30}H_{26}O_{12}$	577.1346	577.1350	0.69	451.1026, 425.0873, 407.0768, 289.0715	3 dpi
18	theasinensin A	5.71	polymerized catechin derivatives	$[M-2H]^{2-}$	$C_{44}H_{34}O_{22}$	913.1463	913.1446	−1.86	743.1231, 591.1140, 169.0142, 125.0243	1 dpi
19	unknown	5.72	unknown	$[M-2H]^{2-}$	$C_{26}H_{43}O_{36}$	930.1456	930.1450	−0.65	ND	1, 3 dpi
20	(E)GC-(E)CG dimer	5.86	polymerized catechin derivatives	$[M-H]^-$	$C_{37}H_{30}O_{17}$	745.1405	745.1399	−0.81	593.1121, 423.0715, 169.0145, 125.0249	1, 3 dpi
21	theasinensin D	5.90	polymerized catechin derivatives	$[M-H]^-$	$C_{44}H_{34}O_{22}$	913.1463	913.1435	−3.07	423.0709, 285.0396, 169.0140, 125.0241	3, 6 dpi
22	procyanidin trimer	6.00	polymerized catechin derivatives	$[M-H]^-$	$C_{45}H_{38}O_{18}$	865.1980	865.1980	0.00	739.1656, 713.1496, 695.1396, 413.0866, 245.0472	3 dpi
23	carthamidin diglucoside	6.03	flavonol glycosides	$[M-H]^-$	$C_{27}H_{32}O_{16}$	611.1612	611.1628	2.62	491.1191, 449.1241, 397.0771	1, 3 dpi
24	isovitexin glucoside	6.03	flavone glycosides	$[M+H]^+$	$C_{27}H_{30}O_{15}$	595.1663	595.1656	−1.18	433.1122, 313.0711	1 dpi
25	(E)GC+(E)C+(E)C isomer 3	6.10	polymerized catechin derivatives	$[M-H]^-$	$C_{45}H_{38}O_{19}$	881.1929	881.1922	−0.79	423.0712, 305.0658, 287.0557, 125.0240	3 dpi
26	unknown	6.10	unknown	$[M-H]^-$	$C_{38}H_{32}O_{19}$	791.1460	791.1453	−0.88	ND	1 dpi
27	diglucopyranosyl trihydroxyflavanone	6.12	flavanone glycosides	$[M-H]^-$	$C_{27}H_{32}O_{15}$	595.1663	595.1667	0.67	475.1249, 433.1353, 313.0934	1, 3 dpi
28	epicatechin *	6.23	flavanols	$[M-H]^-$	$C_{15}H_{14}O_6$	289.0712	289.0722	3.46	245.0820, 203.0710, 123.0451	1 dpi
29	unknown	6.30	unknown	$[M-H]^-$	$C_{19}H_{30}O_8$	385.1862	385.1854	−2.08	153.0916	1 dpi
30	epigallocatechin gallate *	6.35	flavanols	$[M-H]^-$	$C_{22}H_{18}O_{11}$	457.0771	457.0788	3.72	305.0669, 169.0152, 125.0248	1, 3, 6 dpi
31	theacitrin C	6.36	polymerized catechin derivatives	$[M-H]^-$	$C_{44}H_{32}O_{22}$	911.1307	911.1317	1.10	453.0459, 231.0278	1, 3, 6 dpi
32	unknown	6.37	unknown	$[M-H]^-$	$C_{13}H_{18}O_4$	227.0344	227.0347	1.32	ND	1, 3 dpi
33	unknown	6.38	unknown	$[M-H]^-$	$C_{19}H_{32}O_8$	387.2019	387.2006	−3.36	ND	1, 3 dpi
34	unknown	6.38	unknown	$[M-H]^-$	$C_{38}H_{36}O_{19}$	795.1773	795.1773	0.00	ND	1 dpi
35	unknown	6.50	unknown	$[M-H]^-$	$C_{38}H_{32}O_{19}$	791.1460	791.1457	−0.38	ND	1 dpi
36	(E)CG+(E)C isomer 1	6.70	polymerized catechin derivatives	$[M-H]^-$	$C_{37}H_{30}O_{16}$	729.1456	729.1457	0.14	577.1154, 407.0759, 289.0709, 125.0242	3 dpi
37	unknown	6.72	unknown	$[M-H]^-$	$C_{38}H_{32}O_{19}$	791.1460	791.1459	−0.13	ND	1 dpi
38	theasinensin F	6.75	polymerized catechin derivatives	$[M-2H]^{2-}$	$C_{44}H_{34}O_{21}$	897.1514	897.1505	−1.00	575.1186	3 dpi
39	(E)CG+(E)C isomer 2	6.80	polymerized catechin derivatives	$[M-H]^-$	$C_{37}H_{30}O_{16}$	729.1456	729.1446	−1.37	577.1193, 407.0758, 289.0710, 125.0249	1, 3 dpi
40	myricetin 3-neohesperidoside	6.96	flavonol glycosides	$[M-H]^-$	$C_{27}H_{30}O_{17}$	625.1404	625.1405	0.16	316.0224	1, 3 dpi
41	myricetin 3′-glucoside	7.15	flavonol glycosides	$[M-H]^-$	$C_{21}H_{20}O_{13}$	479.0826	479.0829	0.63	316.0225	1 dpi
42	unknown	7.35	unknown	$[M-H]^-$	$C_{39}H_{50}O_{25}$	917.2563	917.2569	0.65	ND	1 dpi
43	quercetin 3-*O*-glucosyl rutinoside	7.39	flavonol glycosides	$[M-H]^-$	$C_{33}H_{40}O_{21}$	771.1984	771.1988	0.52	609.1451, 463.0621, 301.0350, 300.0278	1 dpi
44	epigallocatechin 3-(3-*O*-methylgallate) *	7.43	flavanols	$[M-H]^-$	$C_{23}H_{20}O_{11}$	471.0927	471.0932	1.06	287.0556, 269.0453	1, 6 dpi
45	digalloylprocyanidin dimer	7.53	polymerized catechin derivatives	$[M-H]^-$	$C_{44}H_{34}O_{20}$	881.1565	881.1568	0.34	729.1407, 711.1342, 169.0141	3 dpi
46	rutin *	7.73	flavonol glycosides	$[M-H]^-$	$C_{27}H_{30}O_{16}$	609.1456	609.1457	0.16	533.1294, 300.0277, 271.0251, 255.0299, 243.0298	1, 3, 6 dpi
47	kaempferol 3-*O*-galactosyl rutinoside	7.75	flavonol glycosides	$[M-H]^-$	$C_{33}H_{40}O_{20}$	755.2035	755.2030	−0.66	533.1287, 285.0403	1 dpi
48	epicatechin 3-*O*-gallate *	7.85	flavanols	$[M-H]^-$	$C_{22}H_{18}O_{10}$	441.0822	441.0833	2.49	331.0458, 289.0720, 169.0146, 125.0247	3 dpi
49	tricetin	7.92	flavones	$[M+H]^+$	$C_{15}H_{10}O_7$	303.0505	303.0504	−0.33	285.0421	1, 6 dpi
50	quercetin 3-*O*-galactoside	7.93	flavonol glycosides	$[M-H]^-$	$C_{21}H_{20}O_{12}$	463.0877	463.0881	0.86	300.0276	1 dpi
51	kaempferol 3-*O*-glucosyl rutinoside	8.02	flavonol glycosides	$[M-H]^-$	$C_{33}H_{40}O_{20}$	755.2035	755.2036	0.13	285.0404	1 dpi
52	quercetin 3-*O*-glucoside *	8.04	flavonol glycosides	$[M-H]^-$	$C_{21}H_{20}O_{12}$	463.0877	463.0872	−1.08	301.0344, 300.0272	1, 3 dpi
53	capilliposide I isomer	8.17	flavonol glycosides	$[M-H]^-$	$C_{48}H_{56}O_{27}$	1063.2931	1063.2919	−1.13	917.2332, 755.1826, 609.1454, 377.0873, 301.0343, 300.0270	3 dpi
54	unknown	8.39	unknown	$[M-H]^-$	$C_{24}H_{42}O_{11}$	505.2649	505.2645	−0.79	551.2706	1, 3 dpi
55	kaempferol 3-*O*-rutinoside *	8.46	flavonol glycosides	$[M-H]^-$	$C_{27}H_{30}O_{15}$	593.1506	593.1504	−0.34	1187.3071, 285.0398	1, 3 dpi
56	unknown	8.58	unknown	$[M-H]^-$	$C_{20}H_{36}O_9$	419.2281	419.2269	−2.86	355.1054	1 dpi
57	epiafzelechin 3-gallate	8.94	flavanols	$[M-H]^-$	$C_{22}H_{18}O_9$	425.0873	425.0877	0.94	273.0763, 255.0652	6 dpi

Table 1. *Cont.*

No.	Metabolite Assignment	RT (min)	Metabolite Class	Adducts	Formula	Theoretical *m*/*z*	Measured *m*/*z*	PPM Error	MS/MS Fragments	Time Point
58	sarmenoside II isomer 1	10.31	flavonol glycosides	$[M-H]^-$	$C_{42}H_{46}O_{22}$	901.2402	901.2404	0.22	755.1822, 609.1461, 301.0351, 300.0274	1, 3 dpi
59	theaflavin *	10.59	polymerized catechin derivatives	$[M-H]^-$	$C_{29}H_{24}O_{12}$	563.1190	563.1193	0.53	425.0883, 269.0456, 137.0246	1 dpi
60	sarmenoside II isomer 2	10.91	flavonol glycosides	$[M-H]^-$	$C_{42}H_{46}O_{22}$	901.2402	901.2402	0.00	755.1823, 609.1449, 301.0346, 300.0270	1, 3 dpi
61	theaflavin 3-gallate *	10.97	polymerized catechin derivatives	$[M-H]^-$	$C_{36}H_{28}O_{16}$	715.1299	715.1302	0.42	563.1178, 423.2219, 125.0244	1, 3 dpi
62	theaflavin 3,3′-digallate *	11.17	polymerized catechin derivatives	$[M-H]^-$	$C_{43}H_{32}O_{20}$	867.1409	867.1413	0.46	697.1193, 389.0657, 178.8423	1, 3, 6 dpi
63	theaflavin-3′-gallate *	11.21	polymerized catechin derivatives	$[M-H]^-$	$C_{36}H_{28}O_{16}$	715.1299	715.1287	−1.68	563.1173, 125.0238	1, 3 dpi
64	unknown	13.78	unknown	$[M-H]^-$	$C_{33}H_{54}O_{15}$	689.3384	689.3376	−1.20	671.3262, 653.3162	0 dpi

* Confirmed by comparison with authentic standards. ND, not determined.

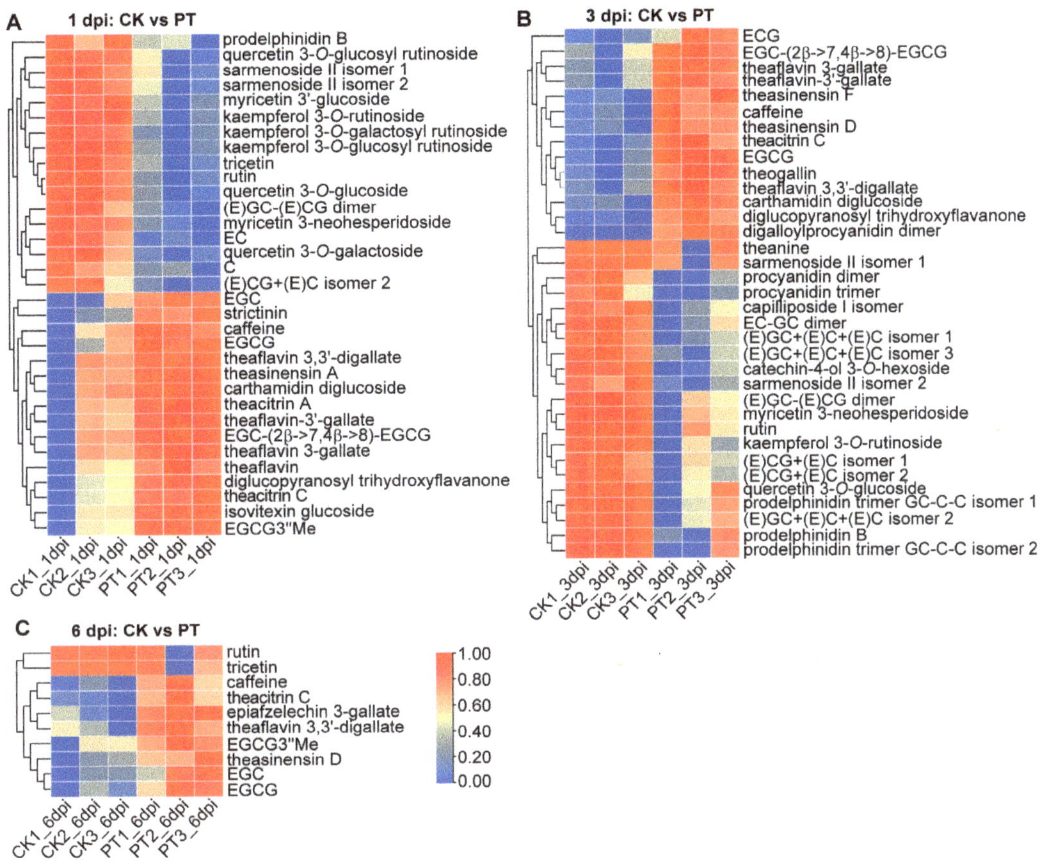

Figure 3. Heatmaps of annotated metabolites showing significant differences between fungal-inoculated and mock-inoculated tea leaves at 1 dpi (**A**), 3 dpi (**B**) and 6 dpi (**C**). Red and blue colors signify increased and decreased levels of metabolites. PT, tea leaves inoculated with *P. theae*. CK, tea leaves inoculated with the mock control.

In addition to alterations in catechin metabolism, the concentrations of various proanthocyanidin oligomers, including prodelphinidin B, procyanidin dimers, EC-GC dimer, (E)GC-(E)CG dimer and (E)GC-(E)C-(E)C trimer, decreased in tea leaves. This decline could be attributed to the depletion of their precursors, EC and C. There was also a marked reduction in the biosynthesis of flavonol glycosides, such as several derivatives of quercetin, kaempferol and myricetin, which are typically abundant in tea leaves. On the contrary, the concentrations of several oxidative polymerized products of catechins, namely theaflavins (theaflavin, theaflavin 3-gallate, theaflavin 3′-gallate and theaflavin 3, 3′-digallate), theasinensins (theasinensin A, D and F) and theacitrins (theacitrin A and C), increased alongside the elevated levels of their precursors, EGCG and EGC (Figure 3). There are two potential explanations for this accumulation. Firstly, these oxidative polymers might possess additional antifungal or antioxidant properties that directly contribute to the stress response against fungal infection. Alternatively, fungal infection could trigger oxidative stress in tea leaves through the enhanced activities of enzymes like peroxidase and polyphenol oxidase. Such oxidative stress may result in increased catechin polymerization. Although theaflavins, theasinensins and theacitrins are typically found in higher quantities during the fermentation of oolong and black teas, it is worth noting that a study detected increased levels of theaflavins in fresh tea leaves infested by tea green leafhoppers [36–39]. Nonetheless, the induction of oxidative catechin polymers as a response to biotic stresses has received little exploration. Understanding whether these polymers function directly as defensive compounds or are formed indirectly due to oxidative stress would enhance our comprehension of the metabolic defenses employed by tea plants.

To analyze the variation trends of catechin monomers and polymers during the course of infection, we plotted the relative contents of these compounds in fungal-infected tea samples (Figure 4). EGCG consistently showed a significant increase at 1 dpi ($p < 0.05$) and maintained higher levels until 6 dpi. The level of EC was significantly lower at 1 dpi compared to 0 dpi ($p < 0.05$). C levels also tended downward, albeit the change was not statistically significant. As for oxidized catechin polymers, the aforementioned theaflavins, theasinensins and theacitrins all showed a marked surge at 1 dpi and primarily remained elevated at subsequent time points despite some fluctuations.

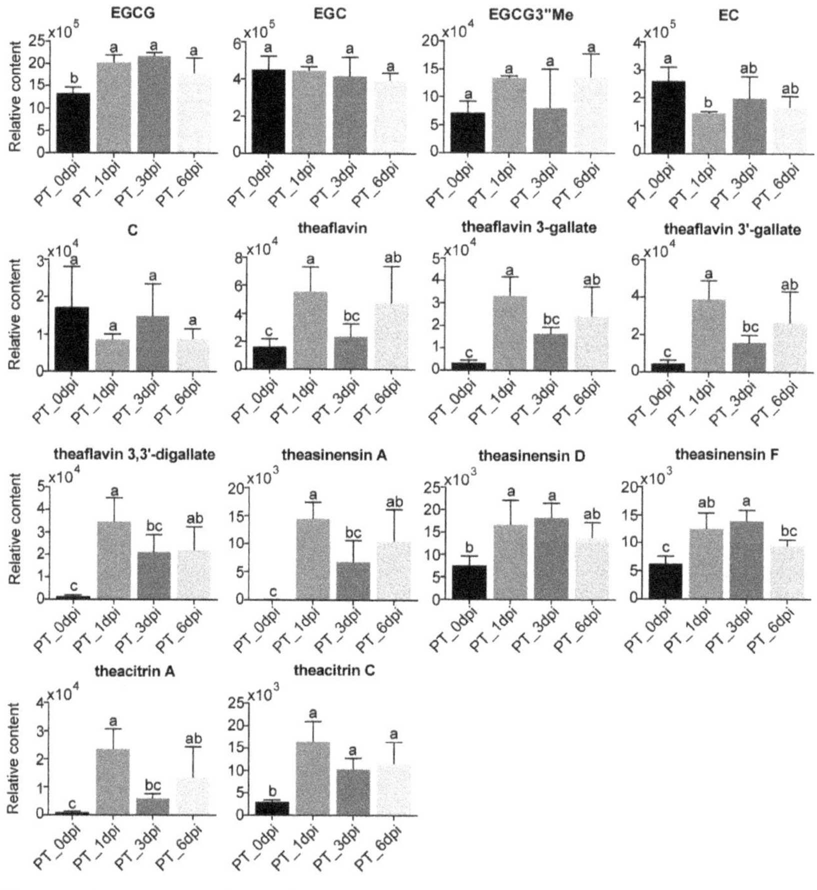

Figure 4. Quantitative analysis of catechins and polymerized catechin derivatives extracted from metabolomics data. The *y*-axis denotes peak area values obtained through normalization to the "normalize to all compounds function" in Progenesis QI. The relative abundance of each metabolite is presented as mean ± SD ($n = 3$). Different letters denote significant differences in metabolite contents at different time points, as determined by Fisher's least significant difference test at $p < 0.05$.

Collectively, the results show that tea plants strategically reconfigure their flavonoid profiles in response to infection by prioritizing the production of potent antimicrobial flavanols (e.g., EGCG) while reallocating resources away from potentially less bioactive compounds (e.g., EC, C and flavonol glycosides). This orchestrated adjustment likely strengthens the defense against fungal pathogens. Our work offers unique perspectives into the intricate flavonoid-mediated responses of tea plants to control infection.

3.4. Transcriptomic Analysis of Tea Plant Response to *P. theae* Infection

To further understand the global transcriptome responses modulated by *P. theae* infection, RNA sequencing was performed on *P. theae*-inoculated tea leaves at 0, 1, 3 and 6 dpi. Detailed information on RNA sequencing is provided in Table S2. Overall, 330,641,626 clean reads were generated from 12 samples, and these reads were mapped to the reference genome of "Suchazao", with alignment rates ranging between 86.00% and 89.61%. PCA analysis of the transcriptome data demonstrated clear clustering of biological replicates within each time point, providing evidence of *P. theae*-induced changes in gene expression (Figure 5A). The first two PCs collectively accounted for 64.6% of the total variance. Samples at 1 dpi were positioned at the far left of the score plot, whereas samples at 0 dpi were located at the far right along the first PC. This is concordant with the metabolomics data showing that the most drastic variation was induced at 1 dpi. A total of 29,037 transcripts were detected in all samples (Table S3). After filtering by FDR < 0.05 and absolute $\log_2$ (fold change) > 1, we identified 5967 DEGs (4547 up and 1420 down), 1485 DEGs (1269 up and 216 down) and 2408 DEGs (1708 up and 700 down) in PT_0 dpi vs. PT_1 dpi, PT_0 dpi vs. PT_3 dpi and PT_0 dpi vs. PT_6 dpi comparisons, respectively, suggesting that *P. theae* infection caused substantial transcriptional reprogramming in the tea plants (Figure 5B). In all comparisons, a higher number of genes were up-regulated compared to those that were down-regulated, which is consistent with the changes at the metabolomic level. In addition, many DEGs were unique to each time point, with 2845, 419 and 322 genes uniquely up-regulated and 1001, 88 and 278 genes uniquely down-regulated at 1, 3 and 6 dpi, respectively (Figure 5C). The reliability of the RNA-seq data was validated through qRT-PCR analysis, which confirmed the expression patterns of 12 randomly selected DEGs in the phenylpropanoid, flavonoid, shikimate and lignin pathways (Figure S3).

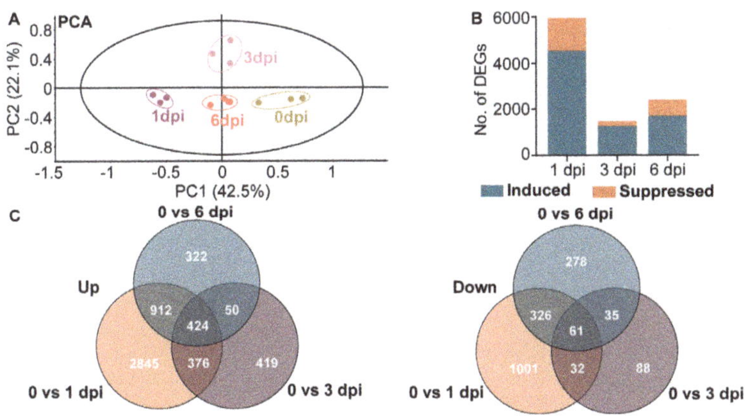

Figure 5. Time course analysis of tea transcriptome reprogramming in response to gray blight infection. (**A**) Principal component analysis of fungal-inoculated tea leaves at 0, 1, 3 and 6 dpi. (**B**) Number of differentially expressed genes (DEGs) at each time point. Teal and orange columns indicate up- and down-regulated genes, respectively. (**C**) Venn diagrams illustrating the number of DEGs up- or down-regulated due to fungal treatment over the time course.

GO classification of DEGs was performed to categorize the unigenes into biological processes, molecular functions and cellular components (Figure S4). Within the biological process category, the DEGs were found to be abundant in GO terms such as "cellular process", "single organism process", "metabolic process" and "response to stimulus". The molecular function category mainly comprised DEGs associated with "binding" and "catalytic activity". In the cellular component category, highly represented groups included "cell", "cell part", "organelle" and "membrane". We conducted KEGG pathway enrichment analysis on DEGs to investigate the primary metabolic pathways involved in tea plant response. The top 10 enriched up-regulated and down-regulated pathways at 1, 3 and 6 dpi are listed in Figure 6. Key enriched pathways included phenylpropanoid biosynthesis, flavonoid biosynthesis, photosynthesis, amino acid metabolism, starch and sucrose metabolism, plant–pathogen interaction and plant hormone signal transduction. Three pathways—"phenylpropanoid biosynthesis", "flavonoid biosynthesis" and "photosynthesis-antenna proteins"—were commonly enriched across all time points. Previous studies have demonstrated that pathogen-infected plant leaves often experience a decrease in photosynthetic rates due to disturbances in the photosynthetic machinery [40]. In line with this, our studies also found that *P. theae* infection, particularly at 1 dpi, disrupted pathways related to photosynthesis, such as "photosynthesis-antenna proteins", "carbon fixation in photosynthetic organisms" and "porphyrin and chlorophyll metabolism" (Table S4). The disruption of photosynthesis may have broader implications for plant defense, as the chloroplast plays a pivotal role in fortifying plant immunity through the synthesis of a diverse array of molecules, hormones and proteins that possess antimicrobial properties [41].

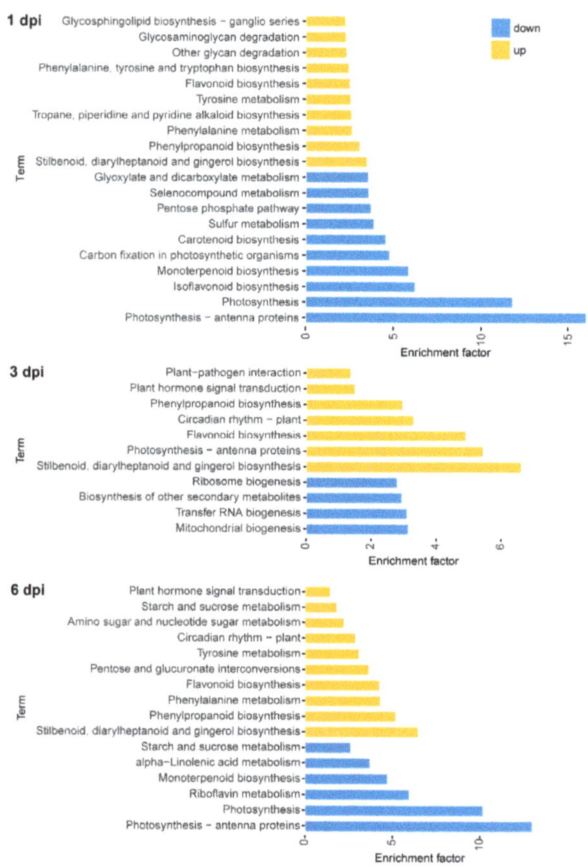

Figure 6. KEGG pathway enrichment analysis of DEGs following fungal treatment. The up-regulated pathways are in yellow, while the down-regulated pathways are in blue.

3.5. Transcriptomic Rewiring of Phenylpropanoid and Flavonoid Metabolism in Tea Plants in Response to P. theae Infection

KEGG pathways for phenylpropanoid and flavonoid biosynthesis were significantly enriched following infection (Figure 6). This correlates well with the elevated flavonoid levels from the metabolomics data, indicating the importance of these pathways in tea plant response to gray blight disease. Similar enrichment of phenylpropanoid and flavonoid biosynthesis was also observed in a prior study involving the interaction between the tea cultivar "Suchazao" and a pathogenic *Pseudopestalotiopsis* isolate [18]. Therefore, we closely examined the transcriptional profiles of genes involved in phenylpropanoid and flavonoid metabolism in tea leaves.

The role of phenylpropanoids as inducible antimicrobial compounds and signaling molecules in plant–pathogen interactions is well established [42]. Similarly, we observed pronounced increases in the transcription levels of genes encoding phenylalanine ammonia-lyase (PAL), cinnamate 4-hydroxylase (C4H) and 4-coumarate-CoA ligase (4CL) within the general phenylpropanoid pathway (Figure 7A). PAL, which catalyzes the initial committed step critical for regulating phenylpropanoid metabolism, showed up-regulation of all six transcripts at 1 dpi, with sustained high expression at late infection stages. C4H and 4CL function sequentially downstream to produce *p*-coumaroyl CoA, an essential precursor for the biosynthesis of various phenolic compounds, such as flavonoids and lignin monomers [43]. They showed similar expression patterns to PAL (Figure 7A,B).

With only some exceptions, genes in the flavonoid pathway were predominantly up-regulated in response to the pathogen at 1 dpi (Figure 7A). Metabolite profiling demonstrated differential accumulation of flavanols and flavonol glycosides. However, differential expression of genes responsible for directing metabolic flux towards specific branches of flavanols or flavonols was not evident. This disconnect between metabolite and transcriptional changes hints at the possibility of additional post-transcriptional controls in regulating flavonoid biosynthesis in tea plants challenged by *P. theae*.

Also being generated via the phenylpropanoid pathway, lignin comprises a highly branched polymer of monolignols that functions as a fundamental building block of plant cell walls. During the early defense response to pathogen infection (1 and 3 dpi), most lignin biosynthetic genes in tea leaves were up-regulated, though there were some instances of down-regulation for certain transcripts (Figure 7A). This is consistent with the well-documented role of lignin in enhancing plant tolerance to various biotic and abiotic stresses across plant species [44]. Aside from serving as a barricade for pests and pathogens, mounting evidence shows that cell wall lignin dynamically adjusts its concentration and composition in response to environmental stimuli, providing protection in a wide range of contexts [44–46]. The resilience of lignin against microbial degradation suggests that increased lignification may fortify the infected cells and restrict pathogen spread [45]. It should be noted that variations in lignin accumulation under fungal infection were not investigated in our study, and future qualitative and quantitative analyses of lignin dynamics will provide additional insights. Intriguingly, peroxidase and laccase transcripts universally showed up-regulation at 1 dpi (Figure 7A). These enzymes are actively involved in the polymerization of monolignols into lignin polymers [47]. Isolation and biochemical analysis of individual isoenzymes in the future experiments will disentangle their roles in lignin formation and pathogen protection.

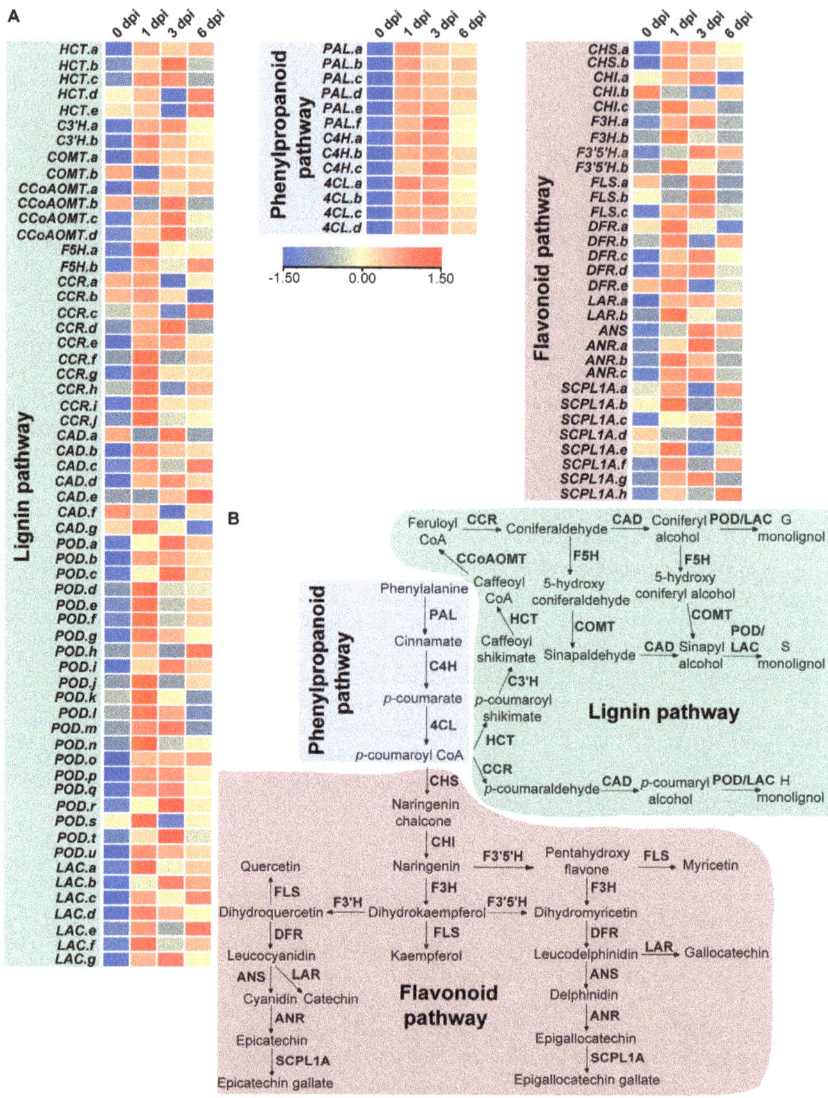

Figure 7. Expression profiles of genes involved in phenylpropanoid, lignin and flavonoid pathways in tea leaves after fungal treatment. (**A**) Heatmap comparison of the TPM values of related genes in three pathways based on RNA-seq data. (**B**) Simplified diagram of phenylpropanoid, lignin and flavonoid pathways. PAL, phenylalanine ammonia lyase; C4H, cinnamate 4-hydroxylase; 4CL, 4-coumarate-CoA ligase; CHS, chalcone synthase; CHI, chalcone isomerase; F3H, flavanone 3-hydroxylase; F3′H, flavanone 3′-hydroxylase; F3′5′H, flavanone 3′,5′-hydroxylase; FLS, flavonol synthase; DFR, dihydroflavonol 4-reductase; LAR, leucoanthocyanidin reductase; ANS, anthocyanidin synthase; ANR, anthocyanidin reductase; SCPL1A, type 1A serine carboxypeptidase-like acyltransferases; CCR, cinnamoyl-CoA reductase; HCT, hydroxycinnamoyl-CoA:shikimate/quinate hydroxycinnamoyl transferase; C3′H, *p*-coumaroyl quinate/shikimate 3′-hydroxylase; CCoAOMT, caffeoyl-CoA *O*-methyltransferase; CAD, cinnamyl alcohol dehydrogenase; COMT, caffeic acid O-methyltransferase; F5H, ferulate 5-hydroxylase; PER, peroxidase; LAC, laccase.

4. Conclusions

Through integrated "omics" analyses, this study uncovered the dynamic global transcriptomic and metabolomic changes in tea plants challenged with *P. theae*. By employing untargeted metabolomics coupled with multivariate testing, we identified 64 differentially accumulated metabolites between infected and control groups across multiple time points, highlighting the significant impact of this pathogen on tea plant metabolism. Rapid and substantial metabolic rewiring was observed at 1 dpi, being characterized by enhanced synthesis of caffeine, EGCG and oxidative catechin polymers, as well as a reduction in EC, proanthocyanidins and flavonol glycosides. Combined analysis of metabolomic and transcriptomic data implicates the involvement of phenylpropanoid, flavonoid and lignin pathways in tea plant resistance. Notably, we reveal for the first time the differential modulation of different branches of the flavonoid biosynthetic pathway in the tea plant in response to fungal challenge. These findings shed light on the complex and dynamic interactions between the tea plant and the causal agent of gray blight. Furthermore, our study enables the identification of key defensive genes, metabolites and pathways with potential for disease resistance breeding. Given that resistance can vary among tea cultivars, future comparative studies exploring varieties with varying degrees of susceptibility will provide deeper insights into the resistance mechanisms and aid in the development of more effective strategies for disease management and crop improvement.

Supplementary Materials: The following supporting information can be downloaded at: https://www.mdpi.com/article/10.3390/metabo13111122/s1, Figure S1: representative total ion chromatograms of mock-inoculated and fungal-inoculated tea leaves at 1 dpi in the negative ionization mode; Figure S2: OPLS-DA analysis of metabolomics datasets; Figure S3: qRT-PCR validation of RNA-seq DEGs; Figure S4: gene ontology (GO) enrichment analysis of differentially expressed genes in gray blight-infected tea plants at 1, 3 and 6 dpi; Table S1: list of primers for qRT-PCR; Table S2: the quality of RNA-seq data; Table S3: gene detected in all samples; Table S4: KEGG enrichment analysis.

Author Contributions: Conceptualization, X.Y. and Z.W.; methodology, Z.D., X.W. and C.Z.; formal analysis, S.Z.; investigation, S.Z. and X.Y.; data curation, S.Z.; writing—original draft preparation, S.Z. and X.Y.; writing—review and editing, X.Y. and Z.W.; supervision, X.Y. and Z.W.; funding acquisition, X.Y. All authors have read and agreed to the published version of the manuscript.

Funding: This research was funded by the Fujian Agriculture and Forestry University (FAFU) Construction Project for Technological Innovation and Service System of Tea Industry Chain (K1520005A02) and Fundamental Research Project of Fujian Provincial Research Institute for Public Welfare, China (Grant No. 2023R1027002).

Institutional Review Board Statement: Not applicable.

Informed Consent Statement: Not applicable.

Data Availability Statement: The data presented in this study are available on request from the corresponding author.

Acknowledgments: We would like to thank Hongli Hu of Fujian Agriculture and Forestry University for providing the strain used in our study.

Conflicts of Interest: The authors declare no conflict of interest.

References

1. Food and Agriculture Organization Statistics. FAOSTAT (Online). 2021. Available online: http://www.fao.org/faostat/en/#data/QC (accessed on 16 June 2022).
2. Pandey, A.K.; Sinniah, G.D.; Babu, A.; Tanti, A. How the global tea industry copes with fungal diseases-challenges and opportunities. *Plant Dis.* **2021**, *105*, 1868–1879. [CrossRef]
3. Wang, S.; Mi, X.; Wu, Z.; Zhang, L.; Wei, C. Characterization and pathogenicity of *Pestalotiopsis*-like species associated wth gray blight disease on *Camellia sinensis* in Anhui Province, China. *Plant Dis.* **2019**, *103*, 2786–2797. [CrossRef] [PubMed]
4. Maharachchikumbura, S.S.N.; Guo, L.-D.; Cai, L.; Chukeatirote, E.; Wu, W.P.; Sun, X.; Crous, P.W.; Bhat, D.J.; McKenzie, E.H.C.; Bahkali, A.H.; et al. A multi-locus backbone tree for *Pestalotiopsis*, with a polyphasic characterization of 14 new species. *Fungal Divers.* **2012**, *56*, 95–129. [CrossRef]

5. Chen, Y.; Zeng, L.; Shu, N.; Jiang, M.; Wang, H.; Huang, Y.; Tong, H. *Pestalotiopsis*-like species causing gray blight disease on *Camellia sinensis* in China. *Plant Dis.* **2018**, *102*, 98–106. [CrossRef] [PubMed]
6. Takeda, Y. Genetic analysis of tea gray blight resistance in tea plants. *Jpn. Agric. Res. Q.* **2002**, *36*, 143–150. [CrossRef]
7. Joshi, S.; Sanjay, R.; Baby, U.; Mandal, A.K. Molecular characterization of *Pestalotiopsis* spp. associated with tea (*Camellia sinensis*) in southern India using RAPD and ISSR markers. *Indian J. Biotechnol.* **2009**, *8*, 377–383.
8. Liu, F.; Hou, L.; Raza, M.; Cai, L. *Pestalotiopsis* and allied genera from *Camellia*, with description of 11 new species from China. *Sci. Rep.* **2017**, *7*, 866. [CrossRef]
9. Pandey, A.K.; Hubbali, M.; Vandana; Dutta, P.; Babu, A. Characterization and identification of fungicide insensitive *Pestalotiopsis*-like species pathogenic to tea crop in India. *World J. Microbiol. Biotechnol.* **2022**, *39*, 34. [CrossRef]
10. Yamada, K.; Sonoda, R.; Ishikawa, K. Population genetic structure of QoI-resistant *Pestalotiopsis longiseta* isolates causing tea gray blight. *Plant Dis.* **2016**, *100*, 1686–1691. [CrossRef] [PubMed]
11. Nishad, R.; Ahmed, T.; Rahman, V.J.; Kareem, A. Modulation of plant defense system in response to microbial interactions. *Front. Microbiol.* **2020**, *11*, 1298. [CrossRef]
12. Lindsay, W.P.; Lamb, C.J.; Dixon, R.A. Microbial recognition and activation of plant defense systems. *Trends Microbiol.* **1993**, *1*, 181–187. [CrossRef]
13. González-Lamothe, R.; Mitchell, G.; Gattuso, M.; Diarra, M.S.; Malouin, F.; Bouarab, K. Plant antimicrobial agents and their effects on plant and human pathogens. *Int. J. Mol. Sci.* **2009**, *10*, 3400–3419. [CrossRef]
14. Zeng, L.; Zhou, X.; Liao, Y.; Yang, Z. Roles of specialized metabolites in biological function and environmental adaptability of tea plant (*Camellia sinensis*) as a metabolite studying model. *J. Adv. Res.* **2021**, *34*, 159–171. [CrossRef] [PubMed]
15. Friedman, M. Overview of antibacterial, antitoxin, antiviral, and antifungal activities of tea flavonoids and teas. *Mol. Nutr. Food Res.* **2007**, *51*, 116–134. [CrossRef]
16. Yang, Y.; Zhang, T. Antimicrobial activities of tea polyphenol on phytopathogens: A review. *Molecules* **2019**, *24*, 816. [CrossRef] [PubMed]
17. Wang, L.; Wang, Y.; Cao, H.; Hao, X.; Zeng, J.; Yang, Y.; Wang, X. Transcriptome Analysis of an Anthracnose-Resistant Tea Plant Cultivar Reveals Genes Associated with Resistance to Colletotrichum camelliae. *PLoS ONE* **2016**, *11*, e0148535. [CrossRef]
18. Wang, S.; Liu, L.; Mi, X.; Zhao, S.; An, Y.; Xia, X.; Guo, R.; Wei, C. Multi-omics analysis to visualize the dynamic roles of defense genes in the response of tea plants to gray blight. *Plant J.* **2021**, *106*, 862–875. [CrossRef]
19. Zheng, S.; Chen, R.; Wang, Z.; Liu, J.; Cai, Y.; Peng, M.; Zhang, T.; Li, Y.; Wang, B.; Bao, J.; et al. High-quality genome assembly of *Pseudopestalotiopsis theae*, the pathogenic fungus causing tea gray blight. *Plant Dis.* **2021**, *105*, 3723–3726. [CrossRef]
20. Yu, X.; Xiao, J.; Chen, S.; Yu, Y.; Ma, J.; Lin, Y.; Li, R.; Lin, J.; Fu, Z.; Zhou, Q.; et al. Metabolite signatures of diverse *Camellia sinensis* tea populations. *Nat. Commun.* **2020**, *11*, 5586. [CrossRef] [PubMed]
21. Montenegro-Burke, J.R.; Guijas, C.; Siuzdak, G. METLIN: A tandem mass spectral library of standards. *Methods Mol. Biol.* **2020**, *2104*, 149–163. [CrossRef]
22. Horai, H.; Arita, M.; Kanaya, S.; Nihei, Y.; Ikeda, T.; Suwa, K.; Ojima, Y.; Tanaka, K.; Tanaka, S.; Aoshima, K.; et al. MassBank: A public repository for sharing mass spectral data for life sciences. *J. Mass Spectrom.* **2010**, *45*, 703–714. [CrossRef] [PubMed]
23. Wishart, D.S.; Guo, A.; Oler, E.; Wang, F.; Anjum, A.; Peters, H.; Dizon, R.; Sayeeda, Z.; Tian, S.; Lee, B.L.; et al. HMDB 5.0: The Human Metabolome Database for 2022. *Nucleic Acids Res.* **2022**, *50*, D622–D631. [CrossRef] [PubMed]
24. Patro, R.; Duggal, G.; Love, M.I.; Irizarry, R.A.; Kingsford, C. Salmon provides fast and bias-aware quantification of transcript expression. *Nat. Methods* **2017**, *14*, 417–419. [CrossRef] [PubMed]
25. Yu, G.; Wang, L.G.; Han, Y.; He, Q.Y. clusterProfiler: An R package for comparing biological themes among gene clusters. *Omics A J. Integr. Biol.* **2012**, *16*, 284–287. [CrossRef] [PubMed]
26. Mao, X.; Cai, T.; Olyarchuk, J.G.; Wei, L. Automated genome annotation and pathway identification using the KEGG Orthology (KO) as a controlled vocabulary. *Bioinformatics* **2005**, *21*, 3787–3793. [CrossRef]
27. Zhao, X.; Chen, S.; Wang, S.; Shan, W.; Wang, X.; Lin, Y.; Su, F.; Yang, Z.; Yu, X. Defensive responses of tea plants (*Camellia sinensis*) against tea green leafhopper attack: A multi-omics study. *Front. Plant Sci.* **2020**, *10*, 1705. [CrossRef]
28. Livak, K.J.; Schmittgen, T.D. Analysis of relative gene expression data using real-time quantitative PCR and the 2(-Delta Delta C(T)) Method. *Methods* **2001**, *25*, 402–408. [CrossRef]
29. Chen, C.; Chen, H.; Zhang, Y.; Thomas, H.R.; Frank, M.H.; He, Y.; Xia, R. TBtools: An integrative toolkit developed for interactive analyses of big biological data. *Mol. Plant* **2020**, *13*, 1194–1202. [CrossRef]
30. Shen, Y.; Liu, N.; Li, C.; Wang, X.; Xu, X.; Chen, W.; Xing, G.; Zheng, W. The early response during the interaction of fungal phytopathogen and host plant. *Open Biol.* **2017**, *7*, 170057. [CrossRef]
31. Wiklund, S.; Johansson, E.; Sjöström, L.; Mellerowicz, E.; Edlund, U.; Shockcor, J.; Gottfries, J.; Moritz, T.; Trygg, J. Visualization of GC/TOF-MS-based metabolomics data for identification of biochemically interesting compounds using OPLS class models. *Anal. Chem.* **2008**, *80*, 115–122. [CrossRef]
32. Kim, Y.-S.; Choi, Y.E.; Sano, H. Plant vaccination: Stimulation of defense system by caffeine production in planta. *Plant Signal. Behav.* **2010**, *5*, 489–493. [CrossRef] [PubMed]
33. Zhang, Z. Antifungal activity of caffeine against fungal pathogens of tea plant. *J. Nanjing Agric. Univ.* **2010**, *2*, 63–67.
34. Ramaroson, M.-L.; Koutouan, C.; Helesbeux, J.-J.; Clerc, V.; Latifa, H.; Geoffriau, E.; Briard, M. Role of phenylpropanoids and flavonoids in plant resistance to pests and diseases. *Molecules* **2022**, *27*, 8371. [CrossRef]

35. Chen, Z.M.; Sun, X.L.; Dong, W.X. Genetics and chemistry of the resistance of tea plant to pests. In *Global Tea Breeding*; Chen, L., Apostolides, Z., Chen, Z.-M., Eds.; Zhejiang University Press: Zhejiang, China, 2012; pp. 343–360.
36. Balentine, D.A.; Wiseman, S.A.; Bouwens, L.C. The chemistry of tea flavonoids. *Crit. Rev. Food Sci. Nutr.* **1997**, *37*, 693–704. [CrossRef]
37. Davis, A.L.; Lewis, J.R.; Cai, Y.; Powell, C.; Davis, A.P.; Wilkins, J.P.G.; Pudney, P.; Clifford, M.N. A polyphenolic pigment from black tea. *Phytochemistry* **1997**, *46*, 1397–1402. [CrossRef]
38. Matsuo, Y.; Li, Y.; Watarumi, S.; Tanaka, T.; Kouno, I. Production and degradation mechanism of theacitrin C, a black tea pigment derived from epigallocatechin-3-O-gallate via a bicyclo[3.2.1]octane-type intermediate. *Tetrahedron* **2011**, *67*, 2051–2059. [CrossRef]
39. Liao, Y.; Yu, Z.; Liu, X.; Zeng, L.; Cheng, S.; Li, J.; Tang, J.; Yang, Z. Effect of major tea insect attack on formation of quality-related nonvolatile specialized metabolites in tea (*Camellia sinensis*) leaves. *J. Agric. Food. Chem.* **2019**, *67*, 6716–6724. [CrossRef]
40. Duan, S.; Jin, J.; Gao, Y.; Jin, C.; Mu, J.; Zhen, W.; Sun, Q.; Xie, C.; Ma, J. Integrated transcriptome and metabolite profiling highlights the role of benzoxazinoids in wheat resistance against Fusarium crown rot. *Crop J.* **2022**, *10*, 407–417. [CrossRef]
41. Lu, Y.; Yao, J. Chloroplasts at the crossroad of photosynthesis, pathogen infection and plant defense. *Int. J. Mol. Sci.* **2018**, *19*, 3900. [CrossRef]
42. Naoumkina, M.A.; Zhao, Q.; Gallego-Giraldo, L.; Dai, X.; Zhao, P.X.; Dixon, R.A. Genome-wide analysis of phenylpropanoid defence pathways. *Mol. Plant Pathol.* **2010**, *11*, 829–846. [CrossRef]
43. Volpi e Silva, N.; Mazzafera, P.; Cesarino, I. Should I stay or should I go: Are chlorogenic acids mobilized towards lignin biosynthesis? *Phytochemistry* **2019**, *166*, 112063. [CrossRef] [PubMed]
44. Yadav, S.; Chattopadhyay, D. Lignin: The building block of defense responses to stress in plants. *J. Plant Growth Regul.* **2023**, *42*, 6652–6666. [CrossRef]
45. Moura, J.C.; Bonine, C.A.; de Oliveira Fernandes Viana, J.; Dornelas, M.C.; Mazzafera, P. Abiotic and biotic stresses and changes in the lignin content and composition in plants. *J. Integr. Plant Biol.* **2010**, *52*, 360–376. [CrossRef] [PubMed]
46. Ithal, N.; Recknor, J.; Nettleton, D.; Maier, T.; Baum, T.J.; Mitchum, M.G. Developmental transcript profiling of cyst nematode feeding cells in soybean roots. *Mol. Plant-Microbe Interact.* **2007**, *20*, 510–525. [CrossRef]
47. Boerjan, W.; Ralph, J.; Baucher, M. Lignin Biosynthesis. *Annu. Rev. Plant Biol.* **2003**, *54*, 519–546. [CrossRef]

metabolites

MDPI

Article

Secondary Metabolites and Antioxidant Activity against Moko Disease as a Defense Mechanism of *Musa* spp. from the Ecuadorian Coast Area

Raluca A. Mihai [1,*], Vanessa A. Terán-Maza [1], Karen A. Portilla-Benalcazar [1], Lissette E. Ramos-Guaytarilla [1], María J. Vizuete-Cabezas [1], Erly J. Melo-Heras [1], Nelson S. Cubi-Insuaste [1] and Rodica D. Catana [2]

1 Army Scientific and Technological Research Center—CICTE, Department of Life Science and Agriculture, Universidad de Las Fuerzas Armadas—ESPE, Av. General Rumiñahui s/n y, Sangolqui 171103, Ecuador; vateran1@espe.edu.ec (V.A.T.-M.); kaportilla@espe.edu.ec (K.A.P.-B.); leramos4@espe.edu.ec (L.E.R.-G.); mjvizuete@espe.edu.ec (M.J.V.-C.); ejmelo@espe.edu.ec (E.J.M.-H.); nscubi@espe.edu.ec (N.S.C.-I.)

2 Developmental Biology Department, Institute of Biology Bucharest of Romanian Academy, 296 Splaiul Independenţei, 060031 Bucharest, Romania; rodica.catana@ibiol.ro

* Correspondence: rmihai@espe.edu.ec

Abstract: The *Musa* spp. represents the most commonly produced, transitioned, and consumed fruit around the globe, with several important applications in the biotechnology, pharmaceutical, and food industries. Moko disease is produced by *Ralstonia solanacearum*—a factor with a high impact on all crops in Ecuador, representing one of the biggest phytosanitary problems. Four of the most common varieties of *Musa* spp. were tested to identify the metabolic reaction of plants facing Moko disease. The phenolic and flavonoid content has been evaluated as a defense system, and the α-diphenyl-α-picrylhydrazyl free-radical-scavenging method (DPPH), free-radical-scavenging activity (ABTS), ferric-reducing antioxidant power (FRAP) assays, and liquid chromatography and mass spectrometry (LC-MS) have been adapted to analyze the active compounds with the antioxidant capacity necessary to counteract the pathogenic attack. Our results indicate that all the studied varieties of *Musa* spp. react in the same way, such that the diseased samples showed a higher accumulation of secondary metabolites with antioxidant capacity compared with the healthy ones, with high active compound synthesis identified during the appearance of Moko disease symptoms. More than 40 compounds and their derivatives (from kaempferol and quercetin glycosides) with protective roles demonstrate the implication of the *Musa* spp. defense system against *R. solanacearum* infection.

Keywords: flavones; kaempferol-7-O-neohesperidoside; LC-MS; phenolic content; *Ralstonia solanacearum* race (phylotype) 2

1. Introduction

The *Musa* spp. (Musaceae family) enjoys increased interest since bananas represent the world's most commonly produced, traded, and consumed fruit, contributing to food security and incomes, and being in fourth place in importance in numerous countries [1]. Ecuador is the largest exporter globally, among Africa, Asia, and Latin America, which produce and sell bananas [2]. Besides its economic importance, the interest in banana plants grew due to their numerous uses in pharmacy, biotechnology, and food [3].

Among the factors that may affect banana manufacture, diseases caused by bacteria have serious effects [4]; *Ralstonia solanacearum* is one of the top 10 bacterial pathogens worldwide [5], characterized by a wide host range and wide geographic distribution, making it one of the most destructive crop pathogens in the world [6]. An alert was issued in February 2023 regarding Moko disease in Ecuador, reported in *Musa* plantations in more than 12 provinces [7].

Moko disease represents one of the phytosanitary problems with a great economic impact on all banana varieties and plantain crops, which affects over 70% of yield losses [6], as reported in a variety of studies regarding the treatment of this disease. Over time, some studies have been concerned about the Moko disease in *Musa* spp. In this way, molecular approaches like RAPD (Random Amplified polymorphic DNA) and rep-PCR-based fingerprinting (Repetitive sequence-based Polymerase Chain Reaction [8], AFLP molecular markers (Amplified Fragment Length Polymorphism) [9], and real-time PCR [10] were used to investigate the genetic relationship between *Ralstonia solanacearum* and *Musa* spp. diseases. Different strategies were tested to control/or eradicate the Moko disease: resident varieties [11–14], banana leachates [15], insects and beneficial microorganisms [16].

R. solanacearum invades the vascular tissues, causing the leaf's death, which reduces the ability to produce fruit. Insects and/or contaminated farming equipment instead of mechanical and soil are the principal transmission vectors [17]. Recent genetic studies showed that bacterial strains can survive more than 25 years, affecting native flora, organic matter in soils, and crop plants in tropical and subtropic zones [18]. Even though there are some good practices (chlorine dioxide) and resistance inducers in banana plants [19] for farmers to reduce the spread of the disease agent, no cure is available for Moko.

Secondary metabolites are directly involved in plant defense; factors like genetic, ontogenic, morphogenetic, and environmental factors may affect their biosynthesis/accumulation [20]. Recent studies have pointed out that banana flavonoids can form complex compounds with extracellular proteins that can damage the cell membrane of bacteria, followed by the release of intracellular compounds [21]. Our study aimed to demonstrate that the antioxidant capacity resulting from the biosynthesis of secondary metabolites may act as a defense mechanism against the Moko attack in *Musa* spp. To achieve this aim, four of the most commonly used *Musa* spp. were investigated in our research, such as *Musa cavendish* Paxton [22]—the most widespread edible variety [23], *Musa paradisiaca* L.—a plantain used as a nutritional and therapeutic source [24], *Musa textilis* Née, an important source of fibers [25] and *Musa acuminata* Colla—the most frequent species [26].

2. Materials and Methods

2.1. Chemicals

To quantify the different secondary metabolites, the following chemicals (Sigma-Aldrich Chemical Co.—St. Louis, MO, USA) were used: 2.2—diphenyl—1—picrylhydrazyl, diammonium salt of 2.2—azinobis (3–ethylbenzothiazoline– 6-sulfonic acid), 2.4.6-tri(2-pyridyl)-triazine, iron (III) chloride hexahydrate and Folin–Ciocalteu reagent, 6-hydroxy-2.5.7.8-tetramethyl chroman-2-carboxylic acid, 3.4.5-trihydroxy benzoic acid, potassium persulfate, iron (III) sulfate heptahydrate, quercetin, and ethanol. All the reagents were purchased from commercial providers.

2.2. Sample Collection and Processing

This study was carried out with plants of the genus *Musa* spp., from farms located in Santo Domingo de los Tsáchilas province (0°05′01.6″ N, 79°26′32.8″ W), Canton La Concordia, km 38 road Santo Domingo—Quinindé. Healthy plants and plants with Moko disease symptomatology, at different stages of disease development, were collected. Each degree of infection (disease score) has been characterized and identified based on the following characteristics implemented by the Government of Kerala in 2014: Stage 1 (initial stage) symptoms are characterized by the yellowing of the central leaf, transitioning from yellow to green, weakening, and eventually breaking at the junction with the petiole. As the infection progresses, it leads to wilting and drying of young leaves, spreading to older ones. Older leaves then exhibit yellow bands with dark margins on their edges. In Stage 2 (intermediate stage), symptoms appear swiftly in the suckers, characterized by the progressive yellowing and wilting of older leaves. Additionally, small suckers exhibit a gradual death of the central leaf, which extends toward the outer leaves. In Stage 3 (advanced stage), symptoms in infected corms become evident through transverse cuts, revealing brown or black bands in the vascular bundles infected by the bacteria. Diseased pseudostems excrete bacterial exudate, while internally, vascular bundles exhibit a light to dark brown color due to blockage by extracellular polymeric substances. Plants without clusters display symptoms characterized by the grouping of vascular bundles near the pseudostem center, with significant blockages primarily observed in the petioles of the initially infected leaves, becoming less apparent toward the periphery. All the collected banana plants were in the vegetative stage, young plants of approximately 1 year old. The samples were transferred in a cooler to the university laboratory in Cantón Rumiñahui.

2.3. Active Ingredients Extraction

To determine the relationship between the antioxidant character and the phytochemical compounds involved in the defense of plants against Moko disease, in the first phase, the impurities were removed from the leaves by washing them with distilled water, and afterwards, the leaves were dried and ground (5 g) and placed in 25 mL of ethanol of HPLC (High Performance Liquid Chromatography) grade 99.5 (*v/v*), overnight, in the dark, at 4 °C.

2.4. Active Ingredients Determination

The Folin–Ciocalteu method was used to determine the phenolic concentration in the banana leaves [27]. Solutions from the ethanolic extracts (0.4 mL), Folin–Ciocalteu reagent (2 mL of 10% (*v*/*v*), and 7.5% Na_2CO_3 (1.6 mL) were incubated at room temperature for 30 min. The 765 nm absorbance was used to measure the concentrations. Gallic acid (0–250 mg/L concentration) was used for the calibration curve (y = 0.0112x + 0.1759, R^2 = 0.9794).

A colorimetric method using aluminum chloride described by Pekal et al. [28] was used for the determination of the flavonoid content. Quercetin was used for the realization of the standard solution. The standard calibration curve was performed at a concentration between 0 and 1.5 mg L^{-1}, obtaining the equation y = 1.4566x + 0.0265 with a correlation factor R^2 = 0.9935. To the crude extracts of the plant sample (1 mL) was added 1.5 mL of the solvent, 100 µL of CH_3COONa (1 M), 100 µL of $AlCl_3$ (10% *v/v*), and distilled water (2.3 mL). The samples are covered for 40 min at room temperature. In the blank sample, the aluminum chloride was not placed. Finally, the samples were measured at 435 nm absorbance.

2.5. Antioxidant Capacity Determination

A protocol established by Sachett et al. (2021) [29] and modified by Thaweesang [27] was used for the determination of the antioxidant capacity through the 2,2-diphenyl-1-picrylhydrazyl (DPPH) radical. A stock solution of 1 µg L^{-1} was produced; 2 mL of stock was added to every laboratory test tube, including the 0.1 mL crude extracts sample. After the test tubes were left to incubate for 30 min at room temperature, the absorbance was measured at 517 nm and the radical scavenging activity was calculated using the following formula: control absorbance—sample absorbance/control absorbance ×100. The calibration curve (y = 158.07x − 1.6766, R^2 = 0.9955) was produced using a Trolox solution (positive control) with a concentration between 0 and 0.625 mM.

The establishment of the antioxidant capacity by 2,2′-azino-bis(3-ethylbenzothiazoline-6-sulfonate) radical cation ($ABTS^{*+}$) free radical scavenging (obtained from the reaction between ABTS (7 mM) and potassium persulfate) was based on a study by Kuskoski et al. [30]. After incubating the solution for 12–24 h, it was diluted in absolute ethanol until 0.7 ± 0.1 at 754 nm absorbance was obtained. The inhibitory capacity was determined after the reaction between the ABTS solution (2 mL) and sample (20 µL) for 7 min at room temperature and in darkness; absolute ethanol was used for the blank. The calibration curve was established between a 0 and 2.5 mM concentration, obtaining the following equation y = 31.995x + 3.9568 with a correlation factor R^2 = 0.9697, using Trolox (like a positive control).

To determine the antioxidant capacity based on the Ferric ion Reducing Antioxidant Power (FRAP) test, a FRAP solution was prepared from acetate buffer 300 mM (100 mL) pH 3.6 + HCl (40 mM) + $FeCl_3 \cdot 6 H_2O$ (20 mM) in the proportion of 100: 10: 10 according to the method of Rajurkar et al. (2011) [30]. The absorbance of the mixture was read at 593 nm. The standard curve (y = 0.3654x − 0.0532, R^2 = 0.9686) was performed by taking different concentrations of $FeSO_4 \cdot 7H_2O$ ranging from 0 to 5 mM. The solution consisting of sample (100 µL), distilled water (300 µL), and FRAP solution (3 mL) previously prepared was covered for half an hour at room temperature. The FRAP value (mmol Fe^{2+} g sample) was obtained by the following formula:

$$\text{FRAP value} = \text{Sample absorbance} - \text{Control absorbance}$$

2.6. LC-MS Determination

High-resolution LC-MS was used to determine the bioactive compounds in the samples of stem and leaves from *M. cavendish* at stage 1 of Moko disease. The plant extractions were established using a modified method based on Tohma et al. [31]. Ethanolic extracts were obtained from lyophilized samples represented by leaf and stem (1 g) and 80% ethanol (20 mL) maintained for 2 h at 30 °C and centrifuged for 10 min at 5000 rpm at 4 °C. Afterwards, the extracts were filtered and a rotary evaporator at 30 °C was used for ethanol removal. The samples were kept in well-covered plastic tubes until analysis was performed at −20 °C.

Identification and detection of different metabolites were performed by HPLC equipment (Vanquish Thermo Fisher Scientific (Waltham, MA, USA)) and Mass Detector (Ion Trap Thermo Fisher Scientific (Waltham, MA, USA)). The samples were eluted through an Accucore Vanquish 150 × 2.1 mm column at 35 °C and a 0.5 mL/min flow rate [32]. A 10 µL volume of 0.1% formic acid was used as a mobile phase to inject into the HPLC. MS scanning was used to identify the compounds from the samples based on the retention time comparison of each peak and the monitoring of the ion pairs in a standard solution [33]. The compounds were identified by comparing them with those of the standard compounds available in databases (PubChem, ChEBI, Metlin, HPLC, and the literature data) using the MZmine 2.53 software [34].

2.7. Statistical Test

The RStudio statistical program (1.3.8 version) was used. Significant differences were analyzed through a two-way ANOVA test ($p < 0.05$). All the experiments were realized in triplicate, with the values being presented as the mean ± SD. The correlation between the secondary metabolites analyzed and the antioxidant power was evaluated by Pearson's correlation coefficient, while the correlation matrix was depicted as a scatter plot matrix.

3. Results

3.1. Active Ingredient Determination

The total phenolic content was evaluated for each disease stage per species. For *M. cavendish*, the highest phenolic content was registered in Stage I (0.507 ± 0.025 mg GAE g dw), Stage II (0.358 ± 0.052 mg GAE g dw), followed by healthy plants (0.27 ± 0.036 mg GAE g dw), while Stage III presented the lowest values. In contrast, for *M. acuminata*, *M. paradisiaca* and *M. textilis*, the highest phenolic content was registered in Stage II (4.114 ± 0.145 mg GAE g dw; 4.604 ± 0.215 mg GAE g dw, and 6.868 ± 0.526 mg GAE g dw, respectively), while the other stages presented a similar tendency. *M. textilis* showed the highest phenolic content of all the evaluated *Musa* spp. (Figure 1).

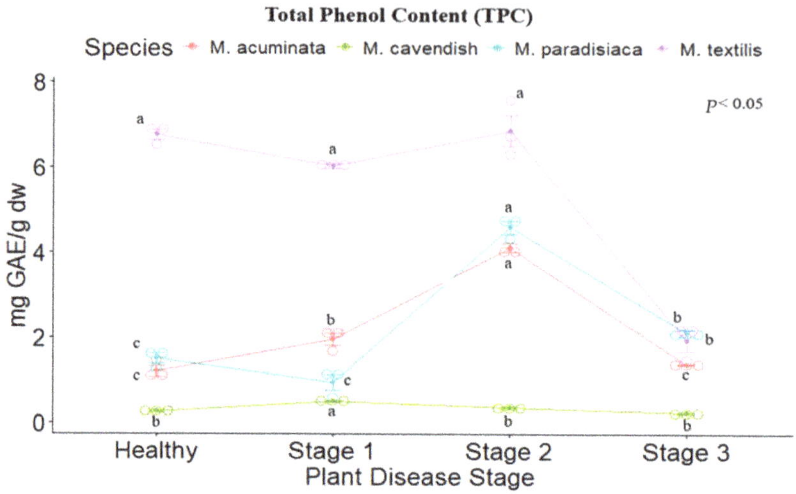

Figure 1. Total phenolic content of the tested *Musa* species. Legend: Stages 1, 2, and 3 represent the Moko disease stages. The plot is represented by the concentration in mg GAE/g dw on the *y*-axis, while the *x*-axis represents the stage for each plant species. Different letters denote significant differences.

M. cavendish showed the highest flavonoid concentration in contrast to the other species, with Stage I presenting the highest value (3.266 ± 0.0295 mg QE/g dw). The other *Musa* spp. show a similar trend to the phenolic content, as the highest value was reported for *M. acuminata, M. paradisiaca,* and *M. textilis*, appearing for Stage II (2.419 ± 0.197, 2.091 ± 0.0937, 1.682 ± 0.154, respectively), followed by the healthy stage, and Stages I and III with similar values. The lowest flavonoid content was reported for *M. textilis* (Figure 2).

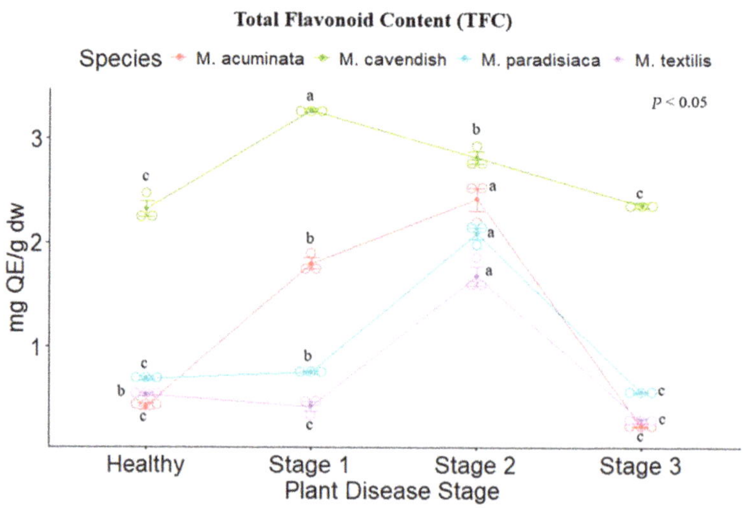

Figure 2. Total flavonoid content of the tested *Musa* species. Legend: Stages 1, 2 and 3 represent the Moko disease stages. The plot is represented by the concentration in mg QE/g dw on the *y*-axis, while the *x*-axis represents the stage for each plant species. Different letters denote significant differences.

3.2. Antioxidant Activity Determination

The antioxidant capacity of the four *Musa* spp. was evaluated through ABTS, DPPH, and FRAP assays. Significant differences were found between the disease stages for each species, following the same trend as their respective results concerning the secondary metabolites analyzed. For the ABTS and DPPH methods, *M. paradisiaca* (ABTS: 62.545 ± 2.524 µmol TEAC (Trolox Equivalent Antioxidant Capacity) g dw, DPPH: 113.997 ± 4.451 µmol TEAC g dw) and *M. textilis* (ABTS: 52.212 ± 2.742 µmol TEAC g dw, DPPH: 225.212 ± 6.708 µmol TEAC g dw) showed the highest values for Stage II. (Figures 3 and 4), while for the FRAP assay. *M. acuminata* (41.803 ± 1.179 µmol Fe^{2+} g dw) and *M. textilis* (67.24 ± 0.284 µmol Fe^{2+} g dw) presented the highest values (Figure 5).

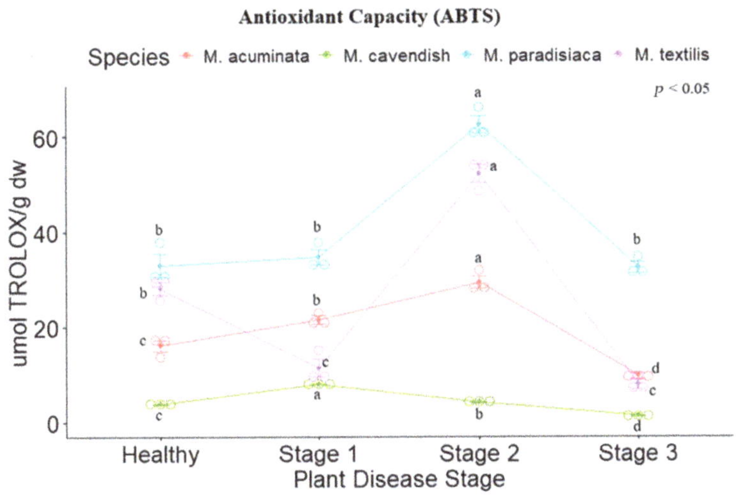

Figure 3. ABTS radical-scavenging assay of the tested *Musa* species. Legend: Stages 1, 2 and 3 represent the Moko disease stages. The plot is represented by the concentration in µmol Trolox/g dw on the *y*-axis, while the *x*-axis represents the stage for each plant species. Different letters denote significant differences.

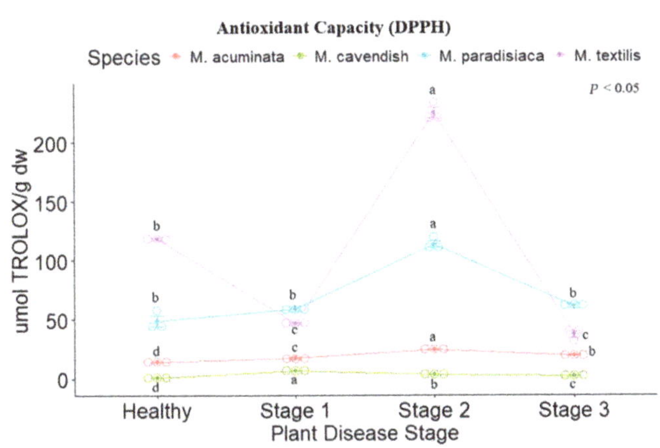

Figure 4. DPPH radical-scavenging assay of the tested *Musa* species. Legend: Stages 1, 2 and 3 represent the Moko disease stages. The plot is represented by the concentration in µmol Trolox/g dw on the *y*-axis, while the *x*-axis represents the stage for each plant species. Different letters denote significant differences.

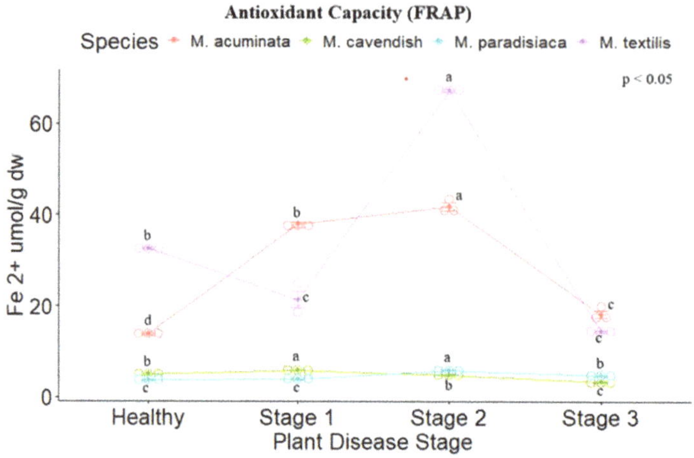

Figure 5. FRAP-reducing potential assay (μmol Fe^{2+} g dw) of the tested *Musa* species. Legend: Stages 1, 2 and 3 represent the Moko disease stages. The plot is represented by the concentration in μmol Fe^{2+}/g dw on the *y*-axis, while the *x*-axis represents the stage for each plant species. Different letters denote significant differences.

The correlation between the two analyzed secondary metabolites was raised for the four *Musa* spp. The contents were correlated with the three methods of antioxidant capacity evaluation used, with each plant showing a higher positive correspondence (R > 0.90) for the total phenolic content and flavonoid content. In contrast, for the phenolic content. *M. textilis* showed a lower correlation, albeit still positive with the ABTS (R = 0.64), DPPH (R = 0.66), and FRAP (R = 0.64) assays (Figure 6).

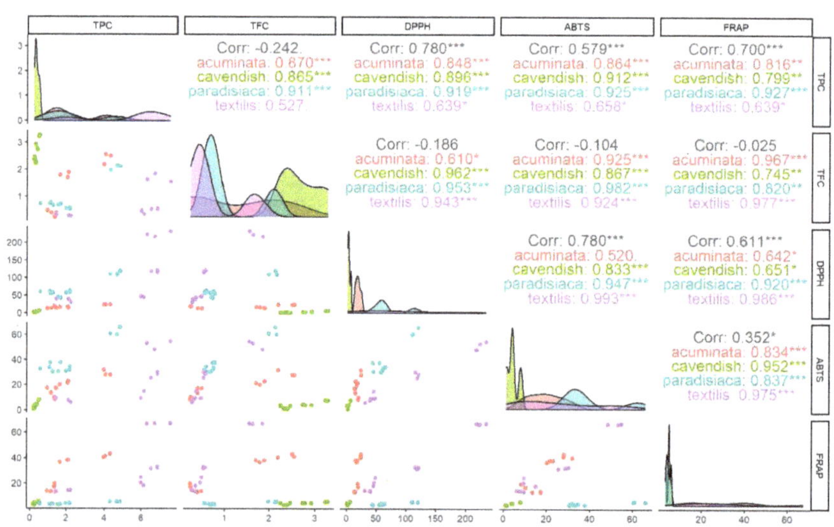

Figure 6. Correlation of flavonoids and total phenolic content with the ABTS, DPPH, and FRAP assays in the *Musa* species at different stages of *R. solanacearum* infection. Correlation values are expressed per species, classified by *** strong correlation, ** medium correlation, and * low correlation.

3.3. LC-MS Determination

LC-MS was used to determine the bioactive compounds in the samples of stem and leaves from *M. cavendish* at Stage 1 of Moko disease. The compounds in the extracts were identified based on their molecular mass and retention time. The phenolic extracts from the stem (34 identified compounds) and leaf (61 identified compounds) samples yielded similar HPLC profiles. However, differences were observed between both samples in terms of the identified phenolic compounds. The leaf and stem samples showed a similar content of phenol precursors such as shikimic acid, caffeoyl alcohol, coumaroyl, and derivatives of amino acids. Flavonoids were exclusively found in the leaf samples, with derivatives of quercetin (spiraeoside, rutin), kaempferol (astragalin, kaempferol-7-O-neohesperidoside), naringenin derivatives, and other flavonoid glycosides. Flavones such as isoorientin and scutellarin 4′-methyl ether had also been identified in the leaf samples. Citric acid, gibberellic acid, and terpenes (e.g., 7-O-methyl rosmanol) were identified mostly in the stem (Table 1, Figures A1 and A2).

Table 1. High-resolution LC-MS analysis of the metabolites identified in *Musa cavendish* at Stage 1 of Moko disease.

		HPLC-MS-NEGATIVE IONS				
ID	**Proposed Compound Identity**	**Molecular Formula**	**Retention Time**	**Molecular Ion**	**Plant Organ**	**Previously Found in *Musa* spp.**
8	Decanoic acid	$C_{10}H_{20}O_2$	1.17	M-H	Both	[35]
25	8-hydroxy-2,7,7,11,15-pentamethyl-5,12,16-trioxapentacyclo [9.8.0]nonadec-13(18)-ene-3,17-dione	$C_{25}H_{40}O_6$	1.15	M-H	Both	
37	Caffeyl alcohol	$C_9H_{10}O_3$	1.15	M-H	Both	
77	Sucrose	$C_{12}H_{22}O_{11}$	1.19	M-H	Both	
92	Citric acid	$C_6H_8O_7$	1.18	M-H	Both	[35]
94	N-Benzoyl-D5-glycine	$C_7H_4D_5NO_3$	1.21	M-H	Both	
98	α, α-Trehalose	$C_{12}H_{22}O_{11}$	1.21	M-H	Both	
102	Shikimic acid	$C_7H_{10}O_5$	1.20	M-H	Both	[35]
124	Glucose, Fructose, Mannose, Galactose	$C_6H_{12}O_6$	1.23	M-H	Both	[36]
131	Isoorientin	$C_{21}H_{20}O_{11}$	1.30	M-H	Both	
187	Kaempferol 7-neohesperidoside	$C_{27}H_{30}O_{15}$	1.56	M-H	Both	[37]
188	Coumaroyl + $C_6H_9O_8$	$C_{15}H_{12}O_8$	1.62	M-H	Both	
192	Isorhamnetin 3-rutinoside	$C_{27}H_{30}O_{15}$	1.59	M-H	Both	[38]
198	Trihydroxyflavone C-hexoside C-pentoside	$C_{23}H_{46}O_7$	1.56	M-H	Both	
211	Geniposide	$C_{17}H_{24}O_{10}$	1.59	M+HCOO	Both	
213	Rutin	$C_{27}H_{30}O_{16}$	1.62	M-H	Both	[39]
215	2-Isopropylmalic acid	$C_7H_{12}O_4$	1.59	M-H	Both	[36]
220	Epicatechin	$C_{15}H_{14}O_6$	1.46	M-H	Both	[40]
251	Isovitexin (4)	$C_{21}H_{20}O_{10}$	1.60	M-H	Both	
252	Spiraeoside, Spiraein, quercetin-4′-glucoside	$C_{21}H_{20}O_{12}$	1.74	M+H	Both	
256	Kaempferol-3-O-glucoside	$C_{21}H_{20}O_{11}$	1.80	M-H	Both	
259	Isoorientin	$C_{19}H_{18}O_{10}$	13.48	M-H	Leaf	
268	Pentose-Hexose + $C_{10}H_{17}$	$C_{15}H_{22}O_{10}$	18.59	M+HCOO	Both	
313	(10E,15E)-9,12,13-trihydroxyoctadeca-10,15-dienoic acid	$C_{18}H_{32}O_5$	21.48	M-H	Both	

Table 1. *Cont.*

		HPLC-MS-NEGATIVE IONS				
ID	**Proposed Compound Identity**	**Molecular Formula**	**Retention Time**	**Molecular Ion**	**Plant Organ**	**Previously Found in *Musa* spp.**
322	Labetalol, 2-hydroxy-5-[1-hydroxy-2-(4-phenylbutan-2-ylamino) ethyl] benzamide	$C_{19}H_{25}N_3O_2$	23.23	M-H	Both	
331	Dienogest, 2-[(8S,13S,14S,17R)-17-hydroxy-13-methyl-3-oxo-1,2,6,7,8,11,12,14,15,16-decahydrocyclopenta[a]phenanthren-17-yl] acetonitrile	$C_{20}H_{25}NO_2$	24.00	M-H	Both	
337	13-HpOTrE, 13S-hydroperoxy-9Z,11E,15Z-octadecatrienoic acid, (9Z,11E,15Z)-13-Hydroperoxy-9,11,15-octadecatrienoic acid	$C_{18}H_{32}O_3$	24.62	M-H	Both	
338	Gibberellate, Gibberellic acid, Gibberellin A3, Gibberellin	$C_{19}H_{22}O_6$	24.46	M-H	Both	
349	N-Acetylneuraminic acid	$C_{11}H_{17}NO_8$	25.25	M-H	Both	
351	3-[4-methyl-1-(2-methylpropanoyl)-3-oxocyclohexyl] butanoic acid	$C_{13}H_{22}O_3$	25.66	M-H	Both	
357	9-Hydroperoxy-10E,12Z-octadecadienoic acid	$C_{20}H_{36}O_3$	26.09	M-H	Both	
364	9-hydroxy-7-(2-hydroxypropan-2-yl)-1,4a-dimethyl-2,3,4,9,10,10a-hexahydrophenanthrene-1-carboxylic acid	$C_{20}H_{30}O_4$	26.50	M+H	Both	
384	2-Hydroxyhippuric acid	$C_9H_9NO_4$	28.27	M-H	Both	
390	Canrenone, (9S,14S)-10,13-dimethylspiro [2,8,9,11,12,14,15,16-octahydro-1H-cyclopenta[a]phenanthrene-17,5′-oxolane]-2′,3-dione	$C_{22}H_{28}O_3$	28.62	M-H	Both	
392	Ajmaline	$C_{20}H_{26}N_2O$	28.55	M-H	Both	
398	Docosanol	$C_{22}H_{46}O$	28.85	M+H	Both	[41]
414	Diacylglycerol 18:3	$C_{39}H_{68}O_5$	29.23	M+HCOO	Both	
421	Medroxyprogesterone	$C_{24}H_{34}O_4$	29.26	M-H	Both	
422	Avocadyne Acetate	$C_{21}H_{36}O_3$	29.51	M+H	Both	
456	7-O-Methylrosmanol	$C_{21}H_{36}O_3$	29.90	M+H	Both	
457	Hydroxylated linoleic acid	$C_{18}H_{32}O_4$	29.93	M-H	Both	
458	Dodecylbenzenesulfonic acid	$C_{18}H_{30}O_3S$	30.02	M+H	Both	
466	Fucosyltransferase V	-	30.31	M+H	Both	
470	Furosemide, 4-chloro-2-(furan-2-ylmethylamino)-5-sulfamoylbenzoic acid	$C_{12}H_{11}ClN_2O_5S$	30.22	M-H	Both	
472	Lysophosphatidylcholine 18:3	$C_{27}H_{50}NO_7P$	30.29	M+HCOO	Both	
474	1-[2-methyl-6-[(2S,3R,4S,5S,6R)-3,4,5-trihydroxy-6-(hydroxymethyl) oxan-2-yl] oxyphenyl] ethanone	$C_{14}H_{20}O_8$	30.42	M-H	Both	
495	Phosphatidylinositol 16:0	$C_{37}H_{71}O_8P$	30.67	M-H	Both	
520	[(4E)-7-acetyloxy-6-hydroxy-2-methyl-10-oxo-2,3,6,7,8,9-hexahydrooxecin-3-yl] (E)-but-2-enoate	$C_{14}H_{20}O_7$	30.87	M+H	Both	
545	Threo-7′-O-Butylresveptero acyclic dimer	$C_{32}H_{38}O_4$	31.19	M-H	Both	

Table 1. *Cont.*

ID	Proposed Compound Identity	Molecular Formula	Retention Time	Molecular Ion	Plant Organ	Previously Found in *Musa* spp.
		HPLC-MS-NEGATIVE IONS				
546	Lysophosphatidylethanolamine 18:2	$C_{40}H_{80}NO_8P$	31.08	M-H	Both	
563	Monogalactosyldiacylglycerol 18:3	$C_{57}H_{102}O_{10}P$	31.24	M+HCOO	Both	
581	1-(9Z,12Z-Octadecadienoyl)-2-hydroxy-sn-glycero-3-phosphoethanolamine	$C_{42}H_{78}NO_8P$	31.56	M-H	Both	
588	Ethylenediaminetetraacetic acid	$C_{10}H_{16}N_2O_8$	31.66	M-H	Both	
620	[5-acetyloxy-3-(hydroxymethyl)-2-oxo-6-propan-2-ylcyclohex-3-en-1-yl] 3-methylpentanoate	$C_{17}H_{26}O_6$	32.15	M+H	Both	
621	Naringenin-7-O-glucoside	$C_{21}H_{22}O_{10}$	32.03	M-H	Both	[39]
633	Hydroxyoctadecadienoic acid	$C_{18}H_{32}O_3$	32.23	M-H1	Both	
661	N-(hexadecanoyl)-1-hydroxyethane-2-amide	$C_{20}H_{41}NO_8$	32.76	M-H	Both	
684	9-Keto-octadecadienoic acid	$C_{18}H_3O_3$	33.18	M-H	Both	
709	Threo-7′-O-Isopropylresvepterol acyclic dimer	$C_{29}H_{32}O_6$	33.87	M-H	Both	
716	Sesamin	$C_{20}H_{18}O_6$	33.96	M+H	Both	
721	Methyl (4aR)-5,6-dihydroxy-1,1-dimethyl-7-isopropyl-2,3,4,9,10,10a-hexahydrophenanthrenanthrene-4a-carboxylate	$C_{23}H_{32}O_4$	34.19	M+Na	Both	
733	1-Acyl-sn-glycero-3-phosphocholine	$C_{26}H_{52}NO_7P$	34.55	M+HCOO	Both	
742	Thymol-beta-D-glucoside	$C_{16}H_{24}O_6$	34.68	M+H	Both	
777	Beta-alanyl-L-histidine	$C_9H_{14}N_4O_3$	40.92	M+H	Both	
807	2,6-Dihydroxybenzoic acid	$C_7H_6O_4$	41.56	M-H	Both	
822	Cystine	$C_6H_{12}N_2O_4S_2$	42.03	M+H	Both	
824	Norethindrone	$C_{20}H_{26}O_2$	42.07	M-H	Both	
839	Scutellarein 4′-methyl ether	$C_{16}H_{12}O_7$	42.95	M+H	Both	
		HPLC-MS-POSITIVE IONS				
505	Candesartan	$C_{24}H_{20}N_6O_3$	1.41	M+H		
568	Adenosine	$C_{10}H_{13}N_5O_4$	1.42	M+		
577	Ceramide	$C_{34}H_{66}NO_3$	1.40	M+H		
722	Isoshaftoside	$C_{21}H_{20}O_{11}$	1.53	-		
728	9-Methoxycamptothecin	$C_{22}H_{20}N_2O_5$	1.51	-		
752	Kaempferol-3-O-rutinoside	$C_{27}H_{30}O_{15}$	1.57	M+H		[42]
758	Glycochenodeoxycholic acid	$C_{26}H_{43}NO_6$	1.57	M+H		
763	Isovitexin	$C_{21}H_{20}O_{10}$	1.57	-		
769	Glycolithocholic acid	$C_{26}H_{43}NO_4$	1.61	M+H		
974	Spiraeoside	$C_{21}H_{20}O_{11}$	10.98	M+H		
980	Kaempferol-7-O-neohesperidoside	$C_{27}H_{30}O_{15}$	11.33	M+H		
984	Selenomethionine	$C_5H_{11}NO_2Se$	11.48	M+H		
985	Phytol	$C_{20}H_{40}O$	11.60	M+H		[40]
986	Orientin	$C_{27}H1_{20}O_{11}$	11.67	M+H		
990	Cis-Nerolidol	$C_{15}H_{26}O$	12.05	M+		
991	Luteolin 4′-O-glucoside	$C_{21}H_{20}O_{11}$	12.11	M-2H		
1004	Vitexin-2′′-O-rhamnoside	$C_{27}H_{30}O_{14}$	12.67	M+H		
1012	Apigenin 7-O-neohesperidoside	$C_{27}H_{30}O_{15}$	12.91	M+H		
1018	5,7-dihydroxy-2-(4-hydroxy-3-methoxyphenyl)-3-[3,4,5-trihydroxy-6-[[(2R,3R,4R,5R,6S)-3,4,5-trihydroxy-6-methyloxan-2-yl] oxymethyl] oxan-2-yl] oxychromen-4-one	$C_{27}H_{30}O_{17}$	13.21	M+H		

Table 1. *Cont.*

ID	Proposed Compound Identity	Molecular Formula	Retention Time	Molecular Ion	Plant Organ	Previously Found in *Musa* spp.
		HPLC-MS-NEGATIVE IONS				
1731	6-Acetoxy-9 benzoyloxy-1,8-dihydroxydihydro-β-agarofuran	$C_{26}H_{30}O_8$	28.93	M+H		
1741	Methyl 4-hydroxy-3,5-dimethoxybenzoate	$C_{10}H_{12}O_5$	29.02	-		
1755	Lauryl diethanolamide	$C_{14}H_{31}NO_2$	29.19	-		
1765	(Z)-9,12,13-trihydroxyoctadec-15-enoic acid	$C_{18}H_{34}O_5$	29.24	M-H		
1788	(1R,9S,10S)-3,4-dihydroxy-11,11-dimethyl-5-(propan-2-yl)-16-oxatetracyclo [7.5.2.0] hexadeca-2(7),3,5-triene-8,15-dione	$C_{19}H_{22}O_5$	29.26	M-H		
2542	Dibutylphthalate	$C_{16}H_{22}O_4$	33.90	M+H		
2543	1-Palmitoylglycerophosphocholine	$C_{24}H_{50}NO_7P$	33.88	M+		
2544	Diacylglycerol trimethylhomoserine	$C_{31}H_{60}NOPS$	33.92	M+H		
2548	1-Hexacosanol	$C_{26}H_{54}O$	33.94	-		
2551	1-Oleoylglycerophosphocholine	$C_{30}H_{60}NO_7P$	34.00	M+		
2614	Beta-Peltatin	$C_{18}H_{23}NO_5$	34.46	M-H		
2624	8-(2-hydroxy-3-methoxy-3-methylbutyl)-7-methoxychromen-2-one	$C_{16}H_{18}O_5$	34.56	M+NH4		

4. Discussion

Banana, as a key crop for food security, is considered the primary food for several developing countries [13]. However, plants from the genus *Musa* spp. are susceptible to some pathogens, including *R. solanacearum* race 2. The wilt disease is determined by bacterial accumulation, which blocks the vessels, stopping the sap flow [43]. Bacterial wilt lowers the yield in banana production [13]; therefore, treatment and management of this bacterial disease involve the achievement of advanced technologies together with research. The understanding of the interaction between *Musa* spp. and pathogens will allow for identifying new strategies to be used to control the diseases without damaging the environment. Physical barriers, vulnerability, and phytoalexin yielding are some of the recognized traits used by *Musa* plants as defense mechanisms against pathogens [44,45]. Although the crops are indeed the most affected, plants possess different mechanisms to protect themselves, such as the development of secondary metabolites to eliminate the pathogen from its system.

In our case, all four of the most commonly used *Musa* spp. investigated (*M. cavendish, M. paradisiaca, M. textilis,* and *M. acuminata*) showed a similar trend in front of infection with *R. solanacearum* race 2. The diseased samples exhibited a higher antioxidant activity compared to the healthy samples, and we attempted to show that this is attributed to the presence of secondary metabolites (phenolic compounds) through the LC-MS analysis to reinforce the idea that secondary metabolites and implicitly antioxidant activity represent a defense mechanism against the pathogen that causes Moko disease. Plants possess physical and chemical barriers involved in their defense; they have a strategy to help their survival when in contact with some biotic or abiotic stresses, implicating the synthesis of secondary metabolites, which is well known as part of the plant immune system. A strong emphasis is placed on antioxidant activity to increase plant protection against the pathogen. The synthesis of metabolites comes from the primary metabolism (glycolysis, Krebs cycle, or shikimate pathway) that depends on the degree of stress to which the plant is subjected and can trigger variance in the levels of secondary metabolites, some of them toxic when stored in plant cells [46]. Synthesized phenolic compounds like simple phenols, flavonols, dihydrochals, and cones phenolic acids are antibiotic compounds that generate a response against pathogens. Phytoalexins, synthesized de novo, manage to inhibit a diversity of microorganisms by their accumulation at the infected site [47].

The antioxidant capacity of plants is related to the defense of both types of antioxidants (enzymatic and non-enzymatic) to escape from the toxic effects of free radicals. Their genetic configuration confers a great capacity to synthesize secondary metabolites under biotic or abiotic stress. Some compounds act like substrates in enzyme-catalyzed detoxification reactions and have central and interrelated functions [48–50].

Likewise, there is a considerable difference between plant developmental stages. The vegetative stage of the samples shows a lower content of secondary metabolites and antioxidant activity. The bioactive compounds produced depend on the environmental conditions to generate an adequate impact on the development of the metabolic pathways associated with their biosynthesis [51]. In the vegetative state, plants use photosynthesis and carbon assimilation for growth, development, and defense, and when exposed to some stress, can alter the storage and synthesis of metabolites. These include the harvest time, exposure to factors such as light, temperature, osmotic potential, nutrition, growth regulators, biotic inducers, and fruit-ripening stage, among others [52].

Based on their elemental role in plant protection against different agents, phenolics are known as antioxidants [53]. Phenolic compounds exhibiting significant antioxidant compounds were identified in each *Musa* spp. The predominant polyphenols in the plant defense mechanism are flavonoids, which may be classified as flavonols, isoflavonols, flavones, flavanones, catechins, and anthocyanidins [54]. New research has shown the involvement of flavonoids in plant protection, playing a significant role in the neutralization of free radicals [55], especially in terms of the highly sensitive antimicrobial effect on pathogens, in which compounds such as naringenin, kaempferol, quercetin, and dihydroquercetin stand out [56]. Recent studies have described the antibacterial effect of phenolic compounds obtained from plantain leaves against some Gram-negative bacteria (*Escherichia coli, Staphylococcus, Pseudomonas)* species [57,58]. The bacterium *Ralstonia solanacearum* race 2 (Smith, 1896) is a Gram-negative bacillus with high genetic variability that affects the vascular system of the plant [59]. The phenolic compounds identified in the current study have shown antibacterial activity toward the infection since this strain is responsible for causing the Moko disease in *Musa* spp., whose direct effect is aggravated by the capacity of its causative agent to remain in the soil for a long time, disabling the immediate replanting of the affected lots [60], highlighting the importance of describing plant defense mechanisms to reach a better understanding of the infection. In our current study, these compounds and their derivatives have been identified, from kaempferol (astragalin, kaempferol-7-O-neohesperidoside) and quercetin glycosides (spiraeoside, rutin), as well as flavones like isoorientin and scutellarein 4′-methyl. Quercetin and its derivatives also play significant roles in plant protection from the effects of UV radiation and/or osmotic stress, in which glycosylated derivatives are involved in osmoregulation [61]. In addition, flavonoids are known for their antioxidant capacity by decreasing ROS levels by inhibiting prooxidant enzymes, cyclooxygenase, and lipoxygenase [62].

Furthermore, signaling molecules have been identified; products of oxidative processes' ROS (7-methyl-rosmanol), supporting the idea of the loss of some acids (carnosic acid) with antibacterial activity against both bacteria types (Gram-positive and Gram-negative) causing the accumulation of oxidized derivatives under oxidative degradation by ROS [63,64]. Microbial attack and the oxidative state of plants mediate the activation of the plant protection mechanisms against stress through different signaling pathways, which conduct the production of various protein and non-protein compounds with roles in protection [65], such as salicylic acid (2-hydroxyhippuric acid), which has also been identified in our samples through LC-MS; salicylic acid levels are known to increase during different type of infections (viruses, fungi, insects, bacteria), while exogenous treatment with salicylic acid improves the protection system of the host [66].

Bananas and plantains are the most commonly consumed as food and are used in medicine around the world, being an attractive source of bioactive compounds. These compounds with desirable biological properties for humans are also implicated in the plant protection strategy against pathogens, as in the case of phenolics, which are a key aspect in *Musa* spp. in the protection mechanism, as it has been found in the present study. A higher content of phenols, flavonoids, and antioxidant capacity was noted in the late stages of the infected samples compared to healthy samples, which showed a lower amount of phenolics and activity.

Identification of the phenolic compounds demonstrated the presence of relevant flavonoids that are involved in the defense mechanism of plants and which are known for their antibacterial activity related to infection by Gram-negative bacteria like *R. solanacearum* responsible for the Moko disease affecting the *Musa* spp. genus.

5. Conclusions

Bananas and plantains belonging to the genus *Musa* are largely consumed all over the world as food staples and for medicinal purposes, being an interesting source of bioactive secondary metabolites. These compounds, with valuable biological properties for humans as antioxidants, also play a pivotal role in the plant's defense mechanism against pathogens. For example, phenolic compounds are essential to the defense strategy of *Musa* spp., as demonstrated in the current study, enhancing the antioxidant profile of these plants. Infected samples showcased a higher concentration of phenols, flavonoids, and antioxidant activity compared to healthy ones, which exhibited lower levels of phenolics and activity. The analysis of the phenolic compounds revealed the presence of significant flavonoids that participate in plant defense mechanisms. The flavonoids, including kaempferol, quercetin, and their glycosides, which we found in the banana samples, have antibacterial properties implicated in the fight against various pathogenic bacteria by interfering with their growth and survival.

Author Contributions: Conceptualization, R.A.M.; methodology, V.A.T.-M., K.A.P.-B. and M.J.V.-C.; formal analysis, V.A.T.-M., K.A.P.-B., M.J.V.-C., L.E.R.-G. and N.S.C.-I.; investigation, V.A.T.-M., K.A.P.-B., M.J.V.-C., L.E.R.-G. and N.S.C.-I.; writing—original draft preparation, R.A.M. and R.D.C.; writing—review and editing, R.A.M., E.J.M.-H. and R.D.C.; supervision, R.A.M.; project administration, R.A.M.; funding acquisition, R.A.M. All authors have read and agreed to the published version of the manuscript.

Funding: This work was supported by the Universidad de Las Fuerzas Armadas-ESPE, grant number CV-GNP-0066-2020, and the Institute of Biology Bucharest, Romanian Academy, grant number RO1567-IBB08/2023.

Institutional Review Board Statement: Not applicable.

Informed Consent Statement: Not applicable.

Data Availability Statement: The original contributions presented in the study are included in the article, further inquiries can be directed to the corresponding author due to privacy.

Acknowledgments: The authors are grateful for the financial support of the Society Liaison Management Unit at the University of the Armed Forces-ESPE and for the valuable help with plant collection from Agrocalidad Santo Domingo and INIAP Santo Domingo.

Conflicts of Interest: The authors declare no conflicts of interest. The funders had no role in the design of the study; in the collection, analyses, or interpretation of data; in the writing of the manuscript; or in the decision to publish the results.

Appendix A

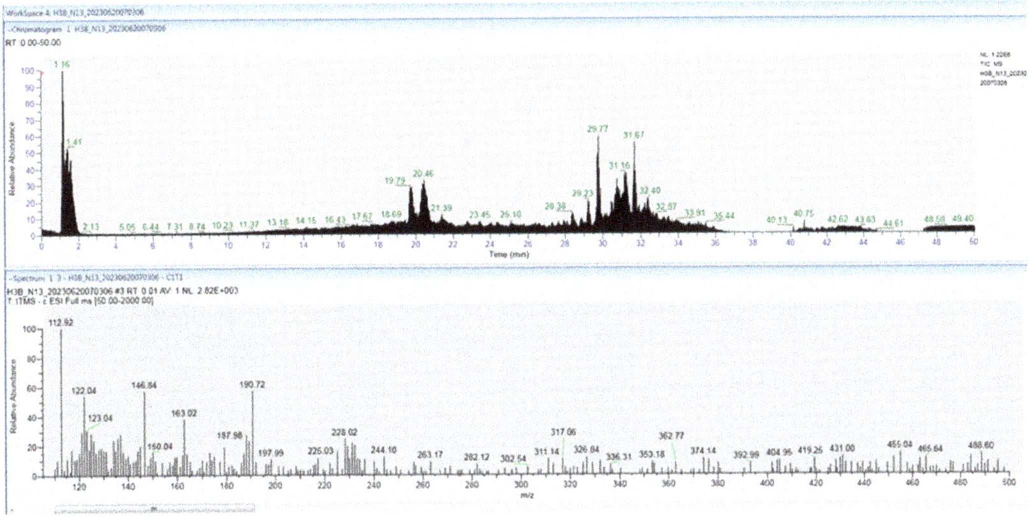

Figure A1. LC-MS chromatogram of the compounds identified in *Musa cavendish* leaf samples at Stage 1 of disease. The *y*-axis depicts the relative absorbance, while the *x*-axis represents the retention time (rt) in the upper chromatogram, and the identified molar mass is represented on the lower chromatogram for each sample.

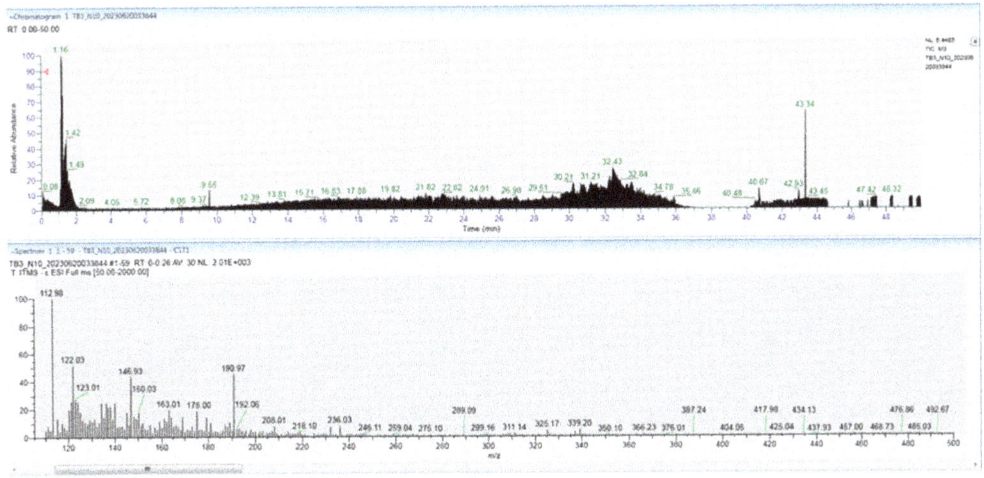

Figure A2. LC-MS chromatogram of the compounds identified in *Musa cavendish* stem samples at Stage 1 of the disease. The *y*-axis depicts the relative absorbance, while the *x*-axis represents the retention time (rt) in the upper chromatogram, and the identified molar mass is represented on the lower chromatogram for each sample.

References

1. International Network for the Improvement of Banana and Plantain, International Plant Genetic Resources Institute. 2000. Available online: https://cgspace.cgiar.org/handle/10568/105424?show=full (accessed on 15 June 2023).
2. FAO. *Banana Market Review—Preliminary Results 2022*; FAO: Rome, Italy, 2022.

3. Lopes, S.; Vanz Borges, C.; de Sousa Cardoso, S.M.; de Almeida Pereira da Rocha, M.F.; Maraschin, M. Banana (*Musa* spp.) as a Source of Bioactive Compounds for Health Promotion. In *Handbook of Banana Production, Postharvest Science, Processing Technology, and Nutrition*, 1st ed.; Siddiq, M., Ahmed, J., Lobo, M.G., Eds.; John Wiley & Sons Ltd.: Hoboken, NJ, USA, 2020.
4. Blomme, G.; Dita, M.; Jacobsen, K.S.; Pérez, V.L.; Molina, A.; Ocimati, W.; Poussier, S.; Prior, P. Bacterial Diseases of Bananas and Enset: Current State of Knowledge and Integrated Approaches Toward Sustainable Management. *Front. Plant Sci.* **2017**, *20*, 1290. [CrossRef] [PubMed]
5. Mansfield, J.; Genin, S.; Magori, S.; Citovsky, V.; Sriariyanum, M.; Ronald, P.; Dow, M.; Verdier, V.; Beer, S.V.; Machado, M.A.; et al. Top 10 plant pathogenic bacteria in molecular plant pathology. *Mol. Plant Pathol.* **2012**, *13*, 614–629. [CrossRef] [PubMed]
6. Denny, T. Plant pathogenic *Ralstonia* species. In *Plant-Associated Bacteria*; Gnanamanickam, S.S., Ed.; Springer Netherlands: Cham, The Netherlands, 2006; pp. 573–644.
7. Banana Moko: A Silent Threat. 2022. Available online: https://www.ecuadortimes.net/banana-moko-a-silent-threat/ (accessed on 15 June 2023).
8. Thwaites, M.; Eden-Green, S. RAPD and rep PCR-based fingerprinting of vascular bacterial pathogens of *Musa* spp. *Plant Pathol.* **1999**, *48*, 121–128. [CrossRef]
9. Abadie, C.; Baudouin, L.; Daugrois, J.-H.; Dollet, M.; Vuillaume, C.; Wicker, E.; Teycheney, P.-Y. CIRAD invasive species initiatives in the Caribbean Basin. In Proceedings of the 44th Annual Meeting of the Caribbean Food Crop Society, Miami, FL, USA, 13–17 July 2008.
10. Alvarez, E.; Gomez, E.; Mejia, J.; Prior, P. Developing a TaqMan probe to detect, through real-time PCR, *Ralstonia solanacearum* which causes Moko in *Musa* spp. in Colombia. Cirad Ritrop. In Proceedings of the 12th International Conference on Plant Pathogenic Bacteria: Programme, Abstracts, List of Participants, INRA, Université de la Réunion, Saint-Denis, Réunion, 7 June 2010; p. 64. Available online: https://agritrop.cirad.fr/555761/ (accessed on 13 January 2024).
11. Oliveira Silva, S.; de Mello Véras, S.; de Gasparotto, L.; de Matos, A.P.; Maciel Cordeiro, Z.; Boher, B. Evaluation of *Musa* spp. for resistance to Moko disease (*Ralstonia solanacearum*, race 2). *InfoMusa* **2000**, *9*, 19–20.
12. Prior, P.; Wicker, E.; Fegan, M. The *Ralstonia solanacearum* species complex: Genetic diversity, phylogeny and molecular typing of strains with particular attention to emerging strains and bacterial wilts of banana known as moko disease, bugtok disease, and blodd disease. In Proceedings of the II Seminario Internacional Sobre Producción, Comercialización e Industrialización de Plátano, Manlzales, Colombia, 29 August–2 September 2005; p. 49.
13. Tripathi, L.; Ntui, V.O.; Tripathi, J.N. Control of Bacterial Diseases of Banana Using CRISPR/Cas-Based Gene Editing. *Int. J. Mol. Sci.* **2022**, *23*, 3619. [CrossRef] [PubMed]
14. Grajales-Amorocho, M.; Acosta-Minoli, C.; Muñoz-Pizza, D.; Manrique-Arias, O.; Munoz-Loaiza, A. Analysis of Moko disease propagation on plantain (*Musa* AAB Simmonds) through a model based on system dynamics. *Eur. J. Plant Pathol.* **2023**, *168*, 437–445. [CrossRef]
15. Blanco, G.; Linares, B.; Hernández, J.; Maselli, A.; Rincón, A.; Ortega, R.; Medina, E.; Hernández, L.; Morillo, J. Microbiological composition and safety of pseudostems and leaf blades leachates of "Harton" plantain in Yaracuy state. *Agron. Trop.* **2013**, *63*, 111–120.
16. Ricardo, F.Á.S.; Vicente, L.F.P. Strategic tactics for the integrated management of pests and diseases in banana. *Braz. J. Anim. Environ. Res.* **2021**, *4*, 4973–5000. [CrossRef]
17. Sequeira, L. Bacterial wilt: The missing element in international banana improvement programs. In *Bacterial Wilt Disease: Molecular and Ecological Aspects*; Prior, P., Elphinstone, A.C.J., Eds.; Springer-Verlag: Berlin/Heidelberg, Germany, 1998; Volume 6, p. 14.
18. Álvarez, E.; Pantoja, A.; Gañán, L.; Ceballos, G. *Current Status of Moko Disease and Black Sigatoka in Latin America and the Caribbean, and Options for Managing Them*; Centro Internacional de Agricultura Tropical (CIAT): Palmira, Colombia; Food and Agriculture Organization of the United Nations (FAO): Rome, Italy, 2015; 40p.
19. Ramirez, G.J.G.O.; Munoz, A.M.; Patino, H.L.F.; Morales, O.J.G. Banana Moko disease management with resistance inducers and chlorine dioxide. *Agron. Colomb.* **2015**, *33*, 194–202. [CrossRef]
20. Yang, L.; Wen, K.S.; Ruan, X.; Zhao, Y.X.; Wei, F.; Wang, Q. Response of Plant Secondary Metabolites to Environmental Factors. *Molecules* **2018**, *23*, 762. [CrossRef]
21. Isroni, M.; EnggarPaskariani, A.; Setiawan, F.; Rakhmawati, A. Banana Peel (*Musa paradisiaca*) Extract's Potency as an Antibacterial *Ralstonia solanacearum* cause Tomato's Desease. *Proceeding Biol. Educ. Conf.* **2018**, *15*, 844–847.
22. Ploetz, R.C.; Kepler, A.K.; Daniells, J.; Nelson, S.C. Banana and plantain—An overview with emphasis on Pacific island cultivars. In *Species Profiles for Pacific Island Agroforestry*; Elevitch, C.R., Ed.; Permanent Agricultural Resources: Holualoa, HI, USA, 2007.
23. Peakland Heritage. 19 July 2002. Available online: https://web.archive.org/web/20160314051757/http://www.peaklandheritage.org.uk/index.asp?peakkey=01001021 (accessed on 18 May 2024).
24. Ajijolakewu, K.A.; Ayoola, A.S.; Agbabiaka, T.O.; Zakariyah, F.R.; Ahmed, N.R.; Oyedele, O.J.; Sani, A. A review of the ethnomedicinal, antimicrobial, and phytochemical properties of *Musa paradisiaca* (plantain). *Bull. Natl. Res. Cent.* **2021**, *45*, 86. [CrossRef]
25. Horry, J.; Ortiz, R.; Arnaud, E.; Crouch, J.H.; Ferris, R.S.B.; Jones, D.R.; Mateo, N.; Picq, C.; Vuylsteke, D. Banana and Plantain. In *Biodiversity in Trust Conservation and Use of Plant Genetic Resources in CGIAR Centres*; Fuccillo, D., Sears, L., Stapleton, P., Eds.; Cambridge University Press: Cambridge, MA, USA, 1997; pp. 67–81.

26. Daniells, J.; Jenny, C.; Karamura, D.; Tomekpe, K. *Musalogue (A Catalogue of Musa Germplasm): Diversity in the Genus Musa*; International Plant Genetic Resources Institute: Rome, Italy, 2001; 207p.
27. Thaweesang, S. Antioxidant activity and total phenolic compounds of fresh and blanching banana blossom (*Musa* ABB CV.Kluai "Namwa") in Thailand. *IOP Conf. Series: Mat. Sci. Eng.* **2019**, *639*, 012047. [CrossRef]
28. Pękal, A.; Pyrzynska, K. Evaluation of aluminium complexation reaction for flavonoid content assay. *Food Anal. Met.* **2014**, *7*, 1776–1782. [CrossRef]
29. Sachett, A.; Gallas-Lopes, M.; Conterato, G.M.M.; Herrmann, A.; Piato, A. Antioxidant Activity by DPPH Assay: In Vitro Protocol. *Protocols* Io. 2021. Available online: https://www.protocols.io/view/antioxidant-activity-by-dpph-assay-in-vitro-protoc-btbpnimn (accessed on 13 January 2024).
30. Kuskoski, E.M.; Asuero, A.G.; Troncoso, A.M.; Mancini-Filho, J.; Fett, R. Aplicación de diversos métodos químicos para determinar actividad antioxidante en pulpa de frutos. *Food Sci. Technol.* **2005**, *25*, 726–732. [CrossRef]
31. Tohma, H.; Köksal, E.; Kılıç, Ö.; Alan, Y.; Yılmaz, M.A.; Gülçin, İ.; Bursal, E.; Alwasel, S.H. RP-HPLC/MS/MS analysis of the phenolic compounds, antioxidant and antimicrobial activities of *Salvia* L. species. *Antioxidants* **2016**, *5*, 38. [CrossRef] [PubMed]
32. Irakli, M.; Skendi, A.; Bouloumpasi, E.; Chatzopoulou, P.; Biliaderis, C.G. LC-MS identification and quantification of phenolic compounds in solid residues from the essential oil industry. *Antioxidants* **2021**, *10*, 2016. [CrossRef] [PubMed]
33. Rajurkar, N.S.; Hande, S.M. Estimation of phytochemical content and antioxidant activity of some selected traditional Indian medicinal plants. *Indian. J. Pharm. Sci.* **2011**, *73*, 146–151. [CrossRef] [PubMed]
34. Luskal, T.; Castillo, S.; Villar-Briones, A.; Orešič, M. MZmine 2: Modular framework for processing, visualizing, and analyzing mass spectrometry-based molecular profile data. *BMC Bioinform.* **2010**, *11*, 395.
35. Cellier, G.; Moreau, A.; Chabirand, A.; Hostachy, B.; Ailloud, F.; Prior, P. A Duplex PCR Assay for the Detection of *Ralstonia solanacearum* Phylotype II Strains in *Musa* spp. *PLoS ONE* **2015**, *10*, e0122182. [CrossRef]
36. Ranjitha, K.; Narayana, C.K.; Roy, T.K. Division of Plant Physiology and Biochemistry, IIHR, Bangalore Aroma profile of fruit juice and wine prepared from Cavendish banana (*Musa* sp., Group AAA) cv. Robusta. *J. Hortl. Sci.* **2013**, *8*, 217–223. [CrossRef]
37. Parijadi, A.A.R.; Yamamoto, K.; Ikram, M.M.M.; Dwivany, F.M.; Wikantika, K.; Putri, S.P.; Fukusaki, E. Metabolome Analysis of Banana (*Musa acuminata*) Treated With Chitosan Coating and Low Temperature Reveals Different Mechanisms Modulating Delayed Ripening. *Front. Sustain. Food Syst.* **2022**, *6*, 835978. [CrossRef]
38. Sonibare, M.; Oresanya, A.I.; Guèye, B.; Abberton, M.; Dsouza, R.; Kuhnert, N. Leaves metabolomic profiling of *Musa acuminata* accessions using UPLC–QTOF–MS/MS and their antioxidant activity. *J. Food Meas. Charact.* **2018**, *12*, 1093–1106. [CrossRef]
39. Matos da Silva, M.; Pereira Alexandre, G.; Magalhães, M.R.; Torres, A.M.; Kato, L.; Costa da Silva, V.; Teixeira de Saboia Morais, S.M.; Garcia Rodriguez, A.; Pacheco Fill, T.; Pereira, A.K.; et al. *Musa* spp. cultivars as a neutralizing source against some toxic activities of Bothrops and Crotalus genus snake venoms. *Toxicon* **2023**, *228*, 107106. [CrossRef] [PubMed]
40. Behiry, S.I.; Okla, M.K.; Alamri, S.A.; EL-Hefny, M.; Salem, M.Z.M.; Alaraidh, I.A.; Salem, A.Z.M. Antifungal and antibacterial activities of *Musa paradisiaca* L. Peel Extract: HPLC analysis of phenolic and flavonoid contents. *Processes* **2019**, *7*, 215. [CrossRef]
41. Waghmare, J.S.; Kurhade, A.H. GC-MS analysis of bioactive components from banana peel (*Musa sapientum* peel). *Eur. J. Exper Biol.* **2014**, *4*, 10–15.
42. Oliveira, C.S.R.; Freire, A.J.D.; Silvestre, N.; Cordeiro, I.C.; Torres, D. Lipophilic extractives from different morphological parts of banana plant "Dwarf Cavendish". *Evtuguin Ind. Crops Prod.* **2006**, *23*, 201–211. [CrossRef]
43. Oresanya, I.O.; Sonibare, M.A.; Gueye, B.; Balogun, F.O.; Adebayo, S.; Ashafa, A.O.T.; Morlock, G. Isolation of flavonoids from *Musa acuminata* Colla (Simili radjah, ABB) and the in vitro inhibitory effects of its leaf and fruit fractions on free radicals, acetylcholinesterase, 15-lipoxygenase, and carbohydrate hydrolyzing enzymes. *J. Food Biochem.* **2020**, *44*, e13137. [CrossRef]
44. Cittan, M.; Çelik, A. Development and validation of an analytical methodology based on Liquid Chromatography–Electrospray Tandem Mass Spectrometry for the simultaneous determination of phenolic compounds in olive leaf extract. *J. Chromatogr. Sci.* **2018**, *56*, 336–343. [CrossRef]
45. Balamurugan, A.; Sakthivel, K.; Gautam, R.K.; Sharma, S.K.; Kumar, A. Ralstonia solanacearum: Biology and its management in solanaceous vegetable crops. In *Rhizosphere Microbes, Microorganisms for Sustainability*; Sharma, S.K., Singh, U.B., Sahu, P.K., Singh, H.V., Sharma, P.K., Eds.; Springer: Singapore, 2020; Volume 23, pp. 259–289.
46. Vaganan, M.M.; Ravi, I.; Nandakumar, A.; Sarumathi, S.; Sundararaju, P.; Mustaffa, M.M. Phenylpropanoid enzymes, phenolic polymers and metabolites as chemical defenses to infection of *Pratylenchus coffeae* in roots of resistant and susceptible bananas (*Musa* spp.). *Indian. J. Exp. Biol.* **2014**, *52*, 252–260.
47. Wang, Z.; Jia, C.H.; Li, J.Y.; Huang, S.Z.; Xu, B.Y.; Jin, Z.Q. Activation of salicylic acid metabolism and signal transduction can enhance resistance to *Fusarium* wilt in banana (*Musa acuminata* L. AAA group, cv. Cavendish). *Funct. Integ. Gen.* **2015**, *15*, 47–62. [CrossRef] [PubMed]
48. Jan, R.; Asaf, S.; Numan, M.L.; Kim, K.-M. Plant Secondary Metabolite Biosynthesis and Transcriptional Regulation in Response to Biotic and Abiotic Stress Conditions. *Agronomy* **2021**, *11*, 968. [CrossRef]
49. Ewané, A.C.; Lepoivre, P.; de Bellaire, L.; Lassois, L. Involvement of phenolic compounds in the susceptibility of bananas to crown rot. A review. *BASE* **2012**, *3*, 393–404.
50. Grene, R. Oxidative Stress and Acclimation Mechanisms in Plants. In *The Arabidopsis Book*; American Society of Plant Biologists: Rockville, MD, USA, 2002; Volume 1, p. e0036. [CrossRef]

51. Kasote, D.M.; Katyare, S.S.; Hegde, M.V.; Bae, H. Significance of Antioxidant Potential of Plants and its Relevance to Therapeutic Applications. *Intern. J. Biol. Sci.* **2015**, *11*, 982–991. [CrossRef] [PubMed]
52. Isah, T. Stress and defense responses in plant secondary metabolites production. *Biol. Res.* **2019**, *52*, 39. [CrossRef] [PubMed]
53. Olędzka, A.J.; Czerwińska, M.E. Role of Plant-Derived Compounds in the Molecular Pathways Related to Inflammation. *Int. J. Mol. Sci.* **2023**, *24*, 4666. [CrossRef] [PubMed]
54. Kytidou, K.; Artola, M.; Overkleeft, H.S.; Aerts, J.M.F.G. Plant Glycosides and Glycosidases: A Treasure-Trove for Therapeutics. *Front. Plant Sci.* **2020**, *11*, 357. [CrossRef] [PubMed]
55. Dias, M.C.; Pinto, D.C.G.A.; Silva, A.M.S. Plant Flavonoids: Chemical Characteristics and Biological Activity. *Molecules* **2021**, *26*, 5377. [CrossRef] [PubMed]
56. Terao, J. Potential Role of Quercetin Glycosides as Anti-Atherosclerotic Food-Derived Factors for Human Health. *Antioxidants* **2023**, *12*, 258. [CrossRef]
57. Ismail, T.N.; Awang, R.; Azmans, R.; Sobhan, M.; Shahidan, W.N.S. Chemical compounds and antimicrobial activity of acetone *Musa acuminate* AA/AAA leaf stalk extracts on selective Gram-negative bacteria. *Malays. J. Anal. Sci.* **2018**, *22*, 957–964.
58. Jouneghani, R.S.; Castro, A.H.F.; Panda, S.K.; Swennen, R.; Luyten, W. Antimicrobial activity of selected banana cultivars against important human pathogens, including candida biofilm. *Foods* **2020**, *9*, 435. [CrossRef] [PubMed]
59. Ghorai, A.K.; Dutta, S.; Barman, R.A. Genetic diversity of *Ralstonia solanacearum* causing vascular bacterial wilt under different agroclimatic regions of West Bengal, India. *PLoS ONE* **2022**, *17*, e0274780. [CrossRef] [PubMed]
60. Mariano, R.L.R.; Silveira, N.S.S.; Michereff, S.J. Bacterial wilt in Brazil: Current Status and Control Methods. In *Bacterial Wilt Disease*; Prior, P., Allen, C., Elphinstone, J., Eds.; Springer: Berlin/Heidelberg, Germany, 1998; pp. 386–393.
61. Jańczak-Pieniążek, M.; Migut, D.; Piechowiak, T.; Buczek, J.; Balawejder, M. The Effect of Exogenous Application of Quercetin Derivative Solutions on the Course of Physiological and Biochemical Processes in Wheat Seedlings. *Int. J. Mol. Sci.* **2021**, *22*, 6882. [CrossRef] [PubMed]
62. An, J.; Kim, S.H.; Bahk, S.; Vuong, U.T.; Nguyen, N.T.; Do, H.L.; Kim, S.H.; Chung, W.S. Naringenin induces pathogen resistance against *Pseudomonas syringae* through the activation of NPR1 in *Arabidopsis*. *Front. Plant Sci.* **2021**, *12*, 672552. [CrossRef] [PubMed]
63. Loussouarn, M.; Krieger-Liszkay, A.; Svilar, L.; Bily, A.; Birtić, S.; Havaux, M. Carnosic Acid and Carnosol, Two Major Antioxidants of Rosemary, Act through Different Mechanisms. *Plant Physiol.* **2017**, *175*, 1381–1394. [CrossRef]
64. Pavić, V.; Jakovljević, M.; Molnar, M.; Jokić, S. Extraction of carnosic acid and carnosol from sage (*Salvia officinalis* L.) leaves by supercritical fluid extraction and their antioxidant and antibacterial activity. *Plants* **2019**, *8*, 16. [CrossRef]
65. War, A.R.; Paulraj, M.G.; War, M.Y.; Ignacimuthu, S. Role of salicylic acid in induction of plant defense system in chickpea (*Cicer arietinum* L.). *Plant Signal Behav.* **2011**, *6*, 1787–1792. [CrossRef]
66. Lefevere, H.; Bauters, L.; Gheysen, G. Salicylic Acid Biosynthesis in Plants. *Front. Plant Sci.* **2020**, *11*, 338. [CrossRef]

metabolites MDPI

Article

Metabolic Reprogramming of Barley in Response to Foliar Application of Dichlorinated Functional Analogues of Salicylic Acid as Priming Agents and Inducers of Plant Defence

Claude Y. Hamany Djande, Paul A. Steenkamp, Lizelle A. Piater, Fidele Tugizimana and Ian A. Dubery *

Research Centre for Plant Metabolomics, Department of Biochemistry, University of Johannesburg, P.O. Box 524, Auckland Park, Johannesburg 2006, South Africa; claudeh@uj.ac.za (C.Y.H.D.); psteenkamp@uj.ac.za (P.A.S.); lpiater@uj.ac.za (L.A.P.); ftugizimana@uj.ac.za (F.T.)
* Correspondence: idubery@uj.ac.za; Tel.: +27-11-5592401

Abstract: Designing innovative biological crop protection strategies to stimulate natural plant immunity is motivated by the growing need for eco-friendly alternatives to conventional biocidal agrochemicals. Salicylic acid (SA) and analogues are known chemical inducers of priming plant immunity against environmental stresses. The aim of the study was to study the metabolic reprogramming in barley plants following an application of three proposed dichlorinated inducers of acquired resistance. 3,5-Dichloroanthranilic acid, 2,6-dichloropyridine-4-carboxylic acid, and 3,5-dichlorosalicylic acid were applied to barley at the third leaf stage of development and harvested at 12, 24, and 36 h post-treatment. Metabolites were extracted using methanol for untargeted metabolomics analyses. Samples were analysed by ultra-high performance liquid chromatography coupled to high-definition mass spectrometry (UHPLC-HDMS). Chemometric methods and bioinformatics tools were used to mine and interpret the generated data. Alterations in the levels of both primary and secondary metabolites were observed. The accumulation of barley-specific metabolites, hordatines, and precursors was observed from 24 h post-treatment. The phenylpropanoid pathway, a marker of induced resistance, was identified among the key mechanisms activated by the treatment with the three inducers. No salicylic acid or SA derivatives were annotated as signatory biomarkers; instead, jasmonic acid precursors and derivatives were found as discriminatory metabolites across treatments. The study highlights differences and similarities in the metabolomes of barley after treatment with the three inducers and points to the triggering chemical changes associated with defence and resistance. This report is the first of its kind, and the knowledge acquired provides deeper insight into the role of dichlorinated small molecules as inducers of plant immunity and can be used in metabolomics-guided plant improvement programmes.

Keywords: antimicrobial metabolites; barley; *Hordeum vulgare*; liquid chromatography; mass spectrometry; metabolomics; multivariate data analysis; secondary metabolites

1. Introduction

Plants are naturally exposed to a plethora of stresses, which severely affect their growth and yield. Upon exposure to biotic stresses, molecular communication with the plant takes place and may lead to a beneficial association or a disease. In this interaction between two biological organisms, the resistance or susceptibility is determined by the ability to ward off/defend against the stress-inducing environment in which each finds itself. Plants have evolved an innate immune system against microbial threats consisting of preformed defensive barriers and constitutive chemical deterrents and inducible defences based on newly synthesised defence proteins and antimicrobial chemicals [1–3]. Resistance responses might not always be effective due to not being timeously triggered or not being launched at the required intensity. Moreover, some pathogens have developed certain strategies to circumvent host defences, thus resulting in disease. Abiotic environmental stressors can also have a negative impact on the resistance or tolerance of plants towards a specific strain of pathogen [4], necessitating the use of antimicrobial chemicals for disease control. As a result of public concern over the risks associated with pesticide usage, alternative control strategies are receiving much attention in order to limit unfavourable hazardous environmental side effects.

Induced resistance (IR) can be described as an altered metabolic state that is activated once the host plant's immune system is triggered or activated by a pathogen attack or other biotic stresses [5]. Fundamentally, as part of IR, systemic acquired resistance (SAR) is known to depend on salicylic acid (SA) and induced systemic resistance (ISR) relies on beneficial microbes and hormones, such as jasmonic acid (JA) and ethylene (ET). However, there is an interconnection involving more participatory mechanisms than SAR and ISR in IR, creating a multi-layered network defining plant immunity [6–9]. The mediation of SAR by the accumulation of SA has previously been demonstrated [5,10–14]. SAR is often effective against a wide spectrum of pathogens and is regarded as the most agronomically important type of plant immunity [15]. The activation of plant immune mechanisms can be initiated by (a non-virulent) pathogen attack but also by treatment with natural or synthetic compounds, such as SA and synthetic analogues [16,17]. Studies on IR elicited by chemicals have revealed previously unknown features of the plant defence response, including defence priming [18].

In addition to the mentioned IR associated with innate immunity, plants can also acquire immunity upon treatment with certain biotic and abiotic stimuli, a phenomenon mediated by preconditioning or priming for inducible defence [19,20]. Immune priming or potentiation enables faster and/or stronger induction of inducible defences following a successful attack by pathogens. As priming events boost multigenic basal resistance, the ensuing protection against disease can be more resilient than race-specific resistance, which is based on single resistance genes. Regardless of the fact that priming rarely provides complete disease protection [21], the application of priming-inducing agents is increasingly considered for exploitation in integrated pest and disease management [22,23].

SA performs a crucial role as an endogenous signalling molecule that activates various aspects of plant defence. The phytohormone also performs important roles in growth and development, respiratory pathways, and regulation of redox homeostasis [24–26]. The exogenous application of SA and related synthetic analogues activating the plant defence system has been reported in several studies [27–31]. Although very efficient in inducing plant resistance, the phytotoxicity and rapid glycosylation of SA result in reduced efficacy and have prevented its use as a plant protection molecule. Hence, synthetic chemical analogues of SA, capable of mimicking various functions thereof, represent an attractive alternative/substitute to the use of conventional biocidal agrochemicals [23,32]. Substitution of SA with the electron-withdrawing element chlorine has displayed an increase in SA activity [33]. Mono- and dichloro-substituted SA derivatives, such as 3,5-DCSA and 2,6-dichloroisonicotinic acid (2,6-DCINA), were found to successfully increase tobacco plant resistance to pathogen attack through the induction and accumulation of pathogenesis-related (PR) proteins [31,34]. It was further established that mono or multiple substitution(s) at positions three and/or five on SA were more active at inducing the PR-1a protein than those at positions two, four, and six, and SA itself [33]. Relatedly, a screening of 60,000 unique chemicals for inducers of pathogen-responsive reporter genes in Arabidopsis seedlings led to the discovery of 114 synthetic elicitor candidates, including 3,5-dichloroanthranilic acid (3,5-DCAA). Upon treatment with 3,5-DCAA, Arabidopsis effectively develops resistance to virulent isolates of the oomycete *Hyaloperonospora parasitica* and *Pseudomonas syringae* DC3000 [28].

Adaptive metabolic reprogramming in plants is often a result of the exposure to a single or a multitude of external stimuli or environmental pressures. Such variation can be observed in a wide variety of primary and secondary metabolites, including ionic inorganic compounds, hydrophilic carbohydrates, amino acids, organic compounds, and compounds linked to hydrophobic lipids. Investigating the metabolic perturbations caused by chemical activators of priming or inducers of IR can significantly contribute to gaining insights into the unique vs. shared features of the host response towards these inducers. In this study, the host was barley (*Hordeum vulgare* L., 'Hessekwa' cultivar), the fourth most important cereal crop in the world. Barley is primarily used for animal feed and human food and alcoholic beverage production. The widely adaptable, short-seasoned, and early maturing crop is farmed as a summer or winter crop in temperate and tropical climates, respectively. In plant science, barley is commonly used as a model plant, particularly when assessing plant resilience to environmental stress [35,36].

Untargeted metabolomics approaches combined with advanced chemometric tools were employed to investigate the metabolic reprogramming in leaves following treatment with synthetic (functional) analogues of SA: 3,5-dichlorosalicylic acid (3,5-DCSA), 3,5-dichloroanthranilic acid (3,5-DCAA), and 2,6-dichloropyridine-4-carboxylic acid (2,6-DCP-4-CA, also known as 2,6-DCINA), in order to evaluate their potential for the induction of a state of enhanced resistance in barley plants. The importance of metabolic reprogramming and priming to improve abiotic stress tolerance in a variety of significant crops is being supported by growing research.

2. Materials and Methods

2.1. Barley Plant Material and Growth Conditions

Barley seeds were collected from the experimental line 'Hessekwa' cultivated in Bredasdorp, in the Western Cape region of South Africa. 'Hessekwa' is a rainfed winter crop, and seeds were provided by the South African Barley Breeding Institute (SABBI). In the current study, seeds were grown as previously described [37,38]. Briefly, the plants were grown in a plant growth room under well-controlled conditions: 12 h fluorescence light (≈85 μmol m^{-2} s^{-2}) and 12 h dark cycle at 22–27 °C. Surfaced-sterilised (70% ethanol) and soaked (in sterile water for 2 h) barley seeds were planted in pasteurised (at 70 °C) soil. The plants were watered twice a week with distilled water containing a water-soluble chemical fertiliser (Multisol 'N', Culterra, Muldersdrift, South Africa). The plants were grown for 21 d (or 16 d post-emergence) before treatment with the inducers. At that time, the seedlings were at the physiological stage 13 according to the Zadoks growth and development scale [37,38].

2.2. Barley Plant Treatment with Priming Inducers

All chemicals, namely, 3,5-dichlorosalicylic acid (3,5-DCSA), 2,6-dichloropyridine-4-carboxylic acid (2,6-DCP-4-CA), and 3,5-dichloroanthranilic acid (3,5-DCAA) were obtained from Merck–Sigma-Aldrich, Johannesburg, South Africa. These inducers were dissolved in dimethylsulphoxide (DMSO, 1 μL/mL; BDH Chemicals, England) mixed with 0.05% of wetting agent (Effekto, Pretoria, South Africa). Control plants received the same concentration of DMSO and wetting agent. Following the foliar application (40 sprays = ±6 mL/pot) of 200 μM of 3,5-DCSA, 2,6-DCP-4-CA, and 3,5-DCAA, treatment groups were kept separately, and leaves (the entire aerial parts of the plants, 1 cm above the surface) were harvested after 12, 24, and 36 h, respectively. The concentration of the inducers and the time interval of the investigation were chosen based on several optimisation studies that indicated no morphological effects. The experimental design included three independent biological replicates for each treatment.

2.3. Metabolite Extraction from Seedlings and Pre-Analytical Sample Preparation

Barley shoot tissues were harvested and snap-frozen in the liquid nitrogen to quench metabolic activity. Metabolites were extracted from the leaves of each replicate with 80% cold aqueous methanol (1:10 *w*/*v* ratio). Following homogenisation for 1 min with an Ultra-Turrax homogeniser, and sonication for 10 s with a probe sonicator (Bandelin Sonopuls, Berlin, Germany), the homogenates were centrifuged at 5100× *g* and 4 °C for 20 min. Supernatants were concentrated to 1 mL using a rotary evaporator at 45 °C and further evaporated to complete dryness in a dry bath pre-heated at 45 °C. The reconstitution of dried extract was completed with 50% UHPLC-grade methanol (Romil, Cambridge, UK) in a 1:10 m/v ratio. In preparation for chromatographic analyses, extracts were filtered through 0.22 μm nylon filters into chromatography vials fitted with 500 μL inserts, capped, and stored at 4 °C.

2.4. Sample Analyses on Mass Spectrometry-Based Analytical Platforms (Ultra-High Performance Liquid Chromatography—High Definition Mass Spectrometry (UHPLC-HDMS))

The Waters Acquity UHPLC hyphenated with a Waters SYNAPT G1 QTOF (quadrupole time-of-flight) mass spectrometer system (Waters Corporation, Milford, MA, USA) was used to analyse the aqueous-methanol extracts. The chromatographic separation of samples was completed using a Waters HSS T3 reverse phase C18 column (150 mm × 2.1 mm × 1.8 μm) thermostatted at 60 °C. The concave gradient elution was carried out at a flow rate of 0.4 mL.min^{-1} using a binary solvent system consisting of water (eluent A) and acetonitrile (eluent B; Romil Pure Chemistry, Cambridge, UK), both of which contained 0.1% formic acid. For the first min of the elution, 95% A and 5% B were kept constant. When the gradient was applied, the chromatographic condition was changed to 10% A and 90% B for 10 s, followed by 5% A and 95% B for 1 min and 50 s, before being restored to the original condition at the end of 28 min. Each sample was injected with a 2 μL volume over the course of a 30 min run. To prevent measurement bias, all sample extracts were randomised. Additionally, pooled quality control (QC) samples were also used to evaluate the stability of the LC-MS system. Blanks consisting of 50% MeOH were used to monitor potential carry-over. Data acquisition was based on three independent biological replicates, and each was analysed in triplicate; thus, $n = 9$.

The TOF MS analyser was used in V-optics mode, and centroid spectral data were acquired using both positive and negative electrospray ionisation (ESI), with a scan range of 50–1200 Da and a scan time of 0.1 s. The cone and desolvation gas flows were at 50 L.h^{-1} and 550 L.h^{-1}, respectively. Nitrogen was used as a nebuliser gas at a flow rate of 700 50 L.h^{-1}. The sampling and extraction cone voltages were 40 V and 4.0 V, respectively, while the capillary voltage was set at 2.5 kV. The desolvation temperature was set at 450 °C, and the source temperature was fixed at 120 °C. Leucine encephalin (50 pg·mL^{-1}, $[M + H]^+ = 556.2771$ and $[M - H]^- = 554.2615$) was used as lock mass, sampled every 15 s, and producing an average intensity of 350 counts per scan. The lock mass serves to correct the centroid mass values in the sample for small deviations from the accurate mass measurement. Both unfragmented and fragmented (using an MS^E method, 10–40 eV) data were acquired. Fragmentation data were used for downstream metabolite structural elucidation and annotation.

2.5. Chemometrics: Data Mining

UHPLC-MS extracted raw data (analysed in positive and negative ionisation modes) was processed using the MarkerLynx XS™ application management tool of the MassLynx XS software, version 4.1 (Waters Corporation, Milford, MA, USA) to generate the corresponding data matrices. For precise peak detection and alignment, the software uses the unique *ApexTrack* (also termed *ApexPeakTrack*) algorithm. A modified Savitzky-Golay smoothing and integration was used prior to the computation of peak intensities. The sample was normalised using the total ion intensities associated with each peak. The processing parameters were set at a mass range of 50–1200 Da, a mass window of 0.05 Da, the intensity threshold (noise elimination level parameter) was set at 100 counts, a mass tolerance of 0.05 Da, and a retention time (Rt) range of 1.5–25.0 min of the chromatograms.

For statistical modelling, generated data matrices were exported to the 'soft independent modelling of class analogy' (SIMCA) software, version 14, equipped with the 'omics skin' function (Sartorius, Umeå, Sweden). As specified in the results section, data were scaled prior to computing chemometric models. As part of the chemometrics models generated, the unsupervised method, principal component analysis (PCA), was applied to reduce the dimensionality of the data and to reveal groupings, trends, and similarities existing between treatments. Moreover, the supervised model, orthogonal projection to latent structures-discriminant analysis (OPLS-DA), was also computed to classify the samples (binary classification), generate descriptive statistics, and provide potential biomarkers. Models were evaluated using the predictive power, Q^2, and the explained cumulative variation in the matrix X, R^2X (cum), also known as the 'goodness of fit' parameter. Additionally, the cross-validated predictive residual analysis of variance (CV-ANOVA) was taken into account to statistically evaluate the accuracy of the OPLS-DA models created. A *p*-value of less than 0.05 suggested a strong model. Moreover, a permutation test was performed on 100 randomly initiated permutations to validate the models. Discriminant features or ions, with both high correlation and covariation, are located at the extreme ends of the generated loading S-plots (e.g., [p(corr) $\geq$ 0.5, $\leq$−0.5, and (p1) $\geq$ 0.1, $\leq$−0.1]) and their statistical significance was assessed on variable importance in projection (VIP) plots which summarise the importance of features. The VIP scores determine the level of significance of each ion in the dataset, and all selected features had VIP scores between 1 and 2.

In the MetaboAnalyst 5.0 (www.metaboanalyst.ca/, accessed on 1 May 2022) [39] platform, dendrogram heatmaps, which allowed the visualisation of the distribution of selected metabolites across the conditions (different time-points and treatments), were constructed. In addition, Metabolomics Pathway Analysis (MetPA), also housed in the MetaboAnalyst bioinformatics software, was used to uncover the key metabolic pathways that define induced responses in barley treated with 3,5-DCAA, 2,6-DCP-4-CA, and 3,5-DCSA. KEGG (Kyoto Encyclopedia of Genes and Genomes, www.genome.jp/kegg/pathway.html, accessed on 10 May 2022) [40] identifiers for each annotated metabolite were used as inputs, and high-quality KEGG metabolic pathways were used as the backend knowledge base.

2.6. Metabolite Annotation

To comprehensively annotate the measured metabolome of barley shoot tissue, molecular networking methods were applied. Briefly, following conversion of 'Waters' (.raw) spectral files into the 'analysis base file' (ABF) format using Reifys Abf converter software [41], the spectral data were uploaded into the Mass Spectrometry-Data Independent AnaLysis (MS-DIAL; http://prime.psc.riken.jp/, accessed on 15 May 2022) [42] software for processing with parameters set at mass accuracy 0.05; minimum peak height 10 amplitude, and the Rt tolerance 0.2 min. The MS-DIAL processed files (GnpsMgf and GnpsTable) were then exported into the Global Natural Product Social Sphere (GNPS; https://gnps.ucsd.edu/, accessed on 3 June 2022) [43] using the WinSCP server for molecular networking. A feature-based molecular networking (FBMN) was computed with a mass tolerance set at 0.05; the minimum pair cosine score was set at 0.6 with a minimum of 4. The search for analogues setting was turned off. Furthermore, to increase the chemical insights that can be obtained from the spectral data, FBMN outputs were combined with outputs from substructural annotation by MS2 latent Dirichlet allocation (MS2LDA), in silico annotation (by Network Annotation Propagation, NAP), and the automated chemical classification (through ClassyFire) into the enhanced molecular networking workflow, MolNetEnhancer. MS2LDA is a tool that decomposes molecular fragmentation data derived from large metabolomics experiments into annotated Mass2Motifs or discovers Mass2Motifs from experimental data, while MolNetEnhancer enables the chemical annotation, visualisation, and discovery of the subtle substructure diversity within molecular families [44,45]. Furthermore, an expert-guided metabolite annotation was performed by manually inspecting individual spectra and mass fragmentation as previously described [37,38]. The metabolite annotation and putative identification was completed and confidently reported herein in accordance with level 2 of the Metabolomics Standards Initiative (MSI) [46].

3. Results

3.1. Chromatographic and Mass Spectrometric Analyses and Molecular Networking Approach to Uncover the Metabolic Space of Barley Leaves Treated with Inducers

Metabolomics studies aim to identify and quantify small molecules involved in metabolic processes. Due to its high throughput and good coverage of metabolites, UHPLC–MS has increasingly been used as a platform of choice for such studies when combined with soft electrospray ionisation (ESI). Reported herein, reverse phase chromatographic separation with mass spectrometric detection revealed differences in peak population (presence or absence) and peak intensities (reduced or increased) of analytes present in methanol-extracted samples, visible on the MS chromatograms. These are time- and treatment-related differences observed between the controls (non-treated) and treated conditions (Figures S1–S3). In addition, a broad coverage of midpolar to nonpolar was also noted with analytes eluting throughout the run time. It is worth pointing out that all three inducers, 3,5-DCAA, 3,5-DCSA, and 2,6-DCP-4-CA, were detected in the samples from treated plants. These inducers were characterised as depicted in Figure 1A–C, showing the extracted ion chromatograms (XIC) of each inducer and the corresponding spectra and structures. A pattern representative of the presence of chlorine atom(s) was observed. The two isotopes of chlorine (^{35}Cl and ^{37}Cl) displayed fragment ions separated by two m/z units. In addition, the inducers were also characterised by the loss of the carboxyl group (neutral loss, NL:45) generating fragments with m/z of 159/161, 145/147, and 160/162 for 3,5-DCAA, 2,6-DCP-4-CA, and 3,5-DCSA, respectively. In some cases, the loss of an HCl (NL:36) molecule was also observed. Furthermore, the presence of inducers or their absorption by barley shoot tissue was demonstrated by an increase in the relative concentration (peak area intensities) over time before dropping to the point of stability. The highest relative concentration in barley was observed at 12 h for 3,5-DCAA and 3,5-DCSA; and at 24 h for 2,6-DCP-4-CA. 3,5-DCAA was the highest in the plants at all the time points under investigation (12 h, 24 h, and 36 h) (Figure 1D).

For metabolome characterisation, a molecular networking (MN) strategy, i.e., MolNetEnhancer, was applied (as described in the methodology section). This allowed a comprehensive exploration and enrichment of chemical annotations, discovering the subtle substructural diversity within molecular families. Furthermore, this MN strategy allowed the visualisation of molecular families with class annotations. Six molecular families were visualised and highlighted in Figure 2. These were lipids and lipid-like molecules, phenylpropanoids and polyketides, organoheterocyclic compounds, benzenoids, organic oxygen compounds, and alkaloids and derivatives. When zooming into these superfamilies, the phenylpropanoids and polyketide superfamily was dominated mainly by flavonoid glycosides. These superfamilies of metabolites served as a template for the annotation of the discriminant metabolites.

3.2. Multivariate Data Analyses: Statistical Description, Evaluation, and Exploration of Changes Observed in the UHPLC-MS Data

Unsupervised chemometrics methods were applied to reveal trends in the datasets obtained from the UHPL-MS analyses. The principal components analysis (PCA) allowed us to summarise information in multidimensional datasets. From the PCA models, some distinct clustering is revealed, indicating treatment- and time-related differences between and within samples (Figures 3 and S4). As a general rule, samples grouped together have more in common (at a metabolome level) than those further apart.

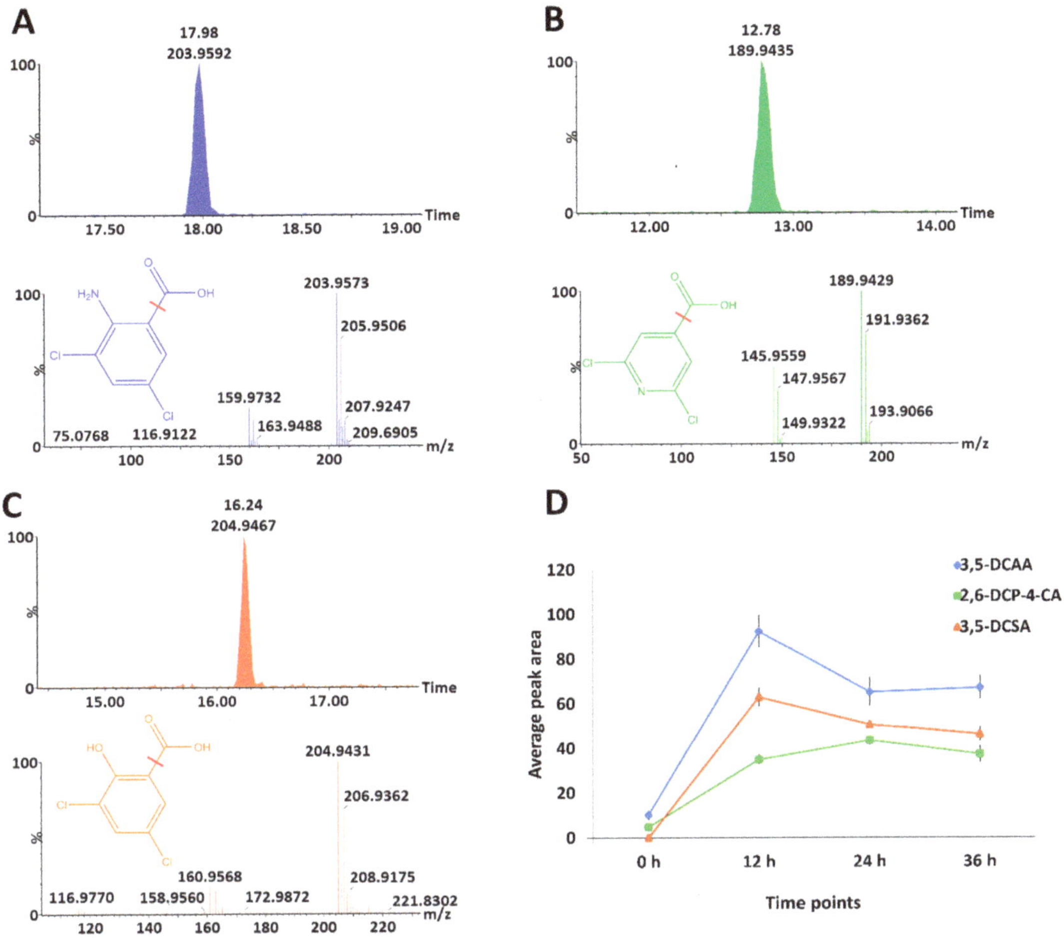

Figure 1. Presence of xenobiotic dichlorinated inducers in barley shoot tissue extracts from different time points. (**A–C**): Extracted ion chromatograms (XIC) and corresponding mass spectrum characteristics of the inducers: 3,5-DCAA, 2,6-DCP-4-CA, and 3,5-DCSA; the structure of each compound is also represented, and the fragmentation site is indicated with the red line. (**D**): Relative concentration of inducers in barley shoot tissue. Error bars indicate the standard deviations of the average peak areas of the samples.

Thus, assessing the scores plots of generated PCA models (Figures 3 and S4), treatment- and time-related sample groupings can be observed, which point to underlying metabolic reprogramming within barley shoot tissue due to the inducer treatments. Except for 3,5-DCSA in the positive ESI mode, controls are separated from the treated, irrespective of the time point. Looking at the controls in each treatment, the time points 12 h and 36 h grouped together and separated from the 24 h, and a slightly similar pattern can also be observed within treated samples (Figure 3C–H). This gives an indication of a possible diurnal effect on the plants. When focusing on each time point (Figure 3I–K), a clear grouping pattern was particularly observed at 24 h between all treatments, showing 3,5-DCSA- and 2,6-DCP-4-CA-treated samples grouped but separated from the control and DCAA treated samples grouping away from the control and other treatments (Figure 3J). This observation drew more attention to the 3,5-DCAA treatment at 24 h. Although able to indicate differences between treatment groups, unsupervised chemometrics are not classification or discrimination methods *per se*, hence less informative in terms of features that explain the differences between sample groups. A supervised method was then used, as described below, to uncover features or metabolites responsible for differential sample groupings.

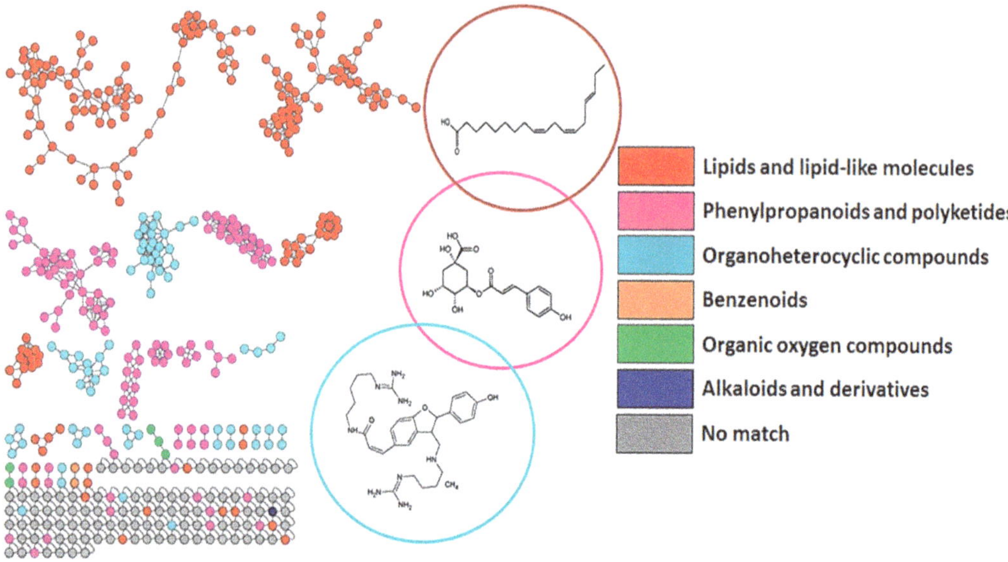

Figure 2. MolNetEnhancer network of ESI(−) spectra obtained from shoot extracts of the barley ('Hessekwa') samples. The enriched molecular network depicts structurally similar nodes as molecular families/clusters, with the annotated metabolites, MS2LDA (MS2 latent Dirichlet allocation) substructures, and Network Annotation Propagation (NAP) annotations, assigned class annotations, represented by coloured nodes and nodes with no class annotation as grey.

3.3. Discriminant Analyses: Treatment- and Time-Related Metabolites and Fold Changes

The supervised learning method, orthogonal projection to latent structure discriminant analysis (OPLS-DA), was performed, and an example is shown in Figure 4. In Figure 4A, the OPLS-DA score plot shows group separation in an OPLS-DA score space. The predictive capability of the computed OPLS-DA model was validated using a permutation test. This consists of comparing the observed R^2 and Q^2 values with randomly permutated ones (n = 100). As seen in the example below, the R^2 and Q^2 values of all computed models were statistically better (or higher) than the 100 permutated ones (Figure 4B). The OPLS loading S-plot allowed the extraction of variables related to each treatment and time point (Figure 4C). Features with both high correlation and covariation were considered, [p(corr) ≥ 0.5, ≤ -0.5, and (p1) ≥ 0.1, ≤ -0.1]. In addition, variable importance in projection (VIP) score plots were used to evaluate the statistical significance of each feature. All selected features had a variable VIP score above 1 (Figure 4D).

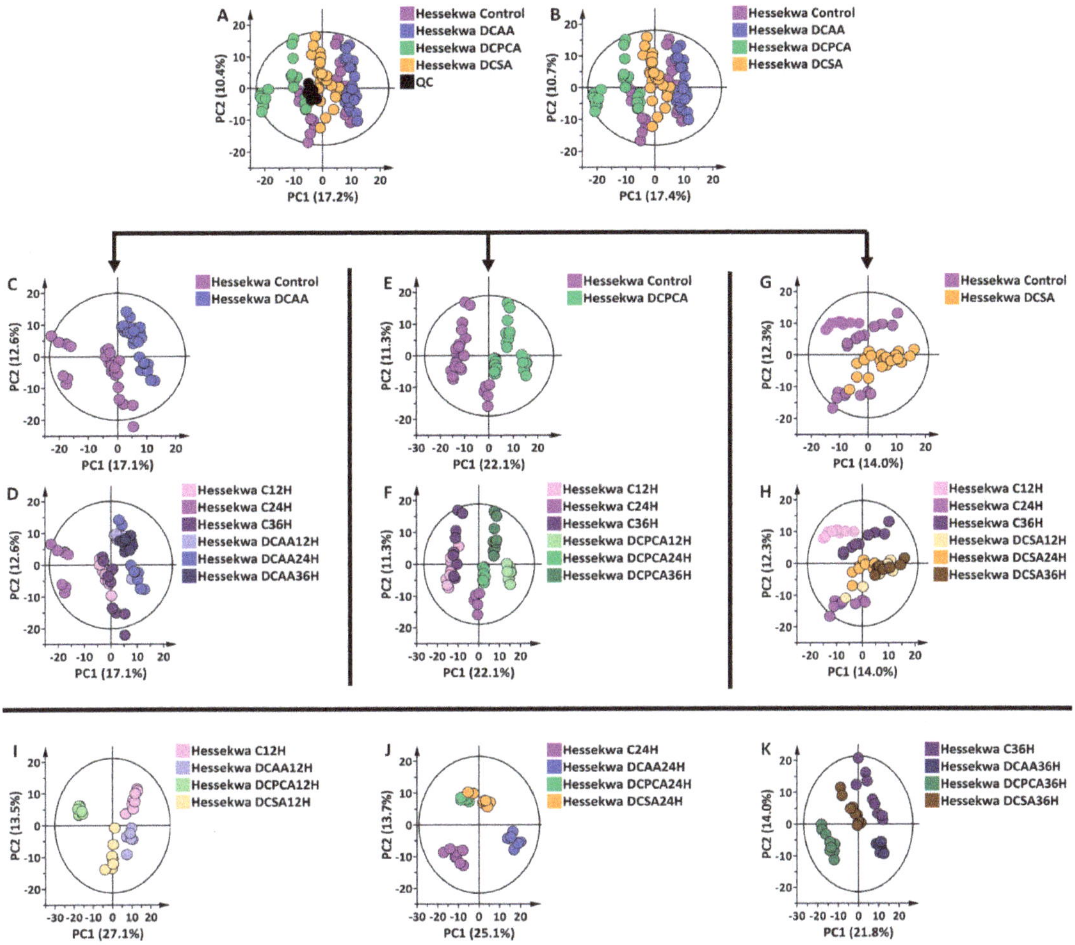

Figure 3. Principal component analysis (PCA) score plots of ESI(−) data from shoot extracts of the 'Hessekwa' cultivar of *Hordeum vulgare*. All data were UV (unit variance) scaled, and the calculated Hoteling's T^2 with a 95% confidence interval is represented by the ellipses present in each PCA score plot. (**A**): a 12-component model of all conditions (including quality controls, QCs), explaining 59.5% variation and predicting 41.6% variation; (**B**) a 12-component model of all conditions (excluding QCs) explaining 60.7% variation, and predicting 42.2% variation. (**C**): a 6-component model of 3,5-DCAA treated and non-treated samples at 12, 24, and 36 h, respectively, explaining 54.6% variation and predicting 37.2% variation; (**D**): same as (**C**) but coloured based on time points; (**E**): a 7-component model of 2,6-DCPCA treated and non-treated samples, explaining 60.6% variation and predicting 41.1% variation; (**F**): same as (**E**) but coloured based on time points. (**G**): a 7-component model of 3,5-DCSA treated and non-treated sample, explaining 53.8% variation and predicting 29.9% variation; (**H**): same as (**G**) but coloured based on time points. (**I**): a 4-component model of all treated and non-treated samples at 12 h, explaining 56.5% variation, and predicting 41.7% variation; (**J**): a 4-component model of all treated and non-treated samples at 24 h, explaining 54.4% variation, and predicting 35.5% variation; (**K**): a 4-component model of all treated and non-treated samples at 36 h, explaining 50.1% variation, and predicting 33.2% variation. (The corresponding set of diagrams for ESI (+) data is presented in Figure S4). As observed in (**A**,**B**), (**C**–**H**) were generated to better visualise the time-dependent changes in each treatment, while (**I**–**K**) highlight the treatment-dependent variation at each selected time point.

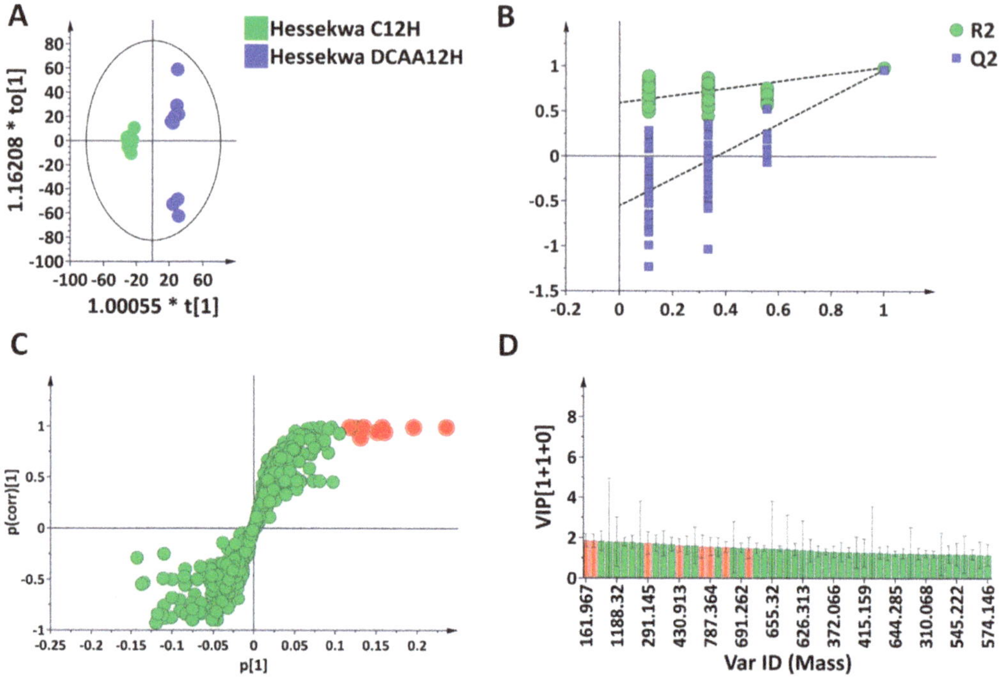

Figure 4. Supervised learning methods for analyses of ultra-high performance liquid chromatography–mass spectrometry (UHPLC–MS) data. Orthogonal partial least squares discriminant analysis (OPLS-DA) modelling and feature selection were performed based on the unique metabolite profiles of 'Hessekwa C12h' (control at 12 h) and 'Hessekwa DCAA 12 h' (treatment at 12 h) leaf extracts in negative ionisation mode. (**A**) OPLS-DA score plots show a clear separation between the two conditions. The model is made of one predictive component and one orthogonal component ($R^2X = 57.5\%$, $R^2Y = 98.9\%$, $Q^2 = 96.1\%$, CV-ANOVA p-value= 4.6539×10^{-9}). (**B**) OPLS-DA validation: Permutation test analysis performed with 100 randomly selected models and showing the above model to be the best among the permutated ones. (**C**) Loading S-plots showing the features responsible for sample clustering, located at the 'outlier' ends of the S-plots with both high correlation and covariation, [p(corr) $\geq$ 0.75 and (p1) $\geq$ 0.1], are highlighted in red. These features are statistically significant candidates as biomarkers related to the DCAA treatment. (**D**) A variable importance in projection (VIP) plot corresponding to the model above and pointing out the mathematical significance of each feature responsible for the discrimination of the treated vs. control conditions. A VIP score >1 is considered as significant in the projection, and the higher the score values, the more significant the features are.

In total, 66 discriminant metabolites belonging to the above-mentioned superfamilies (Figure 2) were annotated/putatively identified as previously described by us [37,38]. These compounds and the corresponding fold changes are presented in Table 1. Compounds with a fold change (FC) > 1 were considered positively correlated to the treatment, and those <1 were negatively correlated. In 3,5-DCAA-treated samples, the highest fold changes were observed with isorhamnetin-3-O-glucoside (FC: 3.49) at 24 h, 4-O-*p*-coumaroylquinic (FC: 5.58), sinapoylagmatine (FC: 3.08), and isovitexin 2″-O-arabinoside (FC: 6.34) at 36 h. In the case of the 2,6-DCPCA treatment, the most positively correlated compounds were mainly fatty acids and derivatives, such as 9,12,13-trihydroxy-10,15-octadecadienoic acid (9,12,13-triHODE) isomer II (FC: 10.73), trihydroxyoctadecenoic acid (triHOME) (FC: 4.59) at 12 h; 9,12,13-triHODE isomer I, 9-oxo-12,13-dihydroxy-10E,15Z-octadecadienoic acid (9-oxo-12,13-diHODE), and a 12-oxophytodienoic acid (OPDA) conjugate and linolenoylglycerol isomer I, at both 12 and 36 h. With the 3,5-DCSA treatment, hordatine A isomer II (FC: 3.49), isovitexin 2″-O-glucoside (FC: 10.97), isovitexin 2″-O-arabinoside (FC: 11.45), and a proline (Pro) derivative, Pro betaine (FC: 15.26), were metabolites with the highest fold change at 36 h. It was also noted that except for valine and proline betaine in extracts from 3,5-DCSA-treated leaves, all amino acids were negatively correlated to all treatments. Subsequently, Venn diagrams were generated for better visualisation of differences across time in individual treatments and at each time point and across treatments.

Table 1. List of all annotated (putatively identified) metabolites extracted from leaves of the barley cultivar 'Hessekwa' treated with 3,5-DCAA, 2,6-DCP-4-CA, and 3,5-DCSA, and harvested at 12, 24, and 36 h post-treatment. The compounds were extracted from the OPLS-DA S-plots, and the fold changes were calculated using a SIMCA software algorithm and are indicated where a metabolite was selected as a significant biomarker. Blue: 3,5-DCAA, Green: 2,6-DCP-4-CA, Orange: 3,5-DCSA, Purple: Control. Different shades of the same colour indicated a specific time point (ranging from lighter to darker, with the former corresponding to 12 h and the latter to 36 h; e.g., 3,5-DCAA 12 h was compared to Control 12 h and metabolites with a FC > 1 were shaded with light blue while those with a FC < 1 were shaded with purple.

	ESI Mode	Compounds	Rt (min)	*m/z*	DCAA Fold Change			DCPCA Fold Change			DCSA Fold Change		
					12 h	24 h	36 h	12 h	24 h	36 h	12 h	24 h	36 h
1	–	*p*-Coumaric acid derivative	0.86	404.103	•	•	•	•	•	•	1.48	1.18	1.16
2	–	4-O-*p*-Coumaroylquinic acid	1.15	337.084	•	•	5.58	•	•	•	•	•	•
3	–	3-Hydroxycoumarin	1.25	161.043	•	•	•	•	0.58	•	•	•	•
4	–	3-O-*p*-Coumaroylquinic acid	3.06	337.112	•	•	•	•	0.83	•	•	0.78	
5	–	Sinapic acid hexose	5.14	385.113	•	0.65	2.18	2.21	•	4.59	1.86	•	2.12
6	–	Dihydroferulic acid 4-O-glucuronide	7.01	371.096	0.87	0.84	1.087	•	•	•	0.82	0.84	•
7	+	Coumaroylputrescine	2.39	235.145	•	•	•	•	1.68	•	•	•	•
8	+	Coumaroylhydroxyagmatine	2.57	293.157	•	1.37	1.27	•	1.45	•	•	1.15	1.09
9	+	Coumaroylagmatine	4.06	277.161	0.35	1.33	1.09	•	1.62	1.17	•	1.34	1.07
10	–	Feruloylhydroxyagmatine	4.49	323.133	•	0.73	•	•	0.84	•	•	0.74	•
11	+	Feruloylagmatine	5.30	307.172	0.612	•	•	•	•	1.14	•	•	•
12	+	Sinapoylagmatine	6.17	337.186	•	•	3.08	•	0.32	•	•	•	•
13	–	Sinapoylhydroxyagmatine	6.30	351.126	•	0.93	0.93	0.88	•	0.89	•	0.92	0.92
14	–	Hordatine B hexose	3.81	787.364	0.818	1.14	1.17	0.74	1.10	0.91	0.78	1.16	1.22
15	–	Hordatine A hexose	4.14	757.353	0.80	•	•	0.63	1.12	0.80	0.75	•	•
16	–	Hordatine C hexose isomer I	4.63	817.376	•	•	•	0.68	1.24	•	0.72	1.31	•
17	–	Hordatine C hexose isomer II	6.26	771.200	•	1.92	•	•	•	•	•	•	•
18	–	Hordatine B isomer I	7.28	579.304	0.63	1.13	1.11	0.86	0.67	•		1.14	•
19	+	Hordatine A Isomer I	7.72	551.304	0.01	1.34	•	0.85	•	•	•	1.32	0.92
20	–	Hordatine A isomer II	7.76	549.294	•	•	•	•	•	2.61	•	•	3.49
21	+	Hordatine B isomer II	7.97	581.319	0.06	1.41	0.82	•	•	•	•	1.50	•
22	+	Hordatine C + 46 isomer I	7.99	655.320	0.83	1.09	1.12	0.87	•	0.88	•	1.17	0.92
23	+	Hordatine C isomer II	8.67	611.330	•	•	•	•	•	•	•	1.38	•
24	–	Isoorientin 7-O-glucoside(Lutonarin)	6.43	609.144	1.45	1.81	1.11	•	1.22	1.13	1.35	1.24	1.15
25	–	Isoorientin 7-O-[6″-sinapoyl]-glucoside	10.53	815.205	1.40	1.26	•	•	0.73	•	•	•	•
26	–	Isovitexin 7,6″-di-*O*-glucoside	8.15	755.205	•	•	0.85	0.53	•	0.55	•	•	0.73
27	+	Isovitexin -7-O-glucoside (Saponarin)	8.39	595.166	0.56	1.21	0.78	0.93	•	0.87	1.06	1.31	0.87
28	–	Isovitexin 7-O-rhamnosylglucoside	8.81	739.208	1.09	0.93	0.89	0.91	•	0.86	•	•	•
29	–	Isovitexin 2″-O-glucoside	9.79	593.150	•	•	•	•	•	•	•	0.01	10.97
30	–	Isovitexin 2″-O-arabinoside	9.93	563.139	•	0.01	6.34	•	•	•	•	0.01	11.45
31	–	Isovitexin	10.59	431.097	0.66	•	0.29	0.40	2.09	0.13	•	2.02	0.43
32	–	Isovitexin 7-O-[6″-sinapoyl]-glucoside	11.42	799.210	1.16	0.77	1.18	1.09	0.65	1.47	•	0.70	•
33	–	Isovitexin 7-O-[X″-feruloyl]-glucoside (Feruloylsaponarin)	11.82	769.200	0.89	0.93	•	0.85	•	•	0.84	•	•
34	–	Apigenin 7-O-arabinosylglucoside	11.90	563.140	•	•	0.82	•	•	•	•	•	0.85
35	–	Apigenin 6-C-arabinoside 8-C-glucoside	8.97	563.140	1.52	•	•	•	•	•	•	•	•
36	–	Isoscoparin 7-O-glucoside	8.99	623.160	•	1.31	0.85	•	1.29	0.78	1.10	1.26	•
37	+	Isoscoparin 2″,6″-di-*O*-glucoside	10.89	787.209	•	1.53	•	•	1.29	•	•	•	•

Table 1. *Cont.*

	ESI Mode	Compounds	Rt (min)	*m/z*	DCAA Fold Change			DCPCA Fold Change			DCSA Fold Change		
					12 h	24 h	36 h	12 h	24 h	36 h	12 h	24 h	36 h
38	−	Isoscoparin 7-O-[6″-sinapoyl]-glucoside	11.53	829.221	•	•	•	0.72	•	•	•	•	•
39	−	Isoscoparin 7-O-[6″-feruloyl]-glucoside	11.95	799.211	•	•	•	•	1.32	0.72	•	1.25	•
40	−	Isorhamnetin-3-O-glucoside	9.74	477.107	•	3.49	•	•	•	•	•	2.43	•
41	−	6-Prenylnaringenin	19.02	339.215	•	0.19	0.68	5.42	0.82	3.60	3.07	0.53	1.80
42	+	Hydroxytryptamine	1.67	177.102	•	2.47	2.50	•	1.31	•	1.84	1.76	1.26
43	−	Coumaroyltryptamine	2.60	289.129	•	1.29	1.55	0.76	1.19	•	1.15	1.14	1.15
44	+	Valine	0.88	118.086	•	•	0.77	•	•	•	1.26	•	•
45	+	Tyrosine	1.14	182.081	0.50	•	0.72	0.70	•	0.58	0.63	•	0.61
46	+	Tyrosine derivatives	1.14	276.107	0.33	1.60	•	•	•	•	•	•	•
47	+	Isoleucine	1.31	132.102	•	•	0.74	•	0.72	0.55	0.61	•	0.69
48	+	Phenylalanine	1.65	166.087	•	0.82	0.58	0.64	0.743	0.44	0.51	0.76	0.52
49	+	Proline betaine	2.05	144.139	•	•	•	•	•	•	•	•	15.26
50	−	Tryptophan	2.43	203.080	•	•	0.84	0.76	0.65	0.65	0.71	0.74	0.63
51	−	Asparaginylglucose	4.24	293.122	•	0.79	•	•	•	•	•	0.81	•
52	−	Citric acid	1.14	191.017	•	0.55	•	•	0.50	•	•	0.49	•
53	−	Isocitric acid	0.93	191.017	•	0.69	•	1.29	0.42	•		0.41	•
54	−	Citric acid derivative	1.39	306.117	•	•	•	•	0.85	•	•	•	•
55	−	Malic acid	0.95	133.012	•	•	0.74	•	0.74	1.22	1.27	0.73	•
56	−	12-Hydroxyjasmonate sulfate	16.61	305.129	•	•	2.93	•	•	•	•	•	•
57	−	Dihydrojasmonic acid (H_2JA) conjugate	16.80	419.173	•	•	•	•	•	2.14	•	•	1.34
58	−	9,12,13,TriHODE isomer I	16.53	327.216	•	0.39	0.70	3.52	•	2.23	1.67	0.61	1.69
59	−	9,12,13,TriHODE Isomer II	16.64	327.216	•	•	•	10.73	•	•	•	•	•
60	−	TriHOME	17.25	329.232	•	0.31	•	4.59	•	•	2.13	0.46	•
61	−	9-Oxo-12,13-diHODE	17.45	325.200	•	0.14	•	6.31	•	3.51	2.41	0.43	2.27
62	−	OPDA conjugate	19.55	309.205	•	0.23	•	3.90	•	2.47	3.73	0.31	•
63	+	Linolenoylglycerol isomer I	20.69	353.267	•	•	2.12	1.93	•	3.08	1.71	•	1.71
64	+	Linolenoylglycerol isomer II	20.98	353.265	•	•	1.70	•	•	2.71	1.64	•	1.70
65	+	Linolenoylglycerol isomer III	21.80	353.263	•	•	1.72	•	•	•	•	•	•
66	+	Linolenoylglycerol isomer VI	22.07	353.263	2.57	•	•	•	•	1.63	2.17	0.70	•

• = annotated in MS chromatograms, (+) ESI positive mode, (−) ESI negative mode.

3.4. Time-Related Differences and Similarities in the Chemical Profiles of Barley Leaves, following Foliar Application of 3,5-DCAA, 2,6-DCP-4-CA and 3,5-DCSA

The Venn diagrams in Figure 5 indicate that there is a wide diversity of metabolites distributed across time in each treatment condition. Starting with extracts from 3,5-DCAA-treated leaves (Figure 5A; Table 1), four metabolites were specifically found as discriminant at 12 h, and these were feruloylagmatine (a precursor in the biosynthesis of hordatines B and C), an isomer of hordatine A, linolenoylglycerol, and the flavonoid apigenin 6-C-arabinoside 8-C-glucoside. At 24 h, specific discriminant metabolites included organic acids citric and isocitric acid, flavonoids, such as isoscoparin 2″,6″-di-O-glucoside, and phenolic- and fatty-derivatives. Twelve discriminant metabolites were found specifically at 36 h post-treatment with 3,5-DCAA. These included some amino acids, a chlorogenic acid and sinapoylagmatine. As part of the flavonoids and fatty acid derivatives, isovitexin 7,6″-di-O-glucoside, apigenin 7-O-arabinosylglucoside, 12-hydroxyjasmonate sulphate, and linolenoylglycerol isomers I and II were noted. In addition to the specific metabolites annotated, some were overlapping across time points. In total, 10 signatory metabolites were found across all time points and were represented on a dendrogram heatmap to show their differential distribution and to highlight the relationship among the sample groups. As seen on the dendrogram heatmap below the Venn diagram in Figure 5A, the common metabolites were mainly the barley-specific hydroxycinnamic acid amides (HCAAs), hordatine B isomer I and II, and the corresponding glycosylated form, hordatine C isomer I, and coumaroylagmatine. In addition, dihydroferulic acid 4-O-glucuronide, lutonarin (luteolin as aglycone), saponarin, isovitexin 7-O-rhamnosylglucoside, and isovitexin 7-O-[6″-sinapoyl]-glucoside (all with apigenin as aglycone) were also found. The heatmap shows an increase in the relative concentration of the HCAAs over time. Lutonarin was also up-regulated at each time point. The dendrogram corresponding to the samples shows treatments branching apart from controls and an interesting association of control samples at 12 and 36 h, both separated from 24 h.

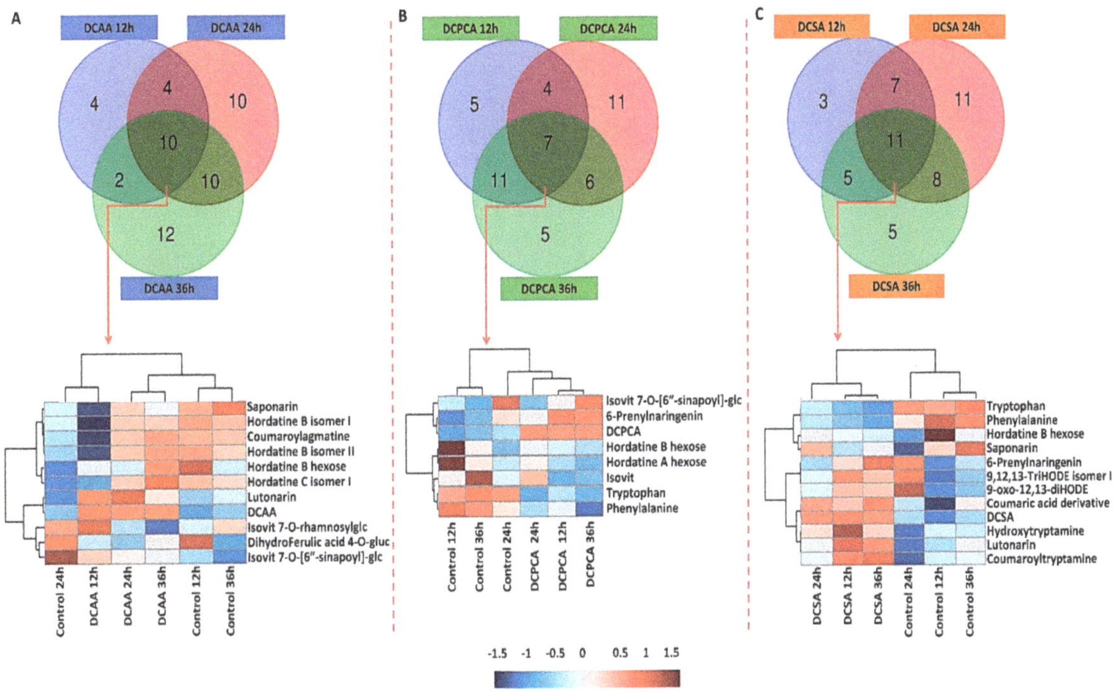

Figure 5. Venn diagrams and dendrogram heatmaps of discriminant metabolites. The distribution of metabolites across time points (12, 24, and 36 h) following foliar application of (**A**) 3,5-DCAA, (**B**) 2,6-DCP-4-CA, and (**C**) 3,5-DCSA is displayed. While Venn diagrams were generated from all the annotated discriminant metabolites (Table 1), the dendrogram heatmaps were generated from shared metabolites (red arrows) across all three-time points to highlight the differences in the relative concentration over time. Metabolites common to the two groups are discussed in the text. The numerical values in the Venn diagram correspond to the number of unique metabolites at each time-point and overlapping metabolites across the different conditions.

Regarding extracts from 2,6-DCP-4-CA-treated leaves, five metabolites of the phenylpropanoids and fatty acid derivatives were found exclusively at 12 h. At 24 h, 11 metabolites were specific to the time point, amongst which were 3-O-*p*-coumaroylquinic acid, coumaroylputrescine, sinapoylagmatine, a feruloyagmatine derivative, coumaroylhydroxyagmatine, 3-hydroxycoumarin, and hydroxytryptamine. Finally, at 36 h, the five specific discriminant metabolites included feruloylagmatine, hordatine A isomer II, dihydrojasmonic acid conjugate, and linolenoylglycerol isomers II and VI. Four metabolites were overlapping across the 12 and 24 h time points and between 12 and 36 h, where 11 metabolites were shared and included coumaroyltryptamine, sinapoylhydroxyagmatine, saponarin, Tyr, and linolenoylglycerol isomer I, to list a few. Coumaroylagmatine, lutonarin, and malic acid were among the six metabolites shared between 24 and 36 h. Seven discriminant metabolites were overlapping across all time points. Looking at the dendrogram heatmap in Figure 5B, hordatine A and B hexose were less abundant at 12 h and 36 h. At these same time points, isovitexin 7-O-[6″-sinapoyl]-glucoside and 6-prenylnaringenin were more abundant. Isovitexin followed the same pattern as hordatine A and B hexose and was more abundant at 24 h. A down-regulation of Trp and Phe at all-time points was also noted. On the dendrogram, the 'treatment' groups were clearly separated from the corresponding controls; however, the treatment at 24 h was closer to the corresponding control.

In extracts from 3,5-DCSA-treated leaves, three metabolites were specifically found at 12 h, namely, hordatine A hexose, isovitexin 7-O-[X″-feruloyl]-glucoside and Val. At 24 h, 11 metabolites were found, specific among which were hordatine B and hordatine C, and asparaginylglucose. Finally, five were found at 36 h and included hordatine A isomer II, isovitexin 7,6″-di-O-glucoside, apigenin 7-O-arabinosylglucoside, dihydrojasmonic acid, and proline betaine. Here, seven metabolites were shared among the 12 and 24 h treatment groups, including hordatine C hexose isomer I, malic acid, and an oxylipin, 12-oxo-phytodienoic acid (OPDA) conjugate. Sinapic acid hexose, linolenoylgycerol isomer I and II, Tyr, and Ile were overlapping between 12 and 36 h. Finally, coumaroylagmatine, coumaroylhydroxyagmatine, sinapoylhydroxyagmatine, hordatine A, and C isomer I was among the eight metabolites shared between 24 and 36 h. Common metabolites to all three-time points included Trp and Phe, both down-regulated at all time points and lutonarin, hydroxytryptamine and coumaroyltryptamine, all up-regulated throughout, as shown on the heatmap. Again, all control samples clustered separately from the treated and control 24 h branches away from control 12 and 36 h, respectively (Figure 5C).

3.5. Treatment-Related Similarities and Differences in the Chemical Profile of Barley Leaves, following Foliar Application of 3,5-DCAA, 2,6-DCP-4-CA, and 3,5-DCSA

For comparison of all treatments across the chosen time points, Venn diagrams were generated and are represented in Figure 6. At 12 h, among the specific metabolites, coumaroyl- and feruloylagmatine and hordatine B were found as discriminant markers in extracts from 3,5-DCAA-treated leaves, Val, and Ile in 3,5-DCSA-treated samples and isocitric acid in 2,6-DCP-4-CA-treated samples. Hordatines A and B hexose, saponarin, isovitexin 7-O-[X″-feruloyl]-glucoside, and Tyr were all found negatively correlated to every treatment. The dendrogram shows a clear separation between the control and the treated samples (Figure 6A).

At 24 h, metabolites responsible for the uniqueness of each condition involve hordatine C hexose and sinapic acid hexose in extracts from 3,5-DCAA-treated leaves, hordatine C in extracts from 3,5-DCSA-treated leaves and coumaroylputrescine, and sinapoylagmatine in extracts from 2,6-DCP-4-CA-treated leaves. On the dendrogram heatmap, shared metabolites to all three treatments were hydroxytryptamine, coumaroyltryptamine, lutonarin, hordatine B, and its hexosylated form up-regulated in extracts from 3,5-DCAA-treated leaves and closely related to the profile of the 3,5-DCAA treatment. Both coumaroylagmatine and coumaroylhydroxyagmatine were up-regulated in all three conditions but seemed to be more associated with the 2,6-DCP-4-CA treatment. Metabolites closely associated with the 3,5-DCSA treatment were a feruloylagmatine derivative and 6-prenylnaringenin, both down-regulated in the treated samples (Figure 6B).

In Section 2.2, the PCA plots revealed an interesting trend at 24 h, where a clearer sample separation was highlighted between the control and all treatments. Keeping such observation in mind, the relative quantification of barley-important metabolites was evaluated at 24 h. These metabolites were the hordatines and the main flavonoids present in the plant, lutonarin, and saponarin. In addition, two alkaloids, coumaroyltryptamine and hydroxytryptanine, previously not annotated in non-stressed plants [38], were also evaluated (Figure S5). An increase in the biosynthesis of the selected metabolites is observed upon priming, especially with 3,5-DCAA treated samples. Among these, saponarin was by far the most abundant metabolite in the extracts, irrespective of the treatments.

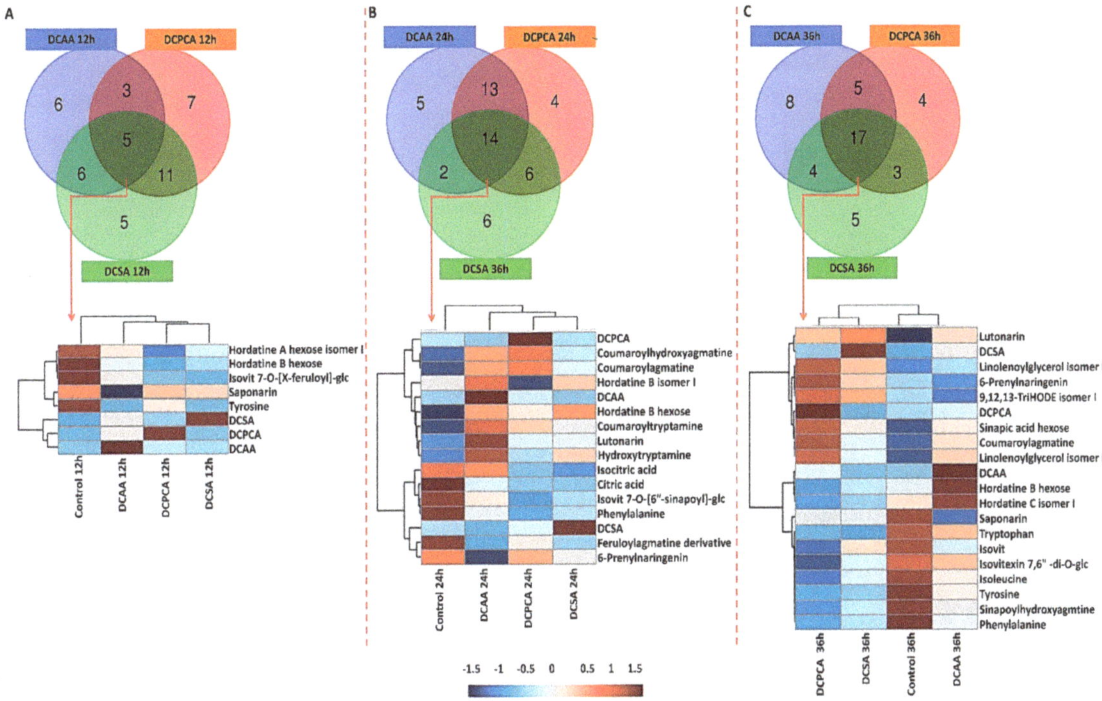

Figure 6. Venn diagrams and dendrogram heatmaps of discriminant metabolites. The distribution of metabolites across treatments (3,5-DCAA, 2,6-DCP-4-CA, and 3,5-DCSA) at (**A**) 12, (**B**) 24, and (**C**) 36 h post-treatment is displayed. While Venn diagrams were generated from all the annotated discriminant metabolites (Table 1), the dendrogram heatmaps were generated from shared metabolites (red arrow) to all three treatments to highlight the differences in the relative concentration at a specific time point. The numerical values in the Venn diagrams correspond to the number of unique metabolites associated with each treatment and the number of overlapping/shared metabolites across the different conditions. Similarly to 12 h, at 36 h, hordatine B was specifically found in extracts from 3,5-DCAA-treated leaves in addition to 12-hydroxyjasmonate sulfate, an isomer of linolenoylglycerol, and other classes of metabolites. Examples of specific metabolites associated with the 3,5-DCSA and 2,6-DCP-4-CA treatments included proline betaine and an OPDA conjugate, respectively. Hordatine B hexose and an isomer of hordatine C were found across all treatments at each time point, and the levels were relatively higher in the 3,5-DCAA samples. Sinapic acid hexose, coumaroylagmatine, 6-prenylnaringenin, 9,12,13-triHODE isomer I, and linolenoyglycerol isomers I and II were all found up-regulated in extracts from 2,6-DCP-4-CA treatments. The relative concentration of sinapoylhydroxyagmatine, saponarin, isovitexin, isovitexin 7,6″-di-O-glucoside, Trp, Leu, and Tyr were down-regulated at 36 h in all treated samples. Here, lutonarin was closely related to 3,5-DCSA and up-regulated in all treated samples. Looking at the dendrogram, similar profiles were observed between DCAA and the control samples and between 2,6-DCP-4-CA and 3,5-DCSA treatments (**C**).

4. Discussion

4.1. Distribution of Metabolite Classes and Metabolic Pathways Analyses for Biological Interpretation

As discussed above, common and specific responses were observed, featuring fluctuations in metabolites grouped as phenolic acids and derivatives, flavonoids, fatty acids and derivatives, amino acids, organic acids, and alkaloids, listed in descending percentage order across all treatments (Figure 7). These classes belonged to the superfamilies identified on the MolNetEnhancer network (Figure 2). While nuanced differences in the % values were noted between the three treatments, the overall patterns were similar, suggesting that the three dichlorinated inducers (3,5-DCAA, 2,6-DCP-4-CA, and 3,5-DCSA) trigger the same type of response in the barley leaves.

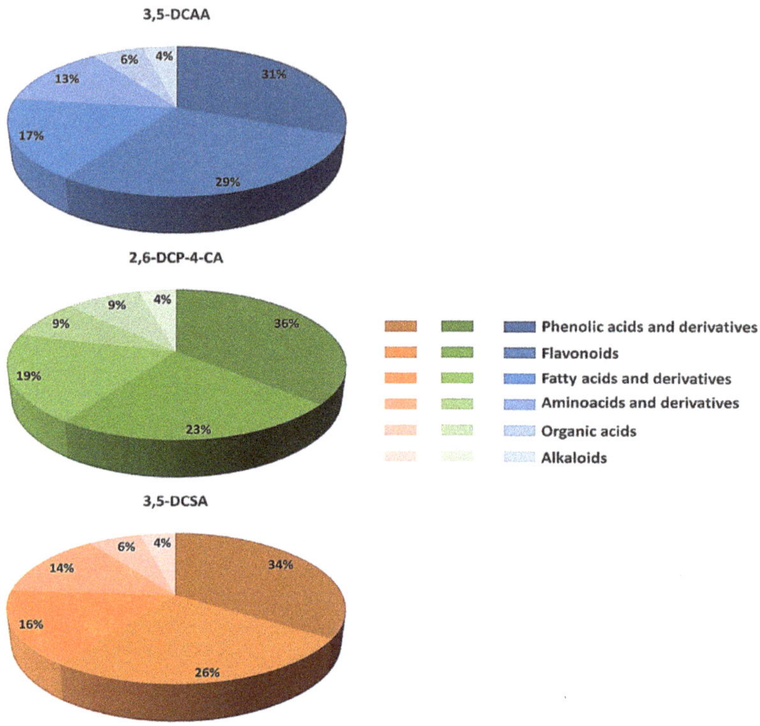

Figure 7. Metabolite class distribution of all annotated discriminant metabolites across all treatments with dichlorinated inducers of acquired resistance. Discriminant metabolites in barley belonged to different classes, as previously identified on the MolNetEnhancer network (Figure 2). Blue: 3,5-DCAA, Green: 2,6-DCP-4-CA, and Orange: 3,5-DCSA.

In the environment, barley plants naturally interact with numerous living organisms, instigating a perturbation in the immune system. SA is a key plant immune hormone that is essential for the development of plant immunity. It was one of the first endogenous plant compounds to be documented as an inducer of SAR [31,47,48] and, eventually, metabolic changes. The barley cultivar, 'Hessekwa' was treated with synthetic functional analogues of SA, namely, 3,5-DCAA, 2,6-DCP-4-CA, and 3,5-DCSA, and metabolic perturbations and associated reprogramming was evident from PCA models (Figure 3) and in Table 1 and Figures 5 and 6.

Functional analyses (i.e., pathway enrichment analysis) using differential (discriminant) metabolites (Figure 8; Table S1) revealed phenylpropanoid biosynthesis, alpha-linolenic acid metabolism, aminoacyl-tRNA biosynthesis, the TCA cycle, Trp metabolism, Phe, Trp, and Tyr catabolism, and the glyoxylate and dicarboxylate metabolism only to name the most significant and/or the most impactful pathways. These pathways are associated with both primary and secondary metabolism and were identified in all treatments with different levels of significance and impact. These differences can be seen, for example, in the 3,5-DCAA and 3,5-DCSA treatments, where the phenylpropanoid biosynthesis was the most significant pathway and in the case of the 2,6-DCP-4-CA treatment, the TCA cycle. Similar to this study, in tobacco and wheat plants, pathogen- and chemical-induced SAR was characterised by the de novo production of phenylpropanoid pathway chemicals in the leaf, regardless of the signalling pathway [49,50].

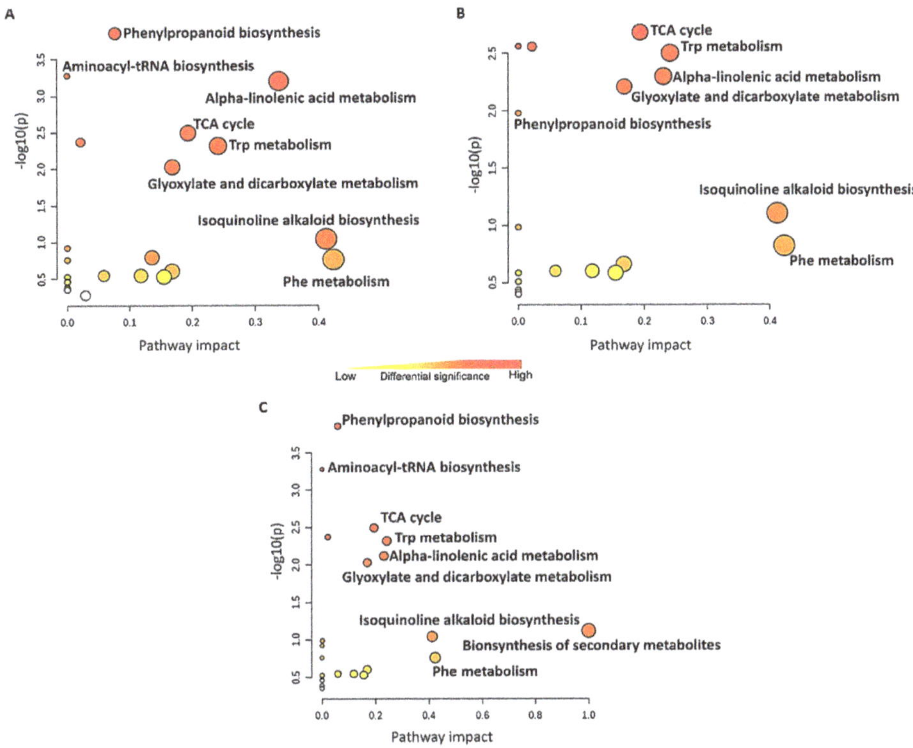

Figure 8. Metabolomic Pathway Analysis (MetPA) as generated by MetaboAnalyst ver. 5.0 with (**A**): 3,5-DCAA, (**B**): 2,6-DCP-4-CA, and (**C**): 3,5-DCSA data. The size of the circle reflects the pathway impact score, while the colour of each circle is based on *p*-values (darker colours imply more significant changes of metabolites in the relevant pathway). TCA: tricarboxylic acid; Trp: tryptophan; Phe: phenylalanine.

4.2. Biological Implication with Regard to Plant Protection

In addition to alterations in transcription, the adaptive responses of plants to environmental cues also involve post-translational protein modifications, metabolite alteration, and/or accumulation. These changes all work together to produce a particular physiological response or phenotype. Depending on the system under investigation, chemical and biological priming agents may each exhibit a level of specificity in their action mechanisms that prepares the plant in a different mode [51]. In this context, it was proposed that metabolites that are strongly affected by chemical or biological inducers (the priming agents) are named 'priming compounds' since their effect on plant metabolism is to trigger the synthesis of 'primed compounds' that assist to counteract the stress. At the same time, these additionally identified metabolites point to defence pathways that the plants deploy to "get ready for the battle" [51].

It has been regarded that a primary metabolism only performs a supportive role in plant defence and that the energy saved by the down-regulation of primary metabolism is diverted and used for defence responses. However, the up-regulation of metabolic pathways associated with primary metabolism may also occur during plant defence, and it was proposed that such up-regulation modulates signal transduction cascades that lead to plant defence responses [52]. Amino acids participate in plant growth and development, signalling processes, and stress responses. In addition to being building blocks for protein biosynthesis, they also perform pivotal roles in several pathways associated with secondary metabolism [53,54]. Free amino acids are essential for SAR in plants, and their presence has previously been attributed to the plant's tolerance [55,56]. The negative correlation between the Tyr, Phe, and Trp levels and the treatments observed in this study might be attributed to the rapid use of the compounds in the biosynthesis of secondary metabolites involved in plant defence, and for de novo protein synthesis as in the case of pathogenesis-related (PR) proteins. For instance, Phe is the substrate of the first committed step in the phenylpropanoid pathway (the most significant pathway to respond to 3,5-DCAA and 3,5-DCSA, Figure 8), which results in the biosynthesis of phenolic compounds, including hydroxycinnamic acids (HCAs) and flavonoids. In addition, amino acids can be converted into precursors or intermediates in the TCA cycle, supporting the mitochondrial metabolism and the production of ATP [57,58]. Consequently, it can be speculated that there is an imbalance in the ratio between amino acid anabolism and catabolism. The level of amino acids produced might not be proportional to the plant's requirement for energy and other processes.

Trp in barley seedlings is a precursor to simple indole alkaloids, gramine, and tryptamine. Coumaroyltryptamine and hydroxytryptamine were found to be discriminatory and positively correlated to the treatments (Figure S5; Table 1). The decrease in Trp could also be linked to the accumulation of tryptamine derivatives which participate in plant protection. In previous studies, tryptamine and derivatives accumulated in the member of the grass family (Poaceae species) following the application of compounds, such JA as and fungal pathogens [59,60]. The reliance of both the constitutive and inducible defence mechanisms on the Trp pathway was previously demonstrated [60].

Commenting on energy, functional analogues of SA are used to activate induced resistance; however, this process consumes a lot of energy and interferes with other metabolic functions. Therefore, balancing the trade-off between defence and growth may be key to a plant's success [20]. Alterations in the production of organic acids, crucial contributors to energy production, were noted, and the TCA cycle was one of the most significant pathways observed, especially in the 2,6-DCP-4-CA treatment. There was a positive correlation between malic acid with 2,6-DCP-4-CA and 3,5-DCSA treatments at 36 and 12 h, respectively. Citric and isocitric acid were consistently found as discriminative metabolites and negatively correlated to all treatments at 24 h. In Pastor et al., 2014, [51], the TCA cycle was activated and potentiated in β-aminobutyric acid (BABA)-primed Arabidopsis plants. Organic acids are essential to all plant species as they constitute good storage for carbon, perform a role in CO_2 fixation and stomatal conductance, help plants deal with excess cations, and are reversibly implicated in the biosynthesis of amino acids and other compounds [61,62]. Specifically, citrate is produced in the TCA cycle from the condensation of oxaloacetate, the end product of a previous turn of the cycle, and acetyl-CoA and provides a bridge between carbohydrate and fatty acid metabolism.

The majority of fatty acid derivatives were positively correlated to the treatments at 12 and 36 h. These compounds are alternative sources of energy for plant growth and development. Of special interest and part of discriminant metabolites are the JA derivatives, such as 12-hydroxy-JA sulphate, only found in the 3,5-DCAA treatment, and dihydro-JA found in both 2,6-DCP-4-CA and 3,5-DCSA treatments. Moreover, a conjugate of OPDA, the precursor of JA, was also increased due to the 2,6-DCPCA and 3,5-DCSA treatments. Jasmonates are by-products of the classical octadecanoid oxylipin pathway deriving from alpha-linolenic acid metabolism, identified here as a significant metabolic pathway [63,64]. Recent studies have elaborated on the direct and active roles of fatty acids and their breakdown products on various defence mechanisms [65]. They are involved in the regulation of plants' basal, effector-triggered, and systemic immunity, and perform an important role in NADPH oxidase activation which result in the biosynthesis of reactive oxygen species (ROS). In addition to regulating plant growth and development, JA is crucial in the resistance against both biotic and abiotic stresses. It acts as a mobile signal for

SAR and is translocated via vasculature [66]. Despite being clearly characterised by the molecules they synthesise, JA, SA, and ET signalling pathways interact both cooperatively and antagonistically in a range of responses [13,66]. While the first reports elaborated on the inhibitory effect of SA on JA actions in tomatoes, positive crosstalk between the two phytohormones has also been highlighted in a number of systems [67,68]. These hormones were not annotated in the study; however, as mentioned above, precursors of JA and JA derivatives were annotated, suggesting the activation of this pathway and a state of alertness (primed state).

Phenolic acids and conjugated derivatives are end products in the phenylpropanoid pathway. The activation of the phenylpropanoid pathway is associated with enhanced resistance to stresses. The chlorogenic acid, 4-O-*p*-coumaroylquinic acid, was found up-regulated in 3,5-DCAA treated plants at 36 h. Sinapic acid hexose accumulated in all treated plants at 12 and 36 h. Pastor et al. [51] hypothesised that phenolics and sinapates are compounds produced in primed plants that support the biosynthesis of metabolites functioning in cell wall reinforcement. Coumaroylagmatine, a well-known precursor of hordatine A and B, was positively correlated to all treatments at 24 and 36 h. Feruloyagmatine and sinapolyagmatine, on the other hand, were only found to be positively correlated to 2,6-DCP-4-CA at 24 h and 3,5-DCAA at 36 h, respectively. Increased production of HCAAs was reported in maize after insect herbivory attack [69,70]. In a study by [58], the HCAA coumaroylputrescine accumulated in barley leaves treated with JA. Here, the positive correlation of the compound to the 2,6-DCP-4-CA treatment at 24 h was also observed. Similarly, to the precursors, hordatine A, B, and C production, was increased mostly at 24 h in both 3,5-DCAA and 3,5-DCSA (Figure S5; Table 1, and hordatine B at 36 h in the case of the 2,6-DCP-4-CA treatment (Table 1). The synchronised increase in the relative content of both hordatines and precursors highlights the role of both the phenylpropanoid- and polyamine pathways in plant immunity. The role of these barley-specific metabolites in antimicrobial defences was previously highlighted [37,38], and the increase in their production can be related to the preconditioning of the plants. From the time point at which the increase in the production of hordatines and precursors occurred, it might be speculated that in the 'Hessekwa' cultivar it takes at least 24 h to induce a positive correlation to the treatment. In terms of specific markers for each treatment in this class of metabolites, a derivative of *p*-coumaric acid, 4-O-*p*-coumaroylquinic acid, and *p*-coumaroylputrescine were only found up-regulated by 3,5-DCSA, 3,5-DCAA, and 2,6-DCP-4-CA, respectively. It might be suggested that each treatment uses different molecules to perform similar functions.

Flavonoids are structurally related compounds performing crucial roles in biotic and abiotic stresses [71], and in transcriptional and growth regulation [72]. The glycosylated derivatives of luteolin and apigenin (lutonarin and saponarin, respectively) are the most dominant flavonoids in barley and were annotated as discriminant metabolites in all treatments at almost all time points. While lutonarin was always up-regulated, saponarin levels increased only at 12 and 24 h in samples from 3,5-DCSA treated plants and at 24 h with 3,5-DCAA treatment. Both compounds are good antioxidants [73] and are proposed as protectants against UV-B radiation [74]. The aglycones have very similar chemical structures and do not differ meaningfully in their antibacterial activity. Likewise, the presence and location of the sugar group(s) in the flavone glucosides do not have a significant effect on the antibacterial activity [75]. Isovitexin 2′-O-glucose was up-regulated only at 24 h and down-regulated at 12 and 36 h, while 6-prenylnaringenin followed the opposite pattern. Here, the diurnal and nocturnal responses might influence the pattern of accumulation since some mechanisms are activated in the presence of light and others in the absence.

Several studies have demonstrated that induction of plant immune and defence responses can be associated with enhanced expression of the phenylalanine ammonia-lyase (*PAL*) gene(s). PAL is the first enzyme in the phenylpropanoid pathway and is responsible for the production of phenolic compounds, such as phenolic acids and derivatives, and flavonoids. The fluctuation of these compounds in the plants is evidence of the plants' reaction to the treatments. Despite the fact that SA is the only specific phenolic molecule that is extensively used as a biochemical marker for SAR, the implication of phenolics in the identification of plant damage and plant signalling during the development of SAR is acknowledged. In wheat, DCPCA significantly increased the phenolic content [42]. In fact, the first artificial substances that were demonstrated to activate SAR were 2,6-DCINA acid and its methyl ester [19,76,77]. In tobacco, 2,6-DCINA and 3,5-DCSA were reported to be efficient at enhancing PR1 protein expression and resistance against the tobacco mosaic virus (TMV) [34,77]. Compared to SA, the dichlorinated 3,5-DCSA was found to be more active at inducing resistance against TMV [32]. 3,5-DCAA has been established as an inducer of transient and rapid resistance to pathogens in Arabidopsis by simultaneously engaging two different branches of the plant defence signalling network: NPR1-dependent and NPR1-independent responses [28]. The interaction of 3,5-DCAA with defence signalling pathways takes place either downstream or independently of SA perception and accumulation. Here, in the absence of SA-related metabolites as discriminant biomarkers, it might be speculated that these treatments in barley, or at least in the 'Hessekwa' cultivar, are not SA-dependent. In priming, multiple layers of induced defence mechanisms can be involved. Enhanced responses developed upon treatment with these inducers were mostly observable through the activation of the TCA cycle, phenylpropanoid pathway, and alpha-linolenic acid metabolism, converting the metabolome of the naïve plants to a primed state. This was attained by the fluctuation of common and specific protective metabolites, such as the accumulation of JA conjugates. Due to the nature of untargeted metabolomics, it is difficult to distinguish between 'priming agents' and 'primed compounds' [51], but the annotated metabolites and the identified metabolic pathways were supportive of a defensive role and its link to priming. The investigation of molecular and biochemical mechanisms in priming events is still ongoing [78], and this study is a contribution in that regard.

5. Conclusions

Timeous activation of defence mechanisms has proven to be of great importance for the plant's survival under environmental stressors. In this context, the fluctuation of several metabolites was observed on the shoot tissue of the barley cultivar 'Hessekwa' using xenobiotic dichlorinated substitutes of salicylic acid, anthranilic acid, and isonicotinic acid (3,5-DCSA, 3,5-DCAA, and 2,6-DCP-4-CA/2,6-DCINA). It is known that even a subtle change in the substitution site of a compound can significantly affect the physicochemical properties thereof and, ultimately, the biological activity; this was reiterated in this study. 3,5-DCAA, 2,6-DCP-4-CA, and 3,5-DCSA were proposed as chemicals with either induced/acquired resistance or priming activity. There is a relationship between the two concepts, and while there is considerable overlap between the phenotypes of the SAR and ISR states, the underlying triggers and mechanisms of action differ. With priming inducers, it has recently become apparent that defence priming should be regarded as an adaptive part of induced resistance and that specific defence mechanisms also depend strongly on the primed state. The question has been raised whether metabolism (and thus also metabolites, i.e., the 'primed compounds') can store and process information regarding induced/imprinted/primed responses to changing environments.

Applying untargeted metabolomics approaches, specific and, interestingly, shared mechanisms were highlighted; notably the activation of the phenylpropanoid pathway, the TCA cycle, and the alpha-linolenic acid metabolism by all three inducers. With regard to metabolites annotated in the study, it is important to note that in some cases, they were either conjugated compounds, substrates, or precursors in the biosynthesis of the main components in plant immunity. For example, no signalling molecules (SA, JA, and ET) were present as discriminant metabolites. Instead, JA derivatives and oxylipins, key components in the phytohormone biosynthesis, were found to be discriminative and were mostly up-regulated in treated samples conditions. The presence of related compounds but not the main actors themselves can be perceived as the plant's readiness to quickly produce more specific defence-related metabolites in response to environmental stress (e.g., pathogen attacks). It was then suggested that plants use metabolic imprints (i.e., the metabolic changes that last beyond recovery from stress events) and priming (e.g., the imprints that function to prepare for upcoming stresses) to integrate diverse environmental stress histories. The lack of SA accumulation in this study led to the speculation that the mechanism of priming by dichlorinated xenobiotics (at least in the 'Hessekwa' cultivar of barley) does not rely on SA and derivatives but rather on JA and derivatives. Hordatines (known phytoanticipins/phytoalexins in barley), their hexosylated conjugates and precursor molecules were also found as important discriminant markers of inducer treatment with a time-lapse of 24 h required to enhance the biosynthesis. In addition, the flavonoid profile was also altered and specific markers were apigenin 6-C-arabinoside 8-C-glucoside, isoscoparin 7-O-[6″-sinapoyl]-glucoside, and isovitexin 2″-O-glucoside, associated with the 3,5-DCAA-, 2,6-DCP-4-CA-, and 3,5-DCSA-treatments, respectively. Although examples of flavanones and flavonols were amongst the discriminant metabolites, the large number of flavones (as glycosidic derivatives of apigenin and luteolin) were especially prevalent. The increase in the production of tryptamine derivatives, precursors in the biosynthesis of various indole alkaloids, pointed to their participation in defence-related mechanisms and their contribution to the preconditioned state.

Metabolomics approaches, combined with chemometric analysis, have proven to be valuable tools in revealing the metabolic perturbation and reprogramming resulting from the above-mentioned treatments. Chemometrics tools allowed the exploration of large datasets and also to discriminate among the different conditions. All three dichlorinated xenobiotic priming inducers were good candidates in perturbing plant metabolic pathways associated with metabolites involved in defence. Although presenting specific markers, the mechanisms of action seem to be similar. Regardless of nuanced differences, their effects appear to work toward the same goal of establishing an enhanced defensive environment.

Supplementary Materials: The following are available online at https://www.mdpi.com/article/10.3390/metabo13050666/s1, Figure S1. Ultra–high performance liquid chromatography–mass spectrometry (UHPLC–MS) base peak intensity (BPI) chromatograms (negative ionisation) of shoot extracts from the 'Hessekwa' cultivar treated with DCAA, DCPCA, and DCSA for 12, 24, and 36 h. (Comparison of time-dependent changes). Figure S2. Ultra–high performance liquid chromatography–mass spectrometry (UHPLC–MS) base peak intensity (BPI) chromatograms (positive ionisation) of shoot extracts from the 'Hessekwa' cultivar treated with DCAA, DCPCA, and DCSA for 12, 24, and 36 h. (Comparison of time-dependent changes). Figure S3. Ultra–high performance liquid chromatography—mass- spectrometry (UHPLC–MS) base peak intensity (BPI) chromatograms (negative ionisation) of shoot extracts from the 'Hessekwa' cultivar treated with DCAA, DCPCA, and DCSA and harvested after 12 h, 24 h, and 36 h. (Comparison according to inducer). Figure S4. Principal component analysis (PCA) score plot models of ESI (+) data from shoot extracts of the 'Hessekwa' cultivar of *Hordeum vulgare*. Figure S5. Relative quantification (based on average peak area values) of selected primed/induced metabolites at 24 h. Hordatine A, B, and C; flavones lutonarin, saponarin, and alkaloids hydroxytryptamine and coumaroyltryptamine. Table S1. Metabolic pathways generated from Metabolomics Pathway Analysis (MetPA) in MetaboAnalyst 5.0 involve selected annotated metabolites. Blue: 3,5-DCAA; Green: 2,6-DCP-4-CA, and Orange: 3,5-DCSA.

Author Contributions: Conceptualisation: I.A.D.; methodology: C.Y.H.D., F.T. and P.A.S.; formal analysis: C.Y.H.D. and P.A.S.; investigation: C.Y.H.D. and I.A.D.; data curation: C.Y.H.D.; writing—original draft preparation: C.Y.H.D.; writing—review and editing: I.A.D., F.T. and L.A.P.; supervision: I.A.D. All authors have read and agreed to the published version of the manuscript.

Funding: This research received no external funding.

Institutional Review Board Statement: Not applicable.

Informed Consent Statement: Not applicable.

Data Availability Statement: The study design information, LC-MS data, data processing and analyses are reported and incorporated into the main text. Raw data, analyses, data processing information, and the meta-data were deposited to the EMBL-EBI metabolomics repository—MetaboLights50, on 12 January 2023 with the identifier MTBLS6874 (www.ebi.ac.uk/metabolights/MTBLS6874).

Acknowledgments: The South African Barley Breeding Institute (SABBI) is thanked for the provision of seeds.

Conflicts of Interest: The authors declare no conflict of interest.

References

1. Tang, D.Z.; Wang, G.X.; Zhou, J.M. Receptor Kinases in Plant-Pathogen Interactions: More than Pattern Recognition. *Plant Cell* **2018**, *29*, 618–637. [CrossRef] [PubMed]
2. Mhlongo, M.I.; Piater, L.A.; Madala, N.E.; Labuschagne, N.; Dubery, I.A. The Chemistry of Plant–Microbe Interactions in the Rhizosphere and the Potential for Metabolomics to Reveal Signaling Related to Defense Priming and Induced Systemic Resistance. *Front. Plant Sci.* **2018**, *9*, 112. [CrossRef] [PubMed]
3. Lu, H.; Wei, T.; Lou, H.; Shu, X.; Chen, Q. A critical review on communication mechanism within plant-endophytic fungi interactions to cope with biotic and abiotic stresses. *J. Fungi* **2021**, *7*, 719. [CrossRef] [PubMed]
4. Prasch, C.M.; Sonnewald, U. Signaling events in plants: Stress factors in combination change the picture. *Environ. Exp. Bot.* **2015**, *114*, 4–14. [CrossRef]
5. Llorens, E.; González-Hernández, A.I.; Scalschi, L.; Fernández-Crespo, E.; Camañes, G.; Vicedo, B.; García-Agustín, P. Priming mediated stress and cross-stress tolerance in plants: Concepts and opportunities. In *Priming-Mediated Stress and Cross-Stress Tolerance in Crop Plants*; Academic Press: Cambridge, MA, USA, 2020; pp. 1–20.
6. Dempsey, D.A.; Klessig, D.F. SOS-too many signals for systemic acquired resistance? *Trends Plant Sci.* **2012**, *17*, 538–545. [CrossRef]
7. Denancé, N.; Sánchez-Vallet, A.; Goffner, D.; Molina, A. Disease resistance or growth: The role of plant hormones in balancing immune responses and fitness costs. *Front. Plant Sci.* **2013**, *4*, 155. [CrossRef]
8. Tugizimana, F.; Mhlongo, M.I.; Piater, L.A.; Dubery, I.A. Metabolomics in plant priming research: The way forward? *Int. J. Mol. Sci.* **2018**, *19*, 1759. [CrossRef]
9. Shine, M.B.; Xiao, X.; Kachroo, P.; Kachroo, A. Signaling mechanisms underlying systemic acquired resistance to microbial pathogens. *Plant Sci.* **2019**, *279*, 81–86. [CrossRef]
10. Conrath, U.; Pieterse, C.M.; Mauch-Mani, B. Priming in plant–pathogen interactions. *Trends Plant Sci.* **2002**, *7*, 210–216. [CrossRef]
11. Pieterse, C.M.J. Prime time for transgenerational defense. *Plant Physiol.* **2012**, *158*, 545. [CrossRef]
12. Fu, Z.Q.; Dong, X. Systemic acquired resistance: Turning local infection into global defense. *Annu. Rev. Plant Biol.* **2013**, *64*, 839–863. [CrossRef]
13. Gao, Q.M.; Kachroo, A.; Kachroo, P. Chemical inducers of systemic immunity in plants. *J. Exp. Bot.* **2014**, *65*, 1849–1855. [CrossRef]
14. Baccelli, I.; Mauch-Mani, B. Beta aminobutyric acid priming of plant defense: The role of ABA and other hormones. *Plant Mol. Biol.* **2016**, *91*, 703–711. [CrossRef]
15. Henry, E.; Yadeta, K.A.; Coaker, G. Recognition of bacterial plant pathogens: Local, systemic and transgenerational immunity. *New Phytol.* **2013**, *199*, 908–915. [CrossRef]
16. Bektas, Y.; Eulgem, T. Synthetic plant defense elicitors. *Front. Plant Sci.* **2015**, *5*, 804. [CrossRef]
17. Merewitz, E. Chemical priming-induced drought stress tolerance in plants. In *Drought Stress Tolerance in Plants*; Springer: Cham, Switzerland, 2016; Volume 1, pp. 77–103.
18. Yi, H.S.; Yang, J.W.; Ryu, C.M. ISR meets SAR outside: Additive action of the endophyte *Bacillus pumilus* INR7 and the chemical inducer, benzothiadiazole, on induced resistance against bacterial spot in field-grown pepper. *Front. Plant Sci.* **2013**, *4*, 122. [CrossRef]
19. Conrath, U.; Beckers, G.J.; Flors, V.; García-Agustín, P.; Jakab, G.; Mauch, F.; Newman, M.A.; Pieterse, C.M.; Poinssot, B.; Pozo, M.J.; et al. Priming: Getting ready for battle. *Mol. Plant Microbe Interact.* **2006**, *19*, 1062–1071. [CrossRef]
20. Buswell, W.; Schwarzenbacher, R.E.; Luna, E.; Sellwood, M.; Chen, B.; Flors, V.; Petriacq, P.; Ton, J. Chemical priming of immunity without costs to plant growth. *New Phytol.* **2018**, *218*, 1205–1216. [CrossRef]
21. Walters, D.R.; Avrova, A.; Bingham, I.J.; Burnett, F.J.; Fountaine, J.; Havis, N.D.; Hoad, S.P.; Hughes, G.; Looseley, M.; Oxley, S.J.; et al. Control of foliar diseases in barley: Towards an integrated approach. *Eur. J. Plant Pathol.* **2012**, *133*, 33–73. [CrossRef]
22. Beckers, G.J.M.; Conrath, U. Priming for stress resistance: From the lab to the field. *Curr. Opin. Plant Biol.* **2007**, *10*, 425–431. [CrossRef]
23. Conrath, U.; Beckers, G.J.M.; Langenbach, C.J.G.; Jaskiewicz, M.R. Priming for enhanced defense. *Annu. Rev. Phytopathol.* **2015**, *53*, 97–119. [CrossRef] [PubMed]
24. Durner, J.; Klessig, D.F. Salicylic acid is a modulator of tobacco and mammalian catalases. *J. Biol. Chem.* **1996**, *271*, 28492–28501. [CrossRef] [PubMed]
25. Rivas-San Vicente, M.; Plasencia, J. Salicylic acid beyond defence: Its role in plant growth and development. *J. Exp. Bot.* **2011**, *62*, 3321–3338. [CrossRef] [PubMed]

26. Maruri-López, I.; Aviles-Baltazar, N.Y.; Buchala, A.; Serrano, M. Intra and extracellular journey of the phytohormone salicylic acid. *Front. Plant Sci.* **2019**, *10*, 423. [CrossRef] [PubMed]
27. Mauch-Mani, B.; Bacelli, I.; Luna, E.; Flors, V. Defense priming: An adaptive part of induced resistance. *Annu. Rev. Plant Biol.* **2017**, *68*, 485–512. [CrossRef] [PubMed]
28. Knoth, C.; Salus, M.S.; Girke, T.; Eulgem, T. The synthetic elicitor 3,5-dichloroanthranilic acid induces NPR1-dependent and NPR1-independent mechanisms of disease resistance in Arabidopsis. *Plant Physiol.* **2009**, *150*, 333–347. [CrossRef] [PubMed]
29. Knoth, C.; Eulgem, T. High-throughput screening of small-molecule libraries for inducers of plant defense responses. In *Plant Chemical Genomics*; Humana Press: Totowa, NJ, USA, 2014; pp. 45–49.
30. Klessig, D.F.; Choi, H.W.; Dempsey, D.M.A. Systemic acquired resistance and salicylic acid: Past, present, and future. *Mol. Plant Microbe Interact.* **2018**, *31*, 871–888. [CrossRef]
31. Zhou, M.; Wang, W. Recent advances in synthetic chemical inducers of plant immunity. *Front. Plant Sci.* **2018**, *9*, 1613. [CrossRef]
32. Faize, L.; Faize, M. Functional analogues of salicylic acid and their use in crop protection. *Agronomy* **2018**, *8*, 5. [CrossRef]
33. Silverman, F.P.; Petracek, P.D.; Heiman, D.F.; Fledderman, C.M.; Warrior, P. Salicylate activity. 3. Structure relationship to systemic acquired resistance. *J. Agri. Food Chem.* **2005**, *53*, 9775–9780. [CrossRef]
34. Conrath, U.; Chen, Z.; Ricigliano, J.R.; Klessig, D.F. Two inducers of plant defense responses, 2, 6-dichloroisonicotinec acid and salicylic acid, inhibit catalase activity in tobacco. *Proc. Natl. Acad. Sci. USA* **1995**, *92*, 7143–7147. [CrossRef]
35. FAOSTAT. Food and Agriculture Organization of the United Nations. Available online: http://www.fao.org/faostat/en/ (accessed on 29 March 2022).
36. Giraldo, P.; Benavente, E.; Manzano-Agugliaro, F.; Gimenez, E. Worldwide Research Trends on Wheat and Barley: A Bibliometric Comparative Analysis. *Agronomy* **2019**, *9*, 352. [CrossRef]
37. Hamany Djande, C.Y.; Piater, L.A.; Steenkamp, P.A.; Tugizimana, F.; Dubery, I.A. A metabolomics approach and chemometric tools for differentiation of barley cultivars and biomarker discovery. *Metabolites* **2021**, *11*, 578. [CrossRef]
38. Hamany Djande, C.Y.; Steenkamp, P.A.; Piater, L.A.; Tugizimana, F.; Dubery, I.A. Hordatines and associated precursors dominate metabolite profiles of barley (*Hordeum vulgare* L.) seedlings: A metabolomics study of five cultivars. *Metabolites* **2022**, *12*, 310. [CrossRef]
39. Pang, Z.; Chong, J.; Zhou, G.; de Lima Morais, D.A.; Chang, L.; Barrette, M.; Gauthier, C.; Jacques, P.É.; Li, S.; Xia, J. MetaboAnalyst 5.0: Narrowing the gap between raw spectra and functional insights. *Nucleic Acids Res.* **2021**, *4*, W388–W396. [CrossRef]
40. Kanehisa, M.; Furumichi, M.; Tanabe, M.; Sato, Y.; Morishima, K. KEGG: New perspectives on genomes, pathways, diseases and drugs. *Nucleic Acids Res.* **2017**, *45*, 353–361. [CrossRef]
41. Abf Converter. Available online: https://www.reifycs.com/AbfConverter/ (accessed on 28 January 2022).
42. Tsugawa, H.; Cajka, T.; Kind, T.; Ma, Y.; Higgins, B.; Ikeda, K.; Kanazawa, M.; VanderGheynst, J.; Fiehn, O.; Arita, M. MS-DIAL: Data-independent MS/MS deconvolution for comprehensive metabolome analysis. *Nat. Methods* **2015**, *12*, 523–526. [CrossRef]
43. Wang, M.; Carver, J.J.; Phelan, V.V.; Sanchez, L.M.; Garg, N.; Peng, Y.; Nguyen, D.D.; Watrous, J.; Kapono, C.A.; Luzzatto-Knaan, T.; et al. Sharing and community curation of mass spectrometry data with Global Natural Products Social Molecular Networking. *Nat. Biotechnol.* **2016**, *34*, 828–837. [CrossRef]
44. Ernst, M.; Kang, K.B.; Caraballo-Rodríguez, A.M.; Nothias, L.F.; Wandy, J.; Chen, C.; Wang, M.; Rogers, S.; Medema, M.H.; Dorrestein, P.C.; et al. MolNetEnhancer: Enhanced molecular networks by integrating metabolome mining and annotation tools. *Metabolites* **2019**, *9*, 144. [CrossRef]
45. Tinte, M.M.; Masike, K.; Steenkamp, P.A.; Huyser, J.; van der Hooft, J.J.; Tugizimana, F. Computational Metabolomics Tools Reveal Metabolic Reconfigurations Underlying the Effects of Biostimulant Seaweed Extracts on Maize Plants under Drought Stress Conditions. *Metabolites* **2022**, *12*, 487. [CrossRef]
46. Sumner, L.W.; Amberg, A.; Barrett, D.; Beale, M.H.; Beger, R.; Daykin, C.A.; Fan, T.W.; Fiehn, O.; Goodacre, R.; Griffin, J.L. Proposed minimum reporting standards for chemical analysis. *Metabolomics* **2007**, *3*, 211–221. [CrossRef] [PubMed]
47. Gozzo, F. Systemic acquired resistance in crop protection: From nature to a chemical approach. *J. Agric. Food Chem.* **2003**, *51*, 4487–4503. [CrossRef] [PubMed]
48. White, P.S. Pattern, process, and natural disturbance in vegetation. *Bot. Rev.* **1979**, *45*, 229–299. [CrossRef]
49. Stadnik, M.J.; Buchenauer, H. Inhibition of phenylalanine ammonia-lyase suppresses the resistance induced by benzothiadiazole in wheat to *Blumeria graminis* f. sp. *tritici*. *Physiol. Mol. Plant Path.* **2000**, *57*, 25–34. [CrossRef]
50. Ramos, O.F.; Smith, C.M.; Fritz, A.K.; Madl, R.L. Salicylic Acid-Mediated Synthetic Elicitors of Systemic Acquired Resistance Administered to Wheat Plants at Jointing Stage Induced Phenolics in Mature Grains. *Crop. Sci.* **2017**, *57*, 3122–3128. [CrossRef]
51. Pastor, V.; Balmer, A.; Gamir, J.; Flors, V.; Mauch-Mani, B. Preparing to fight back: Generation and storage of priming compounds. *Front. Plant Sci.* **2014**, *5*, 295. [CrossRef]
52. Rojas, C.M.; Senthil-Kumar, M.; Tzin, V.; Mysore, K.S. Regulation of primary plant metabolism during plant-pathogen interactions and its contribution to plant defense. *Front. Plant Sci.* **2014**, *5*, 17. [CrossRef]
53. Häusler, R.E.; Ludewig, F.; Krueger, S. Amino acids–a life between metabolism and signaling. *Plant Sci.* **2014**, *229*, 225–237. [CrossRef]
54. Pratelli, R.; Pilot, G. Regulation of amino acid metabolic enzymes and transporters in plants. *J. Exp. Bot.* **2014**, *65*, 5535–5556. [CrossRef]

55. Balmer, D.; de Papajewski, D.V.; Planchamp, C.; Glauser, G.; Mauch-Mani, B. Induced resistance in maize is based on organ-specific defence responses. *Plant J.* **2013**, *74*, 213–225. [CrossRef]
56. Killiny, N.; Hijaz, F. Amino acids implicated in plant defense are higher in *Candidatus liberibacter asiaticus*-tolerant citrus varieties. *Plant Signal. Behav.* **2016**, *11*, e1171449. [CrossRef]
57. Kirma, M.; Araújo, W.L.; Fernie, A.R.; Galili, G. The multifaceted role of aspartate-family amino acids in plant metabolism. *J. Exp. Bot.* **2012**, *63*, 4995–5001. [CrossRef]
58. Hildebrandt, T.M.; Nesi, A.N.; Araújo, W.L.; Braun, H.P. Amino acid catabolism in plants. *Mol. Plant* **2015**, *8*, 1563–1579. [CrossRef]
59. Ogura, Y.; Ishihara, A.; Iwamura, H. Induction of hydroxycinnamic acid amides and tryptophan by jasmonic acid, abscisic acid and osmotic stress in barley leaves. *Z. Naturforsch. C* **2001**, *56*, 193–202. [CrossRef]
60. Ishihara, A.; Kumeda, R.; Hayashi, N.; Yagi, Y.; Sakaguchi, N.; Kokubo, Y.; Ube, N.; Tebayashi, S.I.; Ueno, K. Induced accumulation of tyramine, serotonin, and related amines in response to *Bipolaris sorokiniana* infection in barley. *Biosci. Biotechnol. Biochem.* **2017**, *81*, 1090–1098. [CrossRef]
61. Lopez-Bucio, J.; Nieto-Jacobo, M.F.; Ramırez-Rodrıguez, V.; Herrera-Estrella, L. Organic acid metabolism in plants: From adaptive physiology to transgenic varieties for cultivation in extreme soils. *Plant Sci.* **2000**, *160*, 1–13. [CrossRef]
62. Igamberdiev, A.U.; Eprintsev, A.T. Organic acids: The pools of fixed carbon involved in redox regulation and energy balance in higher plants. *Front. Plant Sci.* **2016**, *7*, 1042. [CrossRef]
63. Gidda, S.K.; Miersch, O.; Levitin, A.; Schmidt, J.; Wasternack, C.; Varin, L. Biochemical and molecular characterization of a hydroxyjasmonate sulfotransferase from *Arabidopsis thaliana*. *J. Biol. Chem.* **2003**, *278*, 17895–17900. [CrossRef]
64. Hamany Djande, C.Y.; Madala, N.E.; Dubery, I.A. Mass spectrometric approaches to study the metabolism of jasmonates: Biotransformation of exogenously supplemented methyl jasmonate by cell suspension cultures of *Moringa oleifera*. In *Jasmonate in Plant Biology*; Humana: New York, NY, USA, 2020; pp. 211–226.
65. Pretorius, C.J.; Zeiss, D.R.; Dubery, I.A. The presence of oxygenated lipids in plant defense in response to biotic stress: A metabolomics appraisal. *Plant Signal. Behav.* **2021**, *16*, 1989215. [CrossRef]
66. Pieterse, C.M.; Leon-Reyes, A.; Van der Ent, S.; Van Wees, S. Networking by small-molecule hormones in plant immunity. *Nat. Chem. Biol.* **2009**, *5*, 308–316. [CrossRef]
67. Pena-Cortés, H.; Albrecht, T.; Prat, S.; Weiler, E.W.; Willmitzer, L. Aspirin prevents wound-induced gene expression in tomato leaves by blocking jasmonic acid biosynthesis. *Planta* **1993**, *191*, 123–128. [CrossRef]
68. Tamaoki, D.; Seo, S.; Yamada, S.; Kano, A.; Miyamoto, A.; Shishido, H.; Miyoshi, S.; Taniguchi, S.; Akimitsu, K.; Gomi, K. Jasmonic acid and salicylic acid activate a common defense system in rice. *Plant Signal. Behav.* **2013**, *8*, e24260. [CrossRef]
69. Marti, G.; Erb, M.; Boccard, J.; Glauser, G.; Doyen, G.R.; Villard, N.; Robert, C.A.M.; Turlings, T.C.; Rudaz, S.; Wolfender, J.L. Metabolomics reveals herbivore-induced metabolites of resistance and susceptibility in maize leaves and roots. *Plant Cell Environ.* **2013**, *36*, 621–639. [CrossRef] [PubMed]
70. Zeiss, D.R.; Piater, L.A.; Dubery, I.A. Hydroxycinnamate amides: Intriguing conjugates of plant protective metabolites. *Trends Plant Sci.* **2021**, *26*, 184–195. [CrossRef] [PubMed]
71. Ramaroson, M.L.; Koutouan, C.; Helesbeux, J.J.; Le Clerc, V.; Hamama, L.; Geoffriau, E.; Briard, M. Role of Phenylpropanoids and Flavonoids in Plant Resistance to Pests and Diseases. *Molecules* **2022**, *27*, 8371. [CrossRef]
72. Mathesius, U. Flavonoid functions in plants and their interactions with other organisms. *Plants* **2018**, *7*, 30. [CrossRef]
73. Benedet, J.A.; Umeda, H.; Shibamoto, T. Antioxidant activity of flavonoids isolated from young green barley leaves toward biological lipid samples. *J. Agric. Food Chem.* **2007**, *55*, 5499–5504. [CrossRef]
74. Markham, K.R.; Mitchell, K.A. The misidentification of the major antioxidant flavonoids in young barley (*Hordeum vulgare*) leaves. *Z. Naturforsch. C* **2003**, *58*, 53–56. [CrossRef]
75. Karpiński, T.M.; Adamczak, A.; Ożarowski, M. Antibacterial activity of apigenin, luteolin, and their C-glucosides. In Proceedings of the 5th International Electronic Conference on Medicinal Chemistry, Online, 1–30 November 2019. [CrossRef]
76. Ward, E.R.; Uknes, S.J.; Williams, S.C.; Dincher, S.S.; Wiederhold, D.L.; Alexander, D.C.; Ahl-Goy, P.; Metraux, J.P.; Ryals, J.A. Coordinate gene activity in response to agents that induce systemic acquired resistance. *Plant Cell* **1991**, *3*, 1085–1094. [CrossRef]
77. Kessmann, H.; Staub, T.; Hofmann, C.; Maetzke, T.; Herzog, J.; Ward, E.; Uknes, S.; Ryals, J. Induction of systemic acquired disease resistance in plants by chemicals. *Ann. Rev. Phytopathol.* **1994**, *32*, 439–459. [CrossRef]
78. Schwachtje, J.; Whitcomb, S.; Firmino, A.A.P.; Zuther, E.; Hincha, D.K.; Kopka, J. Induced, Imprinted, and Primed Responses to Changing Environments: Does Metabolism Store and Process Information? *Front. Plant Sci.* **2019**, *10*, 106. [CrossRef]

metabolites

MDPI

Article

β-Cyclocitral-Mediated Metabolic Changes Optimize Growth and Defense Responses in *Solanum lycopersicum* L.

Shreyas Deshpande and Sirsha Mitra *

Department of Botany, Savitribai Phule Pune University, Pune 411007, India
* Correspondence: smitra@unipune.ac.in

Abstract: β-cyclocitral (βCC) is one of the significant oxidative products of β-carotene. It primes plants for multiple stress acclimation without compromising plant growth. Metabolic reorganization is necessary to maintain a balance between growth and defense. However, the βCC-mediated changes in a plant's metabolic network are unknown. Here, we demonstrate how βCC-induced metabolic changes enable *Solanum lycopersicum* L. (tomato) plants to promote defense and maintain growth under stress. An analysis of early (0–240 min) and late (72 h) changes in the tomato metabolome after βCC-treatment using liquid chromatography and tandem mass spectrometry identified 57 compounds. A principal coordinate analysis suggested that βCC treatment significantly changes the metabolite profile. A variable importance in projection (VIP) analysis revealed 16 and 19 discriminant metabolites from early and late samples, respectively (VIP $\geq$ 1.0). Upregulated metabolites were mainly amino acids and phytophenols. Pathway enrichment analysis showed that βCC treatment influenced amino acid metabolism at early and later times; however, phenylpropanoid and isoquinoline biosynthesis were influenced only at the later time. A 66.6% similarity in the upregulated metabolites of βCC- and simulated-herbivory-treated plants confirmed βCC's role against herbivores. We conclude that βCC steers a temporal separation in amino acids and defense metabolite accumulation that optimizes resource allocation to growth and defense.

Keywords: apocarotenoids; herbivory; liquid chromatography-mass spectrometry; metabolomics; stress signaling

1. Introduction

Environmental stresses are deleterious for the survival of plants. *Solanum lycopersicum* L. (tomato), one of the major food crops in the family Solanaceae, is also challenged by multiple environmental stresses, namely fungal pathogens, herbivores, cold, drought, etc. [1,2]. To counteract environmental challenges, plants optimize their defense strategies via stress sensing, signaling, and reorganizing their metabolic profile under the environmental stimuli. The chloroplast plays a central role in plant defense signaling as it harbors biosynthetic pathways of significant defense metabolites; chloroplast-associated carotenoids are crucial for protecting the photosynthetic apparatus from photo-oxidative damage [3] because the presence of a C40 polyene backbone makes carotenoids susceptible to oxidative cleavage. Cleavage of carotenoids occurs enzymatically by carotenoid cleavage dioxygenase or non-enzymatically by reactive oxygen species (ROS). The oxidative carbonyl products of carotenoids are known as apocarotenoids [4,5]. Apocarotenoids are well known as the precursors of abscisic acid, a phytohormone involved in abiotic stress responses [6]. In recent years, the identification of novel apocarotenoids has established their role in stress signaling [5,7]. Previous studies showed that singlet oxygen ($^1O_2$)-mediated cleavage of β-carotene produces many apocarotenoids, where β-cyclocitral (βCC) is the major apocarotenoid [8].

Interestingly, drought, photooxidative stress, and herbivory increase the level of βCC [9–12]. Studies have shown that exogenous application of βCC can induce $^1O_2$-responsive genes, marker genes for photo-oxidative stress [8]. Recently, we found that βCC treatment triggers the accumulation of transcripts related to both abiotic and biotic stresses. Our study also revealed that exogenous application of βCC primes plants against drought and develops resistance against herbivory [11,13]. These results suggest that βCC causes functional changes in addition to trancriptomic changes. On the other hand, βCC treatment enhances the growth of primary roots and the branching of lateral roots [11]. As defenses are costly, a trade-off between growth and defense is often evident in stress-exposed plants. However, βCC treatment enhances both plant growth and defense. Metabolites can influence plant growth, the branching pattern of roots and shoots, the size, shape, and position of leaves, etc. [14–16]. Therefore, metabolic reorganization is a prerequisite to maintaining a balance between growth and defense. However, βCC-mediated changes in the plant metabolome are yet to be revealed.

In the 'omics' field, metabolomics identifies large-scale changes in the plant metabolome. It is highly challenging, as plant metabolites possess different physical and chemical properties and vary significantly in their concentrations [17]. Metabolomics can be investigated using two main approaches: targeted and untargeted [18,19]. Initially, chromatographic techniques, such as liquid chromatography (LC) and gas chromatography (GC), separate the metabolites, and mass spectroscopy (MS) investigates the quantitative changes in the plant metabolome [20]. The use of nuclear magnetic resonance (NMR) spectroscopy is also evident in metabolomics analysis [21]; however, as the sensitivity of MS is high, it is preferred over NMR spectroscopy for analyzing plant metabolites [22]. The tomato has been considered a model fruit. Its metabolomic profiles have been studied to discriminate different varieties, determine geographical origin, examine fruit development and ripening, and investigate seasonal changes [23]. A comparative metabolomics study of *Ralstonia solanacearum*-infected tomato leaves, stems, and roots revealed the metabolic changes after being infected with *R. solanacearum* [24]. This study identified the metabolites providing plant defense against *R. solanacearum* [24]. However, a comparative metabolomics study of tomato plants after apocarotenoid treatment has yet to be executed.

In light of the reported results and previous findings, we hypothesized that βCC treatment reorganizes plants' metabolic networks to enhance plant growth and defense. Therefore, in the current study, we aimed to reveal how exogenous application of βCC changes the metabolic signature of the tomato plants that enables them to maintain growth under stress using an LC–MS-based approach.

2. Materials and Methods

2.1. Plant Material, Growth Conditions, and Treatments

Seeds of tomato (var. Pusa Ruby) were germinated on cocopeat. 15-day-old seedlings were transferred in pots (9 cm × 9 cm) that contained soil, cocopeat, and vermiculite (5:4:1) and raised for five weeks. These plants were treated with 1 mL of pure βCC (treatment), or water (control) kept in a watch glass in a transparent, closed glass container [11,25]. Tissues were harvested in liquid nitrogen 0, 30, 60, 90, 180, 240 min, and 72 h after treatment. Insect herbivory was simulated in tomato plants by wounding the tomato leaves with a fabric pattern wheel parallel to the mid-vein and applying 20 μL of oral secretion of *Spodoptera litura* larvae diluted with sterile water (1:1) to the wound [26]. Tissues were harvested in liquid nitrogen 72 h after treatment. All harvested tissues were stored at −80°C until further use.

2.2. Extraction of Metabolites and Data Acquisition in LC-QTOF-MS/MS

Leaf samples of tomato plants were pulverized in liquid nitrogen. An amount of 200 mg of tissue was extracted with 1 mL methanol spiked with the internal standard formononetin (10 μg ml^{-1}) by vortexing continuously for 15 min. Further, the extracts were centrifuged at 15,000× *g* at 4 °C for 20 min. The collected supernatant was filtered and subjected to LC-QTOF-MS/MS (Agilent Technologies, Stuttgart, Germany) for analysis.

The extracts were separated on an XDB-C18 column (150 mm × 4.6 mm × 5 μm; Agilent Zorbax-Eclipse) using 0.1% (*v*/*v*) formic acid (solvent A) and acetonitrile with

0.1% formic acid (solvent B) as mobile phases. Compounds were eluted with a solvent gradient profile consisting of 95% A for 1 min followed by a gradient that reached 95% B by 15 min, returned 95% A by 17.5 min, and continued until 20 min. The injection volume was adjusted to 10 μL. The eluted compounds were detected using the centroid mode for both negative and positive ionization modes. The pump limit was 1 min; the draw speed and eject speed were set to 200 μL min^{-1} and 400 μL min^{-1}, respectively. The maximum pressure limit in the column was 800 bar, and the retention time exclusion tolerance was set to (±) 0.2 min. The ion source (dual ESI) was adjusted with a limit of 2 precursors min^{-1}.

2.3. LC-QTOF-MS/MS Data Analysis

Initially, a personal compound database library (PCDL) was prepared by accessing different databases and libraries for the metabolites from Solanaceae plants. The PCDL library and the Mass Hunter Qualitative Analysis BO.07.00 tool (Agilent Technologies) were used to analyze the spectra of the metabolites. The MS/MS data were procured from four replicate samples to assess the biological variations in the control, βCC-, and simulated-herbivory-treated samples. Compounds with an abundance greater than 10,000 counts and a score of more than 75 were considered for further analysis. These compounds were identified with a mass threshold of 7 ppm and a peak distance threshold of 10 ppm in MS/MS mode using the 'find by formula' function in the Mass Hunter software. The molecular ion and daughter ions in MS and MS/MS modes were compared with the reference or predicted spectra available at the Human Metabolome Database (HMDB), MassBank, and Pubchem.

2.4. Statistical Analysis

Principal coordinate analysis (PCoA) of metabolomics data was performed in PAST 3 [27]. A heatmap was prepared with the web tool ClustVis (https://biit.cs.ut.ee/clustvis/; accessed on 1 December 2022). Metaboanalyst 5.0 (https://www.metaboanalyst.ca/; accessed on 1 December 2022) was used for orthogonal partial least square discriminant analysis (OPLS-DA), variable importance in projection (VIP), fold-change analysis, and pathway analysis. Venn diagrams were created using Venny 2.1 (https://bioinfogp.cnb.csic.es/tools/venny/; accessed on 1 December 2022) [28]. Normalized peak areas were analyzed by one-way ANOVA followed by Fisher's least significant difference (LSD) post hoc test. The significance was determined at $p \leq 0.05$.

3. Results

3.1. Application of βCC Altered the Metabolite Profile of Tomato Plants

Previous reports showed that exogenous βCC could improve plants' tolerance to different stresses but did not reduce plant growth [9,11]. Therefore, to investigate how βCC improves plants' tolerance without compromising plant growth, we analyzed the metabolome of tomato plants after βCC treatment and compared it with control plants. Samples were collected at 0, 30, 60, 90, 180, and 240 min after βCC treatment to assess βCC-mediated early changes in the metabolome (Figure S1). Similarly, to track the βCC-induced late metabolic changes, samples were collected at 72 h after βCC treatment (Figure S1). A total of 57 compounds were identified from positive and negative modes across all the time points. Compounds were identified based on the daughter ion spectra at three energy levels 10 eV, 20 eV, and 40 eV (Table S1). Among the identified compounds, 45 and 31 were accumulated at early (0–240 min) and late (72 h) time points, respectively. A heatmap created with the identified compounds showed that the metabolic profiles of βCC-treated samples differed from that of the control (Figure 1a). Principal coordinate analysis (PCoA) showed a total variability of 70.35%, where the maximum variation was captured in coordinates 1 (44.67%) and 2 (25.68%). Interestingly, βCC-treated samples from 0–240 min were grouped separately from the control samples from the same time points (0–240 min). However, the control and βCC-treated samples from 72 h were grouped in the same coordinate in the PCoA plot (Figure 1b). This suggests that βCC-mediated early changes in the metabolite profile were more prominent than later changes.

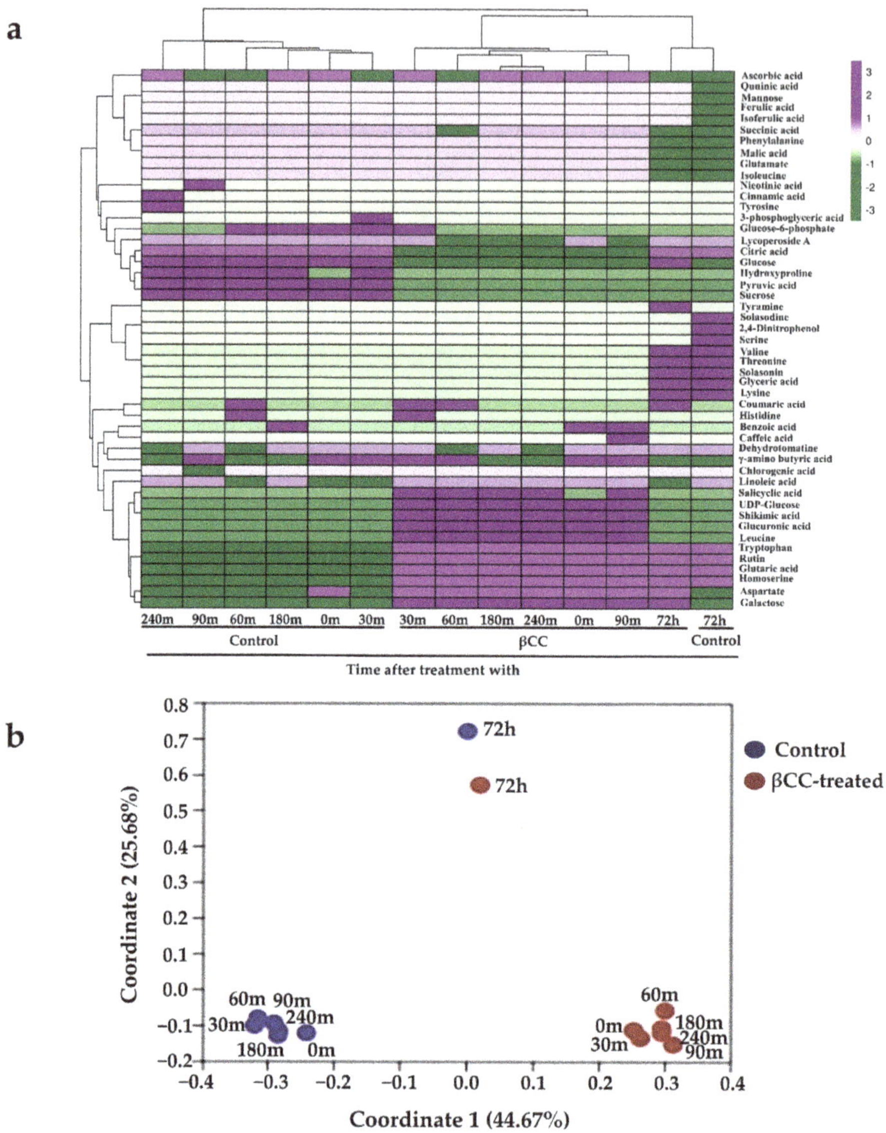

Figure 1. Variation in the metabolites after βCC treatment across different time points. (**a**) The metabolites showed variation in their accumulation pattern across different time points after being treated with βCC. The color scheme at the top-right corner codes for the z-scores (−3 to +3) calculated over binary coordinates of the samples. The average linkage clustering computed between the samples is based on the Manhattan distance and depicted at the top of the heatmap. (**b**) The principal coordinate analysis (PCoA) was performed with the identified metabolites using the Jaccard coefficient and transformation exponent value 2. It showed that the maximum percentage of the variation was captured in coordinates 1 and 2. The values in parentheses show the percentages of the variation.

3.2. Regulation of Metabolites in Early Time Points after βCC Treatment

The PCoA plot indicated a pronounced effect of βCC on metabolite accumulation within 0–240 min after the treatment. To find out the discriminant metabolites that were responsible for separating the control and βCC-treated samples, the peak area of 45 identified compounds from 0–240 min time scale were normalized with the internal standard and analyzed for an orthogonal partial least square discriminant analysis (OPLS-DA) and variable importance in projection (VIP) analysis. OPLS-DA showed that the control and βCC-treated samples were well separated in the score plot (Figure 2a). The data confirmed that the metabolic blend of the control and βCC-treated samples were different. VIP analysis determined the relative contribution of the metabolites to the variation between the control and βCC-treated samples. The analysis identified a total of 16 discriminant compounds (VIP score ≥ 1.0) (Figure 2b).

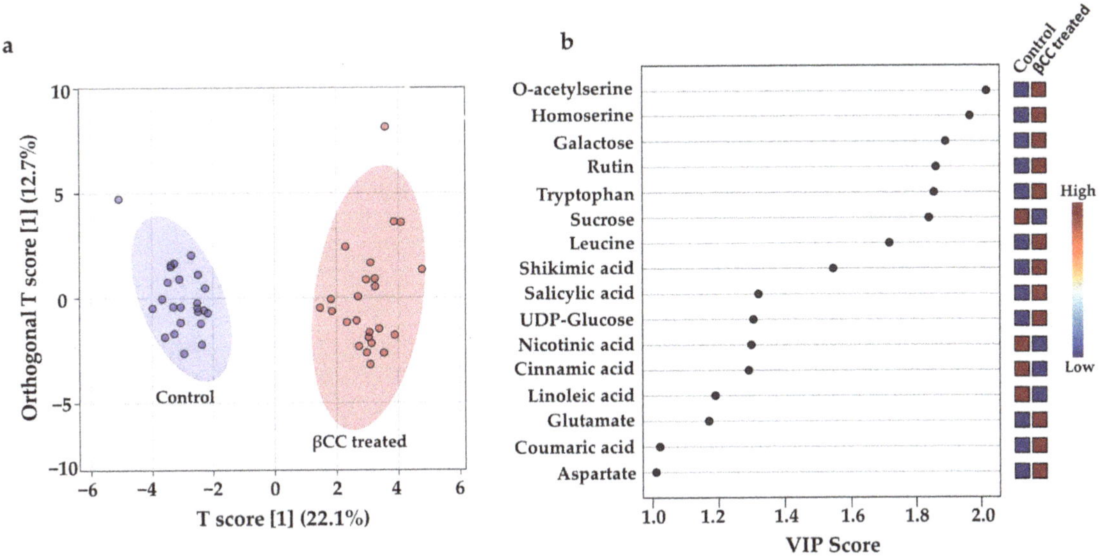

Figure 2. βCC-induced early metabolic changes and discriminant metabolites. To examine the differences in the metabolic profile in tomato plants early after βCC treatment, orthogonal partial least square discriminant analysis (OPLS-DA) and variable importance in projection (VIP) analysis were performed with the normalized peak area of the metabolites identified from 0–240 min after βCC treatment and in control plants. (**a**) An OPLS-DA plot showed the control group clustered to the left region and the βCC-treated group clustered to the right area in the OPLS-DA score plot. The shaded ellipses represent the confidence interval of 95% from OPLS-DA models. (**b**) A variable importance in projection (VIP) plot showed that the metabolites responsible for the significant separation observed between the two sample groups were indicated by a VIP score ≥ 1.0. An increase in VIP score indicates a high contribution of the metabolites to the group separation. The red and blue boxes on the right indicate whether the metabolite concentration was high (red) or low (blue) in the βCC-treated plants compared to control plants.

Further, the fold-change analysis revealed that, out of 16 discriminant compounds, 11 were upregulated ($\log_2 FC \geq 2$), and four were downregulated ($\log_2 FC \leq -2.0$) (Figure S2). Though the $\log_2$FC value of one of the discriminant metabolites (aspartate) was 1.8, its VIP score was 1.01; therefore, it had the most negligible influence in separating the control and βCC-treated sample groups. The normalized peak area of the discriminant metabolites from all the time points was further analyzed by one-way ANOVA and Fisher's LSD post hoc test. Significantly different metabolites were determined at $p \leq 0.05$ (Table S2) and visualized by mapping their regulation in the network of metabolic pathways for each time point (Figure 3a). The analysis showed that the upregulated metabolites were mainly amino acids and their derivatives (i.e., tryptophan, leucine, aspartate, glutamate, homoserine, and o-acetylserine), followed by phytophenols (i.e., shikimate, rutin, and coumaric acid), carbohydrates and their derivatives (i.e., UDP-glucose and galactose), and carboxylic acid (i.e., salicylic acid). Cinnamic acid, linoleic acid, nicotinic acid, and sucrose were significantly downregulated metabolites. Most of the compounds were upregulated, so a pathway enrichment analysis was performed with the upregulated compounds. It revealed that the pathways related to (1) aminoacyl-tRNA biosynthesis (p = 0.0001), (2) glycine, serine, and threonine metabolism (p = 0.001), (3) lysine biosynthesis (p = 0.0010), (4) cysteine and methionine metabolism (p = 0.0028), (5) arginine biosynthesis (p = 0.0059) (6) phenylalanine, tyrosine, and tryptophan biosynthesis (p = 0.0088), (7) alanine, aspartate, and glutamate metabolism (p = 0.0088) (8) galactose metabolism (p = 0.0131), and (9) indole alkaloid biosynthesis (p = 0.0267) were significantly influenced (hypergeometric test; $p \leq 0.05$) by βCC application (Figure 3b). The above data suggest that βCC treatment mainly boosted plants' amino acid metabolism within a few hours of treatment.

3.3. Regulation of Metabolites at a Late Time Point after βCC Treatment

Generally, in plants, accumulation of defense metabolites takes place 3–4 days after exposure to stress [29]. Surprisingly, the PCoA plot indicated that the metabolic blend of βCC-treated samples from 72 h was less different than that of the control samples. This result suggests that quantitative changes were more critical than qualitative changes in later times. The Venn diagram constructed with the metabolites detected 72 h after βCC treatment showed that, among 31 identified compounds, 19 were commonly present (63%) in control and βCC-treated plants, and only eight compounds (25%) were detected explicitly after βCC treatment (Figure 4a). However, the OPLS-DA (Figure 4b) and fold-change analysis (Figure S3) showed that, though the metabolite profile of control and βCC-treated plants were qualitatively similar, they were quantitatively different. Further, we identified the discriminant metabolites from βCC-treated samples after 72 h of treatment by VIP analysis. This revealed 19 discriminant compounds, of which 15 compounds (78%) were significantly upregulated (log2FC $\geq$ 2.0), and four compounds (21%) were significantly downregulated (log2FC ≤ -2.0). The upregulated compounds included mainly phytophenols (i.e., isoferulic acid, ferulic acid, coumaric acid, and quinic acid), followed by amino acids (i.e., homoserine, threonine, valine, tyramine, and aspartate), carbohydrates and their derivatives (i.e., galactose, glucose, and mannose), organic acids (i.e., citric acid and fumaric acid), and steroidal alkaloids (i.e., α-tomatine) (Figure 4c). Significantly downregulated metabolites were serine, solasodine, solasonin, and linoleic acid (Figure 4c). The pathway enrichment analysis of the upregulated discriminant compounds revealed their involvement in (1) glycine, serine, and threonine metabolism (p = 0.0014), (2) lysine biosynthesis (p = 0.0017), (3) aminoacyl-tRNA biosynthesis (p = 0.0038), (4) tyrosine metabolism (p = 0.0056), (5) arginine biosynthesis (p = 0.0071), (6) citrate cycle (p = 0.0088), (7) valine, leucine and isoleucine biosynthesis (p = 0.0106), (8) alanine, aspartate and glutamate metabolism (p = 0.0106), (9) cysteine and methionine metabolism (p = 0.0433), (10) phenylpropanoid biosynthesis (p = 0.0433), and (11) isoquinoline alkaloid biosynthesis (p = 0.0437) (hypergeometric test; $p \leq 0.05$) (Figure 4d). The data suggest that βCC treatment enhanced the pathways related to plant defense at later time points, keeping the amino acid biosynthesis and metabolism boosted.

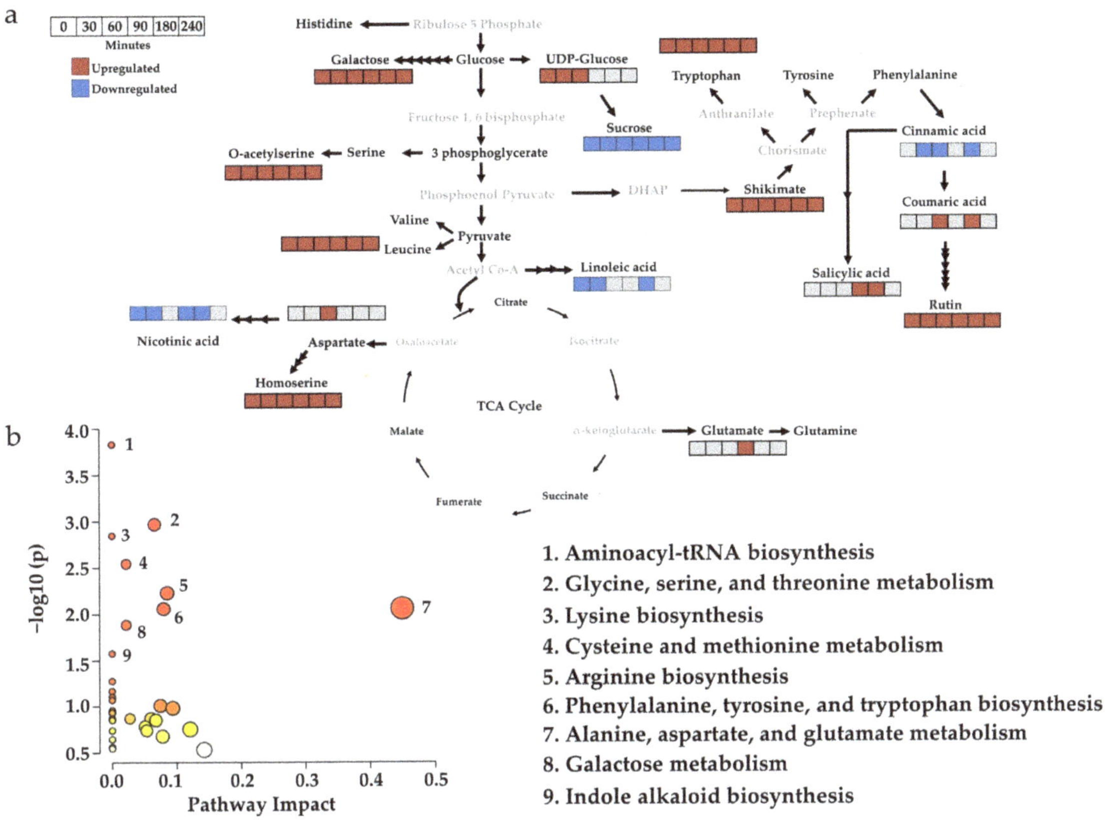

Figure 3. Mapping the relative expression of discriminant metabolites early after βCC treatment and their influence on metabolic pathways. The normalized peak area of the discriminant metabolites from control and βCC-treated samples of different time points (0, 30, 60, 90, 180, and 240 min) were subjected to one-way ANOVA; the significance was determined by Fisher's LSD at $p = 0.05$. (**a**) Significantly up- and down-regulated metabolites from different time points are mapped on the pathways modified from the KEGG database. (**b**) Pathway analysis using the upregulated discriminant metabolites shows altered pathways after βCC treatment (left panel). The x-axis depicts pathway impact values obtained from the pathway topology analysis, and the y-axis depicts the –log of the p values obtained from the pathway enrichment analysis ($p \leq 0.05$). Circular nodes represent the metabolic pathways. The size of the circular nodes positively corresponds with the impact of the proposed pathway based on the pathway topology. The node color, from yellow to orange, shows different levels of significance based on pathway enrichment analysis (yellow–low; orange–high). The most significantly altered pathways are characterized by both a high $-\log(p)$ value and a high impact value. Nine pathways were significantly altered after βCC treatment (right panel). Metabolites in black font were identified in the study, but greys were not. Grey boxes indicate no significant changes in the metabolite accumulation at $p \leq 0.05$.

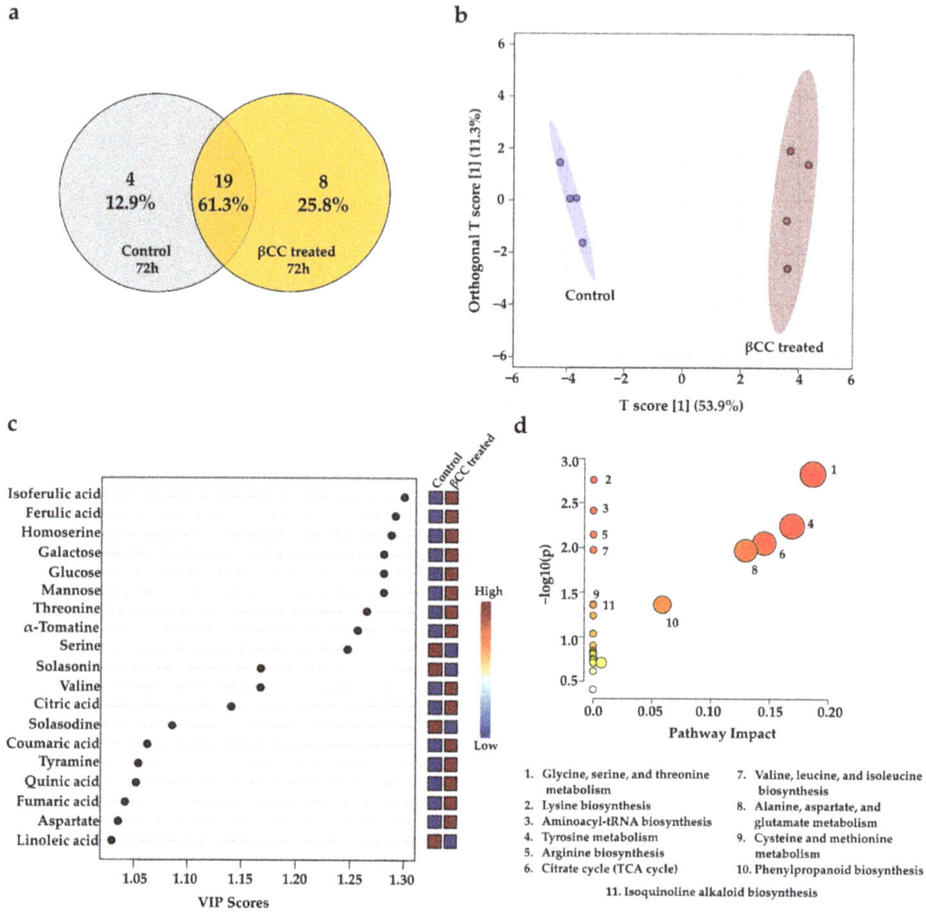

Figure 4. βCC-induced late metabolic changes, discriminant metabolites, and their influence on metabolic pathways. (**a**) A Venn diagram created with the identified metabolites from control and βCC-treated samples showed many metabolites regulated commonly and specifically in them. (**b**) OPLS-DA plot showed distinct clusters of the control group to the left region and the βCC-treated group to the right region of the plot. The shaded ellipses represent the confidence interval of 95% from OPLS-DA models. (**c**) A variable importance in projection (VIP) plot showed that the metabolites responsible for the significant separation observed between these two sample groups were indicated by a VIP score ≥ 1.0. An increase in VIP score indicates a high contribution of the metabolites to the group separation. The red and blue boxes on the right indicate whether the metabolite concentration was high (red) or low (blue) in the βCC-treated plants compared to control plants. (**d**) Pathway analysis using the upregulated discriminant metabolites shows altered pathways after βCC treatment (upper panel). The x-axis depicts pathway impact values obtained from the pathway topology analysis, and the y-axis depicts the –log of the p values obtained from the pathway enrichment analysis ($p \leq 0.05$). Circular nodes represent the metabolic pathways. The size of the circular nodes positively corresponds with the impact of the proposed pathway based on the pathway topology. The node color, from yellow to orange, shows different levels of significance based on pathway enrichment analysis (yellow–low; orange–high). The most significantly altered pathways were characterized by both a high $-\log(p)$ value and a high impact value. A total of 11 pathways were significantly altered after βCC treatment (lower panel).

3.4. βCC Treatment Induces a Similar Metabolic Response as Simulated Herbivory

Previously, we found that βCC treatment was also able to enhance resistance against a generalist herbivore, *Spodoptera littoralis*, in *Arabidopsis thaliana* [13]. To investigate if the discriminant metabolites are also influenced after insect herbivory, we compared the levels of βCC-induced metabolites with simulated-herbivory-treated and control plants after 72 h of treatment. Interestingly, out of the 15 discriminant compounds from βCC-treated plants, ten compounds (66.6%) were significantly upregulated after simulated herbivory compared to the control (Figure 5). The upregulated compounds included ferulic acid, isoferulic acid, coumaric acid, α-tomatine, tyramine, aspartate, citric acid, galactose, glucose, and mannose (Figure 5). In addition, βCC-treated samples accumulated significantly more amounts of quinic acid, threonine, valine, homoserine, and fumaric acid (One-way ANOVA; Fisher's LSD; $p \leq 0.05$) (Figure 5). The data suggest that βCC treatment can induce a similar blend of compounds that are upregulated after insect herbivory with a few more amino acids and phytophenol.

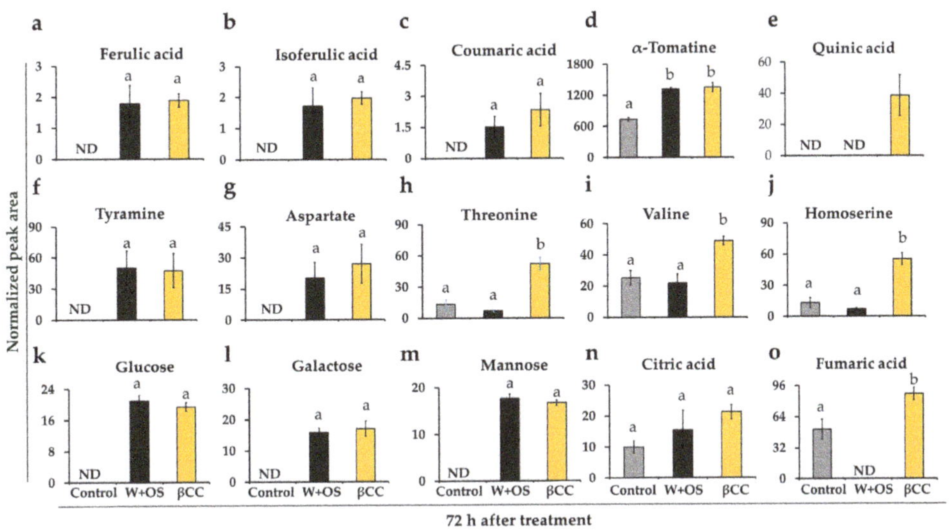

Figure 5. Comparative levels of metabolites after βCC treatment and simulated herbivory. The normalized peak area of the upregulated discriminant metabolites identified after 72 h of βCC treatment was compared with that of control and simulated-herbivory-treated (W+OS) plants. The accumulation of phytophenols, (**a**) ferulic acid ($F_{2,9} = 7.975$; $p_{βCC} = 0.006$, $p_{W+OS} = 0.008$), (**b**) isoferulic acid ($F_{2,9} = 9.012$; $p_{βCC} = 0.0036$, $p_{W+OS} = 0.0078$), (**c**) coumaric acid ($F_{2,9} = 5.554$; $p_{βCC} = 0.014$, $p_{W+OS} = 0.081$), and (**d**) α-tomatine ($F_{2,9} = 37.185$; $p_{βCC \text{ and } W+OS} < 0.001$) were significantly more in βCC- and simulated herbivory-treated plants, but (**e**) quinic acid ($F_{2,9} = 8.672$; $p_{βCC} = 0.0057$, $p_{W+OS} = \text{NA}$) was upregulated only after βCC treatment. Similarly, the levels of amino acids, (**f**) tyramine was also upregulated in βCC- and simulated-herbivory-treated plants ($F_{2,9} = 4.368$; $p_{βCC} = 0.034$, $p_{W+OS} = 0.027$); however, levels of (**g**) aspartate ($F_{2,9} = 4.082$; $p_{βCC} = 0.022$, $p_{W+OS} = 0.069$), (**h**) threonine ($F_{2,9} = 28.458$; $p_{βCC} = 0.0002$, $p_{W+OS} = 0.364$), (**i**) valine ($F_{2,9} = 10.038$; $p_{βCC} = 0.005$, $p_{W+OS} = 0.599$), and (**j**) homoserine ($F_{2,9} = 35.912$; $p_{βCC} < 0.001$, $p_{W+OS} = 0.341$) were upregulated only after βCC treatment. Accumulation of carbohydrates and their derivatives, (**k**) glucose ($F_{2,9} = 137.907$; $p_{βCC \text{ and } W+OS} < 0.001$), (**l**) galactose ($F_{2,9} = 33.337$; $p_{βCC \text{ and } W+OS} < 0.001$), and (**m**) mannose ($F_{2,9} = 260.128$; $p_{βCC} = 0.014$, $p_{βCC \text{ and } W+OS} < 0.001$) were also upregulated after both βCC treatment and simulated herbivory; however, accumulation of organic acids (**n**) citric acid ($F_{2,9} = 1.980$; $p_{βCC} = 0.077$, $p_{W+OS} = 0.356$), and (**o**) fumaric acid ($F_{2,9} = 40.004$; $p_{βCC} = 0.005$, $p_{W+OS} = 0.0006$) was only increased after βCC treatment. The mean normalized peak area (±SE) was analyzed from four replicate plants by one-way ANOVA and Fisher's LSD post hoc test. Different letters indicate significant differences at $p \leq 0.05$.

4. Discussion

Plant growth and productivity are severely affected by environmental stress. However, plants adapt to this unfavorable condition by optimizing their defense strategies. Therefore, the machinery of stress perception, signal transduction, and, ultimately, production of defense responses have been extensively studied. The plant hormones jasmonic acid [30], abscisic acid [31], ethylene [32], and salicylic acid [33], are known as stress signaling molecules; however, the growth hormones auxin, gibberellins, and strigolactones also participate as signaling molecules in plant–environment interactions [34]. In recent years researchers have uncovered apocarotenoids as potential signaling molecules. Studies have shown that βCC, the oxidative product of β-carotene, is a significant apocarotenoid that can induce $^1O_2$-responsive genes, essential for photooxidative stress in *Arabidopsis thaliana* [8]. We found that the exogenous application of βCC reprograms the transcriptome of tomato plants [25] by triggering multiple stress-responsive genes, essential to counteract both abiotic and biotic stresses. We also found that βCC can prime tomato plants against drought and induce resistance against insect herbivores in *A. thaliana* plants [11,13]. However, βCC treatment does not negatively affect plant growth but enhances root growth and lateral branching [10,11]. Generally, plants compromise their photosynthetic ability under stress, and, at the same time, they invest in defense; this reduces the carbon flux towards growth and causes a growth–defense trade-off [35]. Previous studies on tomato metabolomics have revealed that the leaf, stem, and root metabolome present different signatures upon infestation. In addition, metabolomic markers can be used to monitor or predict the performance of plants and their response to environmental stresses [36]. These facts stimulated the hypothesis that βCC specifically upregulates those metabolites that play a dual role in improving plant growth and defense. It is also possible that defense metabolites are upregulated after accomplishing plant growth requirements in βCC-treated plants, which avoids diversion of carbon flux from growth toward defense. To examine this, we compared the metabolome of βCC-treated plants with control plants to reveal βCC-induced metabolic changes in tomato plants. In addition, we also compared βCC-induced metabolic changes with simulated-herbivory-induced changes to reveal βCC's influence on the defense metabolites related to insect herbivory. Our results showed that βCC treatment mainly influenced the metabolism of amino acids and the accumulation of phytophenols.

Traditionally, amino acids are designated as the building blocks of protein; however, they also serve as intermediates for other biosynthetic pathways. Therefore, apart from plant growth and development, they also influence the generation of metabolic energy and signaling processes [37] and confer resistance to both abiotic and biotic stress [38–40]. In a recent study, it was demonstrated that pathogen-inoculated tomato plants' primary metabolic pools were altered [41]. Similarly, βCC treatment induced many primary metabolites very early after the treatment. We found that βCC treatment increased the levels of tryptophan and its precursor shikimate. Shikimate is the common intermediate of the amino acid tryptophan and phenylalanine biosynthetic pathways. As tryptophan is the precursor for auxin and 5-hydroxytryptamine biosynthesis, an increase in the tryptophan level may contribute to plant growth. On the other hand, phenylalanine accumulation remained unaltered after βCC treatment; however, the derivatives of phenylalanine, namely, coumaric acid, rutin, and salicylic acid, were significantly increased (Figure 3a). These metabolites are known for their antioxidant and antipathogenic properties [42–44]. Moreover, rutin and coumaric acid application resulted in increased photosynthesis, chlorophyll content, and shoot growth [45,46]. In addition, the amino acids glutamate and glycine were also increased after βCC treatment. Glutamate is the precursor of chlorophyll tetrapyrrol protoporphyrin IX biosynthesis, where δ-aminolaevulinic acid (ALA) is the major intermediate [47]. Interestingly, the application of ^{14}C-labeled glycine can instantly be incorporated into ALA in dark-grown barley leaves [48]. Moreover, exogenous glycine application can stimulate root hair formation and [49] induce plant growth [50,51]. These findings are consistent with our previous results that showed that βCC treatment could enhance chlorophyll accumulation and root growth in tomato plants [11].

βCC treatment can also enhance the levels of aspartate, which serves as a precursor of many biosynthetic pathways required for growth and defense [52]. For example, it is a precursor for the aspartate oxidase pathway that synthesizes nicotinamide adenine dinucleotide (NAD), an essential component of chlorophyll synthesis [53]. Therefore, presumably, βCC-induced increase in glutamate, glycine, and aspartate accumulation supports increased chlorophyll content in βCC-treated plants. Another vital role of aspartate is to transfer the reduction equivalents from the glycolytic pathway to the mitochondria for ATP generation via the malate–aspartate shuttle [54]. Recent studies showed that mitochondrial components of the malate–aspartate NADH shuttle act as a longevity factor that induces the extension of lifespan in yeast. Therefore, presumably by enhancing the malate–aspartate shuttle, stress-induced excess reducing powers are ameliorated in mitochondria that ultimately helps plant survival during stress. Moreover, the accumulation of aspartate is closely related to stress acclimation. For example, aspartate concentration was increased significantly in drought-exposed *Brassica napus* plants [55]. It is known that out of the eight essential amino acids, four amino acids, namely methionine, threonine, lysine, and isoleucine, are produced from aspartate [56]. An increase in aspartate after βCC treatment is translated into an increase in the levels of homoserine (a common intermediate of methionine, threonine, and lysine), and specifically threonine, but not methionine, lysine, and isoleucine. βCC also induced the accumulation of leucine. An increase in threonine and leucine may be attributed to the critical components of serine/threonine protein kinases [57] and leucine-rich repeat (LRR) proteins [58], respectively, and facilitates the perception of stress signals and protein–protein interactions. An increase in the transcripts of serine/threonine protein kinases and LRR proteins after βCC application [25] suggests the same.

Abiotic stresses, such as drought and salinity, cause osmotic stress in the plant. To maintain osmolarity, plants produce compatible solutes to maintain cell turgor. These non-toxic compounds fall into three categories, namely, amino acids, onium compounds, and sugars/polyols [59]. Proline is one of the significant osmolytes; however, we found proline levels remained unaltered until three days after βCC treatment. Our previous study showed that proline significantly increased after 21 days of βCC treatment. Therefore proline may be upregulated after three days of βCC treatment. However, homoserine is a non-protein amino acid, and increased levels of homoserine can be attributed to the production of the homoserine betain, a known osmolyte in salt stress [60]. Mannose is another important metabolite that also works as an osmolyte and, in addition, enhances antioxidant metabolism and reduces chlorophyll degradation [61].

Accumulation of phytophenols is significantly increased after βCC treatment. They contribute to plant color and protect plants from oxidative stress, pathogen infestation, and herbivore attack [62]. Phytophenols are biosynthesized utilizing amino acids as the precursors; however, a few phenolic compounds are also derived from the shikimic acid pathway [63]. Coumaric acid is one of the phytophenols that is biosynthesized through the shikimate pathway [64] and can be converted into phenolic acids [65,66]. Therefore, an increase in the levels of coumaric acid and its phenolic acids, namely, ferulic acid and its isomer isoferulic acid, suggest a role of βCC in the production of defense metabolites against abiotic and biotic stress. This view is further strengthened by the commonly upregulated defense metabolites in βCC- and simulated-herbivory-treated samples. Interestingly, α-tomatine, a glycoalkaloid specifically present in tomato that deters insect herbivores, is significantly greater in βCC-treated plants than simulated-herbivory-treated plants. Similar trends were evident in the accumulation of quinic acid. Together, this suggests that βCC can upregulate defense metabolites that prime tomato plants against multiple stresses. However, a few metabolites, namely cinnamic acid, nicotinic acid, linoleic acid, solasodin, and solasonin, that are related to plant defense, were downregulated significantly. A high turnover of cinnamic acid to coumaric acid and rutin probably restricts its accumulation. Similarly, the allocation of aspartate towards homoserine production can limit aspartate allocation towards nicotinic acid. Downregulation of linoleic acid after βCC treatment is surprising, as linoleic acid levels are known to be increased within an hour of herbivory and pathogen attack. After being attacked by herbivores, linoleic acid is liberated from cell membranes and is converted to jasmonate, which regulates transcription of the defense genes [67]. In our previous transcriptomic study, we did not find upregulation of any genes from the jasmonate biosynthetic cascade [25]. In the current study, we could not detect jasmonic acids or their derivatives; however, the levels of α-tomatine, a jasmonate-dependent glycoalkaloid [68], increased. Together, these suggest that βCC operates in a jasmonate-independent way in tomato plants. The other glycoalkaloids, solasodine, and solasonine were downregulated after BCC treatment. As α-tomatine is the major glycoalkaloid in tomato plants, downregulation of others may be cost-effective.

5. Conclusions

In the current study, we found that exogenous βCC elicits changes in the metabolome of tomato plants. Interestingly, these changes were more significant at early time points after βCC treatment than later ones. βCC mainly regulates amino acid and phytophenol metabolism. Interestingly, βCC-treated plants precisely upregulated metabolites having a role in improving both growth and defense; moreover, regulation of amino acid and phytophenol metabolism at different times optimized the growth of tomato plants. Therefore, βCC is a promising molecule for inducing resilience against biotic and abiotic stress. In general, most research has focused on the effect of priming molecules on plant phenotypic changes. However, our study sought to reveal the molecular changes that contribute to understanding the molecular mechanisms underlying priming.

Supplementary Materials: The following supporting information can be downloaded at: https://www.mdpi.com/article/10.3390/metabo13030329/s1, Figure S1: Representative base peak chromatogram (BPC) from LC-MS analysis of control, βCC-, and simulated-herbivory-treated samples; Figure S2: Fold-change analysis of the samples from early time points (0–240 min) after βCC treatment compared to control plants; Figure S3: Fold-change analysis of the samples from 72 h after βCC treated plants compared to control plants. Table S1: Compounds identified from control and βCC-treated tomato plants by LC-MS/MS; Table S2: Statistical analysis of the levels of discriminant metabolites identified early after βCC treatment.

Author Contributions: Conceptualization, S.M.; methodology, S.D.; software, S.D. and S.M.; validation, S.D. and S.M.; formal analysis, S.D. and S.M.; investigation, S.M.; resources, S.M.; data curation, S.M. and S.D.; writing—original draft preparation, S.M. and S.D.; writing—review and editing, S.M.; visualization, S.M.; supervision, S.M.; project administration, S.M.; funding acquisition, S.M. All authors have read and agreed to the published version of the manuscript.

Funding: This research was funded by the Ramalingaswami Reentry Fellowship, Department of Biotechnology (DBT), India. Grant number BT/HRD/35/02/2006.

Institutional Review Board Statement: Not applicable.

Informed Consent Statement: Not applicable.

Data Availability Statement: The data presented in this study are available in the main text and supplementary materials.

Acknowledgments: The authors acknowledge Savitribai Phule Pune University for the infrastructure provided for the study, V. T. Barvkar for technical help with LC-MS data acquisition, and S. Raskar, V. Purkar, and Rakesh M. for technical assistance.

Conflicts of Interest: The authors declare no conflict of interest. The funders had no role in the design of the study; in the collection, analyses, or interpretation of data; in the writing of the manuscript, or in the decision to publish the results.

References

1. Thakur, B.; Singh, R.; Nelson, P. Quality attributes of processed tomato products: A review. *Food Rev. Int.* **1996**, *12*, 375–401. [CrossRef]
2. Zhang, Y.; Song, H.; Wang, X.; Zhou, X.; Zhang, K.; Chen, X.; Liu, J.; Han, J.; Wang, A. The roles of different types of trichomes in tomato resistance to cold, drought, whiteflies, and botrytis. *Agronomy* **2020**, *10*, 411. [CrossRef]
3. Nisar, N.; Li, L.; Lu, S.; Khin, N.C.; Pogson, B.J. Carotenoid metabolism in plants. *Mol. Plant* **2015**, *8*, 68–82. [CrossRef] [PubMed]
4. Giuliano, G.; Al-Babili, S.; Von Lintig, J. Carotenoid oxygenases: Cleave it or leave it. *Trends Plant Sci.* **2003**, *8*, 145–149. [CrossRef]
5. Hou, X.; Rivers, J.; León, P.; McQuinn, R.P.; Pogson, B.J. Synthesis and function of apocarotenoid signals in plants. *Trends Plant Sci.* **2016**, *21*, 792–803. [CrossRef]
6. Verslues, P.E.; Agarwal, M.; Katiyar-Agarwal, S.; Zhu, J.; Zhu, J.K. Methods and concepts in quantifying resistance to drought, salt and freezing, abiotic stresses that affect plant water status. *Plant J.* **2006**, *45*, 523–539. [CrossRef]
7. Moreno, J.C.; Mi, J.; Alagoz, Y.; Al-Babili, S. Plant apocarotenoids: From retrograde signaling to interspecific communication. *Plant J.* **2021**, *105*, 351. [CrossRef]
8. Ramel, F.; Birtic, S.; Ginies, C.; Soubigou-Taconnat, L.; Triantaphylidès, C.; Havaux, M. Carotenoid oxidation products are stress signals that mediate gene responses to singlet oxygen in plants. *Proc. Natl. Acad. Sci. USA* **2012**, *109*, 5535–5540. [CrossRef]
9. D'alessandro, S.; Ksas, B.; Havaux, M. Decoding β-cyclocitral-mediated retrograde signaling reveals the role of a detoxification response in plant tolerance to photooxidative stress. *Plant Cell* **2018**, *30*, 2495–2511. [CrossRef]
10. d'Alessandro, S.; Mizokami, Y.; Legeret, B.; Havaux, M. The apocarotenoid β-cyclocitric acid elicits drought tolerance in plants. *Iscience* **2019**, *19*, 461–473. [CrossRef]
11. Deshpande, S.; Manoharan, R.; Mitra, S. Exogenous β-cyclocitral treatment primes tomato plants against drought by inducing tolerance traits, independent of abscisic acid. *Plant Biol.* **2021**, *23*, 170–180. [CrossRef]
12. Prasad, A.; Sedlářová, M.; Kale, R.S.; Pospíšil, P. Lipoxygenase in singlet oxygen generation as a response to wounding: In vivo imaging in Arabidopsis thaliana. *Sci. Rep.* **2017**, *7*, 9831. [CrossRef] [PubMed]
13. Mitra, S.; Estrada-Tejedor, R.; Volke, D.C.; Phillips, M.A.; Gershenzon, J.; Wright, L.P. Negative regulation of plastidial isoprenoid pathway by herbivore-induced β-cyclocitral in Arabidopsis thaliana. *Proc. Natl. Acad. Sci. USA* **2021**, *118*, e2008747118. [CrossRef] [PubMed]
14. Turner, M.F.; Heuberger, A.L.; Kirkwood, J.S.; Collins, C.C.; Wolfrum, E.J.; Broeckling, C.D.; Prenni, J.E.; Jahn, C.E. Non-targeted metabolomics in diverse sorghum breeding lines indicates primary and secondary metabolite profiles are associated with plant biomass accumulation and photosynthesis. *Front. Plant Sci.* **2016**, *7*, 953. [CrossRef]
15. Dickinson, A.J.; Lehner, K.; Mi, J.; Jia, K.-P.; Mijar, M.; Dinneny, J.; Al-Babili, S.; Benfey, P.N. β-Cyclocitral is a conserved root growth regulator. *Proc. Natl. Acad. Sci. USA* **2019**, *116*, 10563–10567. [CrossRef]
16. Reinhardt, D.; Kuhlemeier, C. Plant architecture. *EMBO Rep.* **2002**, *3*, 846–851. [CrossRef] [PubMed]
17. Creydt, M.; Arndt, M.; Hudzik, D.; Fischer, M. Plant metabolomics: Evaluation of different extraction parameters for nontargeted UPLC-ESI-QTOF-mass spectrometry at the example of white Asparagus officinalis. *J. Agric. Food Chem.* **2018**, *66*, 12876–12887. [CrossRef]
18. Piasecka, A.; Kachlicki, P.; Stobiecki, M. Analytical methods for detection of plant metabolomes changes in response to biotic and abiotic stresses. *Int. J. Mol. Sci.* **2019**, *20*, 379. [CrossRef]
19. Shulaev, V.; Cortes, D.; Miller, G.; Mittler, R. Metabolomics for plant stress response. *Physiol. Plant.* **2008**, *132*, 199–208. [CrossRef]
20. Zhao, Y.; Zhao, J.; Zhao, C.; Zhou, H.; Li, Y.; Zhang, J.; Li, L.; Hu, C.; Li, W.; Peng, X. A metabolomics study delineating geographical location-associated primary metabolic changes in the leaves of growing tobacco plants by GC-MS and CE-MS. *Sci. Rep.* **2015**, *5*, 16346. [CrossRef]
21. Zielińska, A.; Siudem, P.; Paradowska, K.; Gralec, M.; Kaźmierski, S.; Wawer, I. Aronia melanocarpa fruits as a rich dietary source of chlorogenic acids and anthocyanins: 1H-NMR, HPLC-DAD, and chemometric studies. *Molecules* **2020**, *25*, 3234. [CrossRef]
22. Jorge, T.F.; Mata, A.T.; António, C. Mass spectrometry as a quantitative tool in plant metabolomics. *Philos. Trans. R. Soc. A Math. Phys. Eng. Sci.* **2016**, *374*, 20150370. [CrossRef]
23. De Vos, R.C.; Hall, R.D.; Moing, A. Metabolomics of a model fruit: Tomato. In *Annual Plant Reviews Online*; Volume 43: Biology of Plant Metabolomics; Wiley-Blacwell Ltd.: Oxford, UK, 2011; Volume 43, pp. 109–155.
24. Zeiss, D.R.; Mhlongo, M.I.; Tugizimana, F.; Steenkamp, P.A.; Dubery, I.A. Metabolomic profiling of the host response of tomato (*Solanum lycopersicum*) following infection by Ralstonia solanacearum. *Int. J. Mol. Sci.* **2019**, *20*, 3945. [CrossRef]
25. Deshpande, S.; Purkar, V.; Mitra, S. β-Cyclocitral, a Master Regulator of Multiple Stress-Responsive Genes in *Solanum lycopersicum* L. Plants. *Plants* **2021**, *10*, 2465. [CrossRef]
26. Mitra, S.; Gershenzon, J. Effects of herbivory on carotenoid biosynthesis and breakdown. *Methods Enzymol.* **2022**, *674*, 497–517.
27. Hammer, Ø.; Harper, D.A.; Ryan, P.D. PAST: Paleontological statistics software package for education and data analysis. *Palaeontol. Electron.* **2001**, *4*, 9.
28. Oliveros, J.C. VENNY. An Interactive Tool for Comparing Lists with Venn Diagrams. 2007. Available online: http://bioinfogp.cnb.csic.es/tools/venny/index.html (accessed on 1 December 2022).
29. Fürstenberg-Hägg, J.; Zagrobelny, M.; Bak, S. Plant defense against insect herbivores. *Int. J. Mol. Sci.* **2013**, *14*, 10242–10297. [CrossRef] [PubMed]

30. Wasternack, C.; Stenzel, I.; Hause, B.; Hause, G.; Kutter, C.; Maucher, H.; Neumerkel, J.; Feussner, I.; Miersch, O. The wound response in tomato–role of jasmonic acid. *J. Plant Physiol.* **2006**, *163*, 297–306. [CrossRef] [PubMed]
31. Fujita, Y.; Fujita, M.; Shinozaki, K.; Yamaguchi-Shinozaki, K. ABA-mediated transcriptional regulation in response to osmotic stress in plants. *J. Plant Res.* **2011**, *124*, 509–525. [CrossRef] [PubMed]
32. Gamalero, E.; Glick, B.R. Ethylene and abiotic stress tolerance in plants. In *Environmental Adaptations and Stress Tolerance of Plants in the Era of Climate Change*; Springer: New York, NY, USA, 2012; pp. 395–412.
33. Loake, G.; Grant, M. Salicylic acid in plant defence—The players and protagonists. *Curr. Opin. Plant Biol.* **2007**, *10*, 466–472. [CrossRef] [PubMed]
34. Verma, V.; Ravindran, P.; Kumar, P.P. Plant hormone-mediated regulation of stress responses. *BMC Plant Biol.* **2016**, *16*, 86. [CrossRef] [PubMed]
35. Herms, D.A.; Mattson, W.J. The dilemma of plants: To grow or defend. *Q. Rev. Biol.* **1992**, *67*, 283–335. [CrossRef]
36. Abreu, A.C.; Fernández, I. NMR metabolomics applied on the discrimination of variables influencing tomato (*Solanum lycopersicum*). *Molecules* **2020**, *25*, 3738. [CrossRef] [PubMed]
37. Häusler, R.E.; Ludewig, F.; Krueger, S. Amino acids–a life between metabolism and signaling. *Plant Sci.* **2014**, *229*, 225–237. [CrossRef] [PubMed]
38. Moe, L.A. Amino acids in the rhizosphere: From plants to microbes. *Am. J. Bot.* **2013**, *100*, 1692–1705. [CrossRef] [PubMed]
39. Watanabe, M.; Balazadeh, S.; Tohge, T.; Erban, A.; Giavalisco, P.; Kopka, J.; Mueller-Roeber, B.; Fernie, A.R.; Hoefgen, R. Comprehensive dissection of spatiotemporal metabolic shifts in primary, secondary, and lipid metabolism during developmental senescence in Arabidopsis. *Plant Physiol.* **2013**, *162*, 1290–1310. [CrossRef]
40. Zeier, J. New insights into the regulation of plant immunity by amino acid metabolic pathways. *Plant Cell Environ.* **2013**, *36*, 2085–2103. [CrossRef]
41. Luna, E.; Flandin, A.; Cassan, C.; Prigent, S.; Chevanne, C.; Kadiri, C.F.; Gibon, Y.; Pétriacq, P. Metabolomics to exploit the primed immune system of tomato fruit. *Metabolites* **2020**, *10*, 96. [CrossRef]
42. Zielinska, D.; Szawara-Nowak, D.; Zielinski, H. Determination of the antioxidant activity of rutin and its contribution to the antioxidant capacity of diversified buckwheat origin material by updated analytical strategies. *Pol. J. Food Nutr. Sci.* **2010**, *60*, 315–321.
43. Liu, X.; Ji, D.; Cui, X.; Zhang, Z.; Li, B.; Xu, Y.; Chen, T.; Tian, S. p-Coumaric acid induces antioxidant capacity and defense responses of sweet cherry fruit to fungal pathogens. *Postharvest Biol. Technol.* **2020**, *169*, 111297. [CrossRef]
44. Verberne, M.C.; Verpoorte, R.; Bol, J.F.; Mercado-Blanco, J.; Linthorst, H.J. Overproduction of salicylic acid in plants by bacterial transgenes enhances pathogen resistance. *Nat. Biotechnol.* **2000**, *18*, 779–783. [CrossRef]
45. Gorni, P.H.; de Lima, G.R.; de Oliveira Pereira, L.M.; Spera, K.D.; de Marcos Lapaz, A.; Pacheco, A.C. Increasing plant performance, fruit production and nutritional value of tomato through foliar applied rutin. *Sci. Hortic.* **2022**, *294*, 110755. [CrossRef]
46. Nkomo, M.; Gokul, A.; Keyster, M.; Klein, A. Exogenous p-coumaric acid improves *Salvia hispanica* L. seedling shoot growth. *Plants* **2019**, *8*, 546. [CrossRef] [PubMed]
47. Willows, R.D. Chlorophyll synthesis. In *The Structure and Function of Plastids*; Springer: Berlin/Heidelberg, Germany, 2007; pp. 295–313.
48. Hendry, G.A.; Stobart, A.K. Glycine metabolism and chlorophyll synthesis in barley leaves. *Phytochemistry* **1977**, *16*, 1567–1570. [CrossRef]
49. Domínguez-May, Á.V.; Carrillo-Pech, M.; Barredo-Pool, F.A.; Martínez-Estévez, M.; Us-Camas, R.Y.; Moreno-Valenzuela, O.A.; Echevarría-Machado, I. A novel effect for glycine on root system growth of habanero pepper. *J. Am. Soc. Hortic. Sci.* **2013**, *138*, 433–442. [CrossRef]
50. Kotinskyi, A.; Salyuk, A.; Zhadan, S. The influence of exogenous glycine on growth and cyanobacteria spirulina platensis (Gom.) Geitl photosynthetic processes. *Biotechnol. Acta* **2018**, *11*, 39–46. [CrossRef]
51. Mohammadipour, N.; Souri, M.K. Beneficial effects of glycine on growth and leaf nutrient concentrations of coriander (*Coriandrum sativum*) plants. *J. Plant Nutr.* **2019**, *42*, 1637–1644. [CrossRef]
52. Han, M.; Zhang, C.; Suglo, P.; Sun, S.; Wang, M.; Su, T. L-Aspartate: An essential metabolite for plant growth and stress acclimation. *Molecules* **2021**, *26*, 1887. [CrossRef]
53. Chai, M.-F.; Chen, Q.-J.; An, R.; Chen, Y.-M.; Chen, J.; Wang, X.-C. NADK2, an Arabidopsis chloroplastic NAD kinase, plays a vital role in both chlorophyll synthesis and chloroplast protection. *Plant Mol. Biol.* **2005**, *59*, 553–564. [CrossRef]
54. Borst, P. The malate–aspartate shuttle (Borst cycle): How it started and developed into a major metabolic pathway. *IUBMB Life* **2020**, *72*, 2241–2259. [CrossRef]
55. Good, A.G.; Zaplachinski, S.T. The effects of drought stress on free amino acid accumulation and protein synthesis in Brassica napus. *Physiol. Plant.* **1994**, *90*, 9–14. [CrossRef]
56. Li, Y.; Wei, H.; Wang, T.; Xu, Q.; Zhang, C.; Fan, X.; Ma, Q.; Chen, N.; Xie, X. Current status on metabolic engineering for the production of l-aspartate family amino acids and derivatives. *Bioresour. Technol.* **2017**, *245*, 1588–1602. [CrossRef] [PubMed]
57. Hardie, D. Plant protein serine/threonine kinases: Classification and functions. *Annu. Rev. Plant Biol.* **1999**, *50*, 97–131. [CrossRef] [PubMed]
58. Bella, J.; Hindle, K.; McEwan, P.; Lovell, S. The leucine-rich repeat structure. *Cell. Mol. Life Sci.* **2008**, *65*, 2307–2333. [CrossRef] [PubMed]

59. Nuccio, M.L.; Rhodes, D.; McNeil, S.D.; Hanson, A.D. Metabolic engineering of plants for osmotic stress resistance. *Curr. Opin. Plant Biol.* **1999**, *2*, 128–134. [CrossRef] [PubMed]
60. Kageyama, H.; Waditee-Sirisattha, R. Osmoprotectant molecules in cyanobacteria: Their basic features, biosynthetic regulations, and potential applications. In *Cyanobacterial Physiology*; Elsevier: Amsterdam, The Netherlands, 2022; pp. 113–123.
61. Zhao, S.; Zeng, W.; Li, Z.; Peng, Y. Mannose regulates water balance, leaf senescence, and genes related to stress tolerance in white clover under osmotic stress. *Biol. Plant.* **2020**, *64*, 406–416. [CrossRef]
62. Pratyusha, S. Phenolic Compounds in the Plant Development and Defense: An Overview. In *Plant Stress Physiology-Perspectives in Agriculture*; Intechopen Limited: London, UK, 2022; pp. 125–140.
63. Santos-Sánchez, N.F.; Salas-Coronado, R.; Hernández-Carlos, B.; Villanueva-Cañongo, C. Shikimic acid pathway in biosynthesis of phenolic compounds. In *Plant Physiological Aspects of Phenolic Compounds*; IntechOpen: London, UK, 2019; Volume 1, pp. 1–15.
64. Herrmann, K.M.; Weaver, L.M. The shikimate pathway. *Annu. Rev. Plant Biol.* **1999**, *50*, 473. [CrossRef]
65. El-Seedi, H.R.; El-Said, A.M.; Khalifa, S.A.; Goransson, U.; Bohlin, L.; Borg-Karlson, A.-K.; Verpoorte, R. Biosynthesis, natural sources, dietary intake, pharmacokinetic properties, and biological activities of hydroxycinnamic acids. *J. Agric. Food Chem.* **2012**, *60*, 10877–10895. [CrossRef] [PubMed]
66. Pei, K.; Ou, J.; Huang, J.; Ou, S. p-Coumaric acid and its conjugates: Dietary sources, pharmacokinetic properties and biological activities. *J. Sci. Food Agric.* **2016**, *96*, 2952–2962. [CrossRef]
67. Conconi, A.; Miquel, M.; Browse, J.A.; Ryan, C.A. Intracellular levels of free linolenic and linoleic acids increase in tomato leaves in response to wounding. *Plant Physiol.* **1996**, *111*, 797–803. [CrossRef]
68. Panda, S.; Jozwiak, A.; Sonawane, P.D.; Szymanski, J.; Kazachkova, Y.; Vainer, A.; Vasuki Kilambi, H.; Almekias-Siegl, E.; Dikaya, V.; Bocobza, S. Steroidal alkaloids defence metabolism and plant growth are modulated by the joint action of gibberellin and jasmonate signalling. *New Phytol.* **2022**, *233*, 1220–1237. [CrossRef] [PubMed]

metabolites

MDPI

Article

Coordinated Regulation of Central Carbon Metabolism in Pyroligneous Acid-Treated Tomato Plants under Aluminum Stress

Raphael Ofoe [1,*], Raymond H. Thomas [2] and Lord Abbey [1,*]

1 Department of Plant, Food, and Environmental Sciences, Faculty of Agriculture, Dalhousie University, 50 Pictou Road, Bible Hill, NS B2N 5E3, Canada
2 Department of Biology, Faculty of Science, Western University 2025E Biological & Geological Sciences Building, 1151 Richmond Street, London, ON N6A 5B7, Canada; rthoma2@uwo.ca
* Correspondence: raphael.ofoe@dal.ca (R.O.); loab07@gmail.com (L.A.)

Abstract: Aluminum (Al) toxicity is a major threat to global crop production in acidic soils, which can be mitigated by natural substances such as pyroligneous acid (PA). However, the effect of PA in regulating plant central carbon metabolism (CCM) under Al stress is unknown. In this study, we investigated the effects of varying PA concentrations (0, 0.25 and 1% PA/ddH_2O (v/v)) on intermediate metabolites involved in CCM in tomato (*Solanum lycopersicum* L., 'Scotia') seedlings under varying Al concentrations (0, 1 and 4 mM $AlCl_3$). A total of 48 differentially expressed metabolites of CCM were identified in the leaves of both control and PA-treated plants under Al stress. Calvin–Benson cycle (CBC) and pentose phosphate pathway (PPP) metabolites were considerably reduced under 4 mM Al stress, irrespective of the PA treatment. Conversely, the PA treatment markedly increased glycolysis and tricarboxylic acid cycle (TCA) metabolites compared to the control. Although glycolysis metabolites in the 0.25% PA-treated plants under Al stress were comparable to the control, the 1% PA-treated plants exhibited the highest accumulation of glycolysis metabolites. Furthermore, all PA treatments increased TCA metabolites under Al stress. Electron transport chain (ETC) metabolites were higher in PA-treated plants alone and under 1 mM, Al but were reduced under a higher Al treatment of 4 mM. Pearson correlation analysis revealed that CBC metabolites had a significantly strong positive ($r = 0.99$; $p < 0.001$) association with PPP metabolites. Additionally, glycolysis metabolites showed a significantly moderate positive association ($r = 0.76$; $p < 0.05$) with TCA metabolites, while ETC metabolites exhibited no association with any of the determined pathways. The coordinated association between CCM pathway metabolites suggests that PA can stimulate changes in plant metabolism to modulate energy production and biosynthesis of organic acids under Al stress conditions.

Keywords: wood vinegar; metabolomics; TCA cycle; carbohydrate metabolism; *Solanum lycopersicum*

1. Introduction

Soil acidity is widespread and accounts for 50% of the global agricultural lands that support up to 80% of world vegetable production [1]. Aluminum (Al) is the most abundant metal in the lithosphere, and its availability is dependent on soil acidity levels [2]. Al exists in soils as non-toxic oxides and Al-silicates, to which several plant roots are exposed. However, when soil pH drops below 5, Al dissociates into trivalent forms that are toxic to most plants, including tomato, even at low concentrations [2]. These sensitive plants exhibit a wide range of Al-induced phytotoxicity, with the inhibition of root growth being the most significant effect [2,3]. Al disrupts root cells, leading to physiological drought by restricting water and nutrient uptake [4]. It also reduces leaf elongation, impairs the photosynthetic ability of plants and instigates the accumulation of reactive oxygen species (ROS) [5,6]. Al-induced buildup of ROS facilitates membrane lipid peroxidation, loss of cell membrane integrity and damage to cellular components including nucleic acids and proteins [7]. Consequently, these phytotoxic effects result in marked yield reduction and total crop failure [4,8].

Generally, plants have evolved cellular and metabolic mechanisms to mitigate Al-induced phytotoxicity. The exclusion and internal detoxification mechanisms of Al tolerance have been well-characterized in plants [2,4]. The exclusion mechanism involves preventing Al entry into root cells by secreting organic acids (OA) to chelate Al within the rhizosphere. The internal tolerance mechanism involves intracellular detoxification of Al and sequestration of the non-toxic Al–OA complexes in the vacuole [4]. These key mechanisms of Al tolerance can be linked to the central carbon metabolism (CCM) and its associated mitochondrial activity, but they are understudied. Furthermore, plant exposure to stress requires a rapid metabolic shift to maintain appropriate metabolism. As a result, several regulatory mechanisms must interact with various metabolic routes to regulate fluxes through the different intermediate pathways associated with CCM [9]. The CCM pathway supplies the fundamental carbon needed to sustain growth and productivity under stress conditions and comprises the Calvin–Benson cycle (CBC), glycolysis, pentose phosphate pathway and tricarboxylic acid (TCA) cycle, followed by oxidative phosphorylation for ATP production through the electron transport chain (ETC) [9–11]. Recently, several studies have demonstrated that Al stress alters metabolic profiles in plant, which may enhance Al tolerance [12–14]. Although most of these studies have focused on the root metabolism of Al tolerance, the effects of Al stress on metabolic changes in aerial parts or plants are poorly understood.

Mitochondrial metabolism plays an important role in Al tolerance by facilitating carbohydrate consumption for OA and cellular ATP biosynthesis [2,8]. Consequently, the glycolytic influx is increased under stress conditions to support not only energy production but also OA generation during Al detoxification. Notably, the TCA cycle produces high malic and citric acids in response to Al stress. These organic acids bind to Al ions, reducing Al toxicity both in the intracellular environment and around the rhizosphere [2]. Therefore, the end products of photosynthesis are highly utilized for OA generation during Al stress [6,12,15], which suggests that Al tolerance is linked to improved photosynthesis. Similarly, leaf metabolism under Al stress represents an important characteristic of plant growth and productivity, and the identification of metabolic alterations modulating photosynthesis and respiration may contribute to improving Al tolerance in plants.

Recent research and development have focused on the use of novel strategies to improve plant growth and resilience due to the multicomplex interaction of various environmental stresses [16]. As a result, the use of biostimulants to enhance crop growth, productivity and tolerance to environmental stresses has attracted the interest of both researchers and farmers [17,18]. Pyroligneous acid (PA) is one widely used biostimulant that is known to improve crop growth, resilience to environmental stresses and productivity [19,20]. PA is a translucent reddish-brown liquid produced through the carbonation of organic biomass in the presence of limited oxygen [19]. It consists of a complex mixture of over 200 water-soluble bioactive compounds, including organic acids, phenolics, sugar and alcohol derivatives [19,21,22]. However, its chemical composition depends on the feedstock, temperature, residence time and heating rate. In agriculture, PA has been demonstrated to promote seed germination, vegetative growth and yield in several crops [3,19,23].

Accumulating evidence has revealed that seed priming with PA treatments enhances plant tolerance to Al and drought stress by regulating genes and proteins involved in energy production and antioxidant defense systems [3,19,23]. Under drought stress, Wang et al. [23] reported that PA stimulates the differential accumulation of proteins involved in carbohydrate metabolism in wheat (*Triticum aestivum* L.) seedlings. This increase in drought tolerance was related to increased activities of enzymes associated to glycolysis and the TCA cycle. Similarly, Ofoe et al. [3] showed that PA enhances peroxidase activities and promotes the expression of auxin response factor and antioxidant genes in primed seedlings under Al stress, thus promoting Al tolerance. These findings suggest that PA can promote normal cellular balance and metabolic processes under stress conditions but require further investigation. Despite the accumulating efforts to study the simulation of plant growth and stress tolerance by PA, the metabolic mechanism underlying plant response to PA and Al stress remains unknown. Furthermore, no studies to date have explored the impact of PA on CCM regulation in plants under Al stress. In this study, we examined the metabolic profile involved in the CCM of PA-treated Scotia tomato seedlings under varying Al stress conditions. This investigation provides metabolic insights governing biochemical pathways that are crucial for understanding how PA mediates Al tolerance in plants.

2. Materials and Methods

2.1. Plant Material and Experimental Conditions

The experiment was conducted from January to March 2022 and repeated from March to May 2022 at the Department of Plant, Food, and Environmental Sciences, Faculty of Agriculture, Dalhousie University, Truro. Scotia tomato seeds were purchased from Halifax seeds (Halifax, Canada), and PA derived from white pine (*Pinus strobus*) biomass was produced by Proton Power Inc (Lenoir City, USA). Details of the PA and its chemical constituents were previously published by Ofoe et al. [21].

The Scotia tomato seeds were initially sterilized with 10% NaClO for 10 min and thoroughly washed 3 times with sterile distilled water. The seeds were further sterilized with 70% ethanol for 5 min and washed thoroughly with sterile distilled water. The sterilized seeds were sown in a Pro-Mix® BX (Premier Tech Horticulture, Rivière-du-Loup, QC, Canada) and grown for four weeks in a growth chamber (Conviron Controlled Environments Ltd., Winnipeg, MB, Canada) with 16/8 h day/night photoperiod, 24/22 °C day/night temperature regime, 300 μmol m^{-2} s^{-1} light intensity and a relative humidity of 70%. Uniform seedlings with an 8-cm root length at the 3rd to 4th true leaf stage were transplanted into a 10.2-cm plastic pot containing 500 g of sterilized sand with an average particle size of 0.5–1.0 mm. The seedlings were maintained with a 25% strength Hoagland nutrient solution (pH = 5.0) at planting for a week under a growth chamber to acclimatize.

2.2. Experimental Treatment and Design

After one week of acclimation, half-strength followed by full-strength nutrient solutions were amended with varying PA and Al concentrations and applied every week. PA treatments were applied to the nutrient solution at 0%, 0.25% and 1% PA/ddH_2O (v/v), with Al ($AlCl_3$) concentrations of 0, 1 and 4 mM. The nutrient solution (pH = 4.5) was renewed every 3 days to maintain adequate moisture content. Throughout the entire study period, the pH of the amended nutrient solution (PA with or without Al) was monitored frequently and adjusted to 4.5 using either sodium hydroxide (NaOH) to increase the pH or HCl to reduce the pH. The study was arranged in a 3 × 3 factorial completely randomized design with five replications.

2.3. Plant Sample Preparation

The fully expanded leaf tissues on the third and fourth petioles from the top (15 leaves per treatment) were collected 40 days after transplanting and immediately frozen in liquid nitrogen. The frozen samples were ground into a fine powder and stored in a −80 °C freezer for central carbon metabolite analyses.

2.4. Metabolite Quantitation Using LC-MRM/MS

Targeted metabolite quantitation was performed at the University of Victoria Genome BC—Proteomics Centre of The Metabolomics Innovation Centre, Canada. Ground samples (50 mg) were mixed with 80% methanol and homogenized with two metal balls on an MM400 mill mixer (Retsch, Haan, Germany) for 3 min at a shaking frequency of 30 Hz. The mixture was then sonicated in an ice-water bath for 5 min and centrifuged at 12,000× g at 5 °C for 20 min. Next, 250 μL of the supernatant was mixed with 150 μL of water and 150 μL of dichloromethane. The mixture was vortexed for 30 s and centrifuged at 21,000× g

for 20 min. Subsequently, 80 μL aliquots of the supernatant were dried under a nitrogen gas flow, and the residues were used for the following assay.

2.4.1. TCA Cycle

A standard solution of all targeted carboxylic acids was prepared using 80% methanol, with concentrations ranging from 200 to 1000 μM. For each sample, 50 μL of both the standard solution and the sample supernatant were mixed with an equal volume of 200 mM 3-nitrophenyl hydrazines (NPH) solution and 150 mM carbodiimide hydrochloride (EDC)-6% pyridine solution. The mixture was then incubated at 30 °C for 40 min. After the reaction, 450 μL of water was added to each solution, and 10 μL of the resulting solution was injected into a C18 liquid chromatography (LC) column (2.1 × 100 mm, 1.8 μm) for quantification of carboxylic acids using ultrahigh LC-multiple reaction monitoring/mass spectroscopy (UPLC-MRM/MS) with (−) ion detection. The UPLC-MRM/MS was performed on an Agilent 1290 UHPLC system coupled to a Sciex 4000 QTRAP MS instrument (AB Sciex, Concord, ON, Canada). The metabolite-dependent parameters used in the UPLC-MRM/MS were based on the procedure described by Han et al. [24].

2.4.2. Glucose and Selected Sugar Phosphates

The dried residue of each sample was mixed with 50 μL of 50% methanol. Then, 50 μL of each serially diluted standard solution of glucose, ribose, ribose-5-phosphate, glucose-6-phosphate and mannose-6-phosphate were mixed with 100 μL of 25 mM 3-amino-9-ethyl carbazole (AEC), 50 μL of 50 mM sodium cyanoborohydride ($NaCBH_3$) and 20 μL of LC/MS grade acetic acid. The mixtures were incubated at 60 °C for 70 min, and 200 μL of water and 300 μL of chloroform were added. Following centrifugation at 12,500× *g* for 5 min, 50 μL of each supernatant was mixed with an equal volume of water, and 10 μL of the resulting solution was injected into a pentafluorophenylpropyl (PFP) LC column (2.1 × 150 mm, 1.7 μm) for UPLC-MRM/MS analysis. The UPLC-MRM/MS analysis was performed on an Agilent 1290 UHPLC system coupled to an Agilent 6495B QQQ instrument (Conquer Scientific, Poway, CA, USA) with positive-ion detection. The metabolite-dependent parameters used in the UPLC-MRM/MS were based on the procedure described by Han et al. [25].

2.4.3. Other Metabolites

An internal standard (IS) solution containing 25 isotope-labelled metabolites, including adenosine diphosphate (ADP), adenosine triphosphate (ATP), fructose-6-phosphate (fructose-6P), fructose-bisphosphate, glycerol-3-phosphate, nicotinamide adenine dinucleotide (NAD), NADH, glucose-1-phosphate, ribose-5-phosphate and others, was prepared in 50% methanol with concentrations ranging from 0.00002 to 10 μM. The dried residue of each sample was dissolved in 100 μL of the IS solution. Then, 10 μL of the resulting solution and standard solutions were injected into a C18 LC column (2.1 × 100 mm, 1.9 μm) for UPLC-MRM/MS analysis with (−) ion detection. The UPLC-MRM/MS analysis was performed on a Waters Acquity UPLC system coupled to a Sciex QTRAP 6500 Plus MS instrument. A tributylamine acetate buffer—acetonitrile/methanol (1:1, *v*/*v*) was used as the mobile phase for gradient elution (10% to 50% B over 25 min) at a flow rate of 0.25 mL/min and a temperature of 60 °C.

2.5. Data Analysis

The concentrations of the detected analytes in the above assays were calculated using internal standard calibration. This involved interpolating the constructed linear regression curves of individual compounds using the analyte-to-internal standard peak area ratios measured from injections of the sample solution. The data analysis of the metabolites was performed using Analyst 1.6.2. A multivariate statistical analysis including two-dimensional (2D) principal component analysis (PCA) and hierarchical clustering was performed to assess the differential metabolism per group. The analysis was performed using XLSTAT version 2022.3 (Addinsoft, New York, NY, USA), and Euclidean distance was utilized for constructing the hierarchical clustering analysis.

3. Results and Discussion

3.1. Overall Metabolic Changes in PA-Treated Plants under Al Stress

Metabolites are the end products of cellular processes, and their levels are influenced by coordinated responses of biological systems to changes in internal and external conditions [26]. Although the metabolic analysis of plant responses to both PA and Al stress is relatively unknown, this study provides a comprehensive understanding of the complex metabolic changes that occur in PA-treated tomato plants under Al stress. Our cluster analysis of both PA-treated and control plants under Al stress revealed two main groupings based on the global metabolic profile (Figure 1B). While the effect of PA was somewhat comparable to the control group, distinct metabolic changes were observed with 1% PA under 1 mM Al stress, but not with 0.25% PA and 1 mM Al alone (Figure 1A,B). Moreover, the clustering of the control group, PA alone, 0.25% PA under 1 mM Al and 1 mM Al alone treatments indicated that the effect of 1 mM Al was less toxic to the tomato plants and resulted in a stable metabolic profile. In contrast, a considerable alteration in metabolic profile was observed in PA-treated and control groups under 4 mM Al stress, with the 1% PA treatment exhibiting the most pronounced effect (Figure 1A,B). These results were expected, as previous studies have shown that high PA and Al concentrations induce phytotoxic effects, while lower concentrations stimulate plant growth and productivity [12,20,27–29].

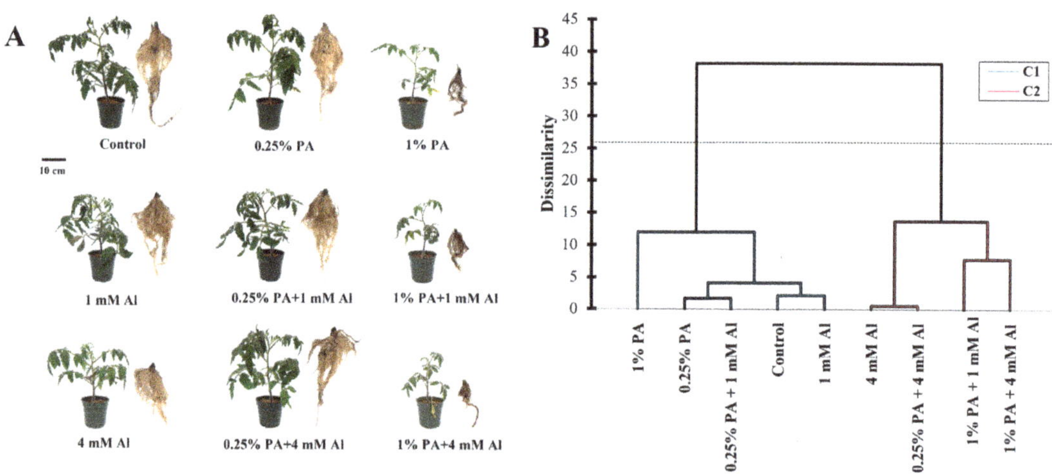

Figure 1. Response of Scotia tomato seedlings to pyroligneous acid (PA) treatment under aluminum stress (Al). (**A**) Morphological effect on tomato growth. (**B**) Cluster analysis of overall central carbon metabolite composition in the leaves.

3.2. Differential Accumulation of Metabolites in PA-Treated Plants under Al Stress

Exposure of plants to adverse environmental conditions, including Al stress, necessitates rapid metabolic reprogramming to stimulate stress tolerance. These metabolic changes are coordinated among diverse pathways to alter fluxes related to the different CCM routes [9]. Table 1 shows the total differential abundance of metabolites involved in specific pathways within the CCM. In total, 48 metabolites were detected (Figure S1; Table S1) and categorized into five groups based on their biological function in the CCM routes: CBC, glycolysis, PPP, TCA cycle and ETC (Table 1). Among these groups, glycolysis was the main differentially expressed metabolic pathway, accounting for approximately 75% of the total identified metabolites (Figure S2). Additionally, the TCA cycle accounted for approximately 25%, while the remaining groups comprised less than 1% of the total differentially expressed metabolites (Figure S2).

Table 1. Total metabolites involved in central carbon metabolic pathways in the leaves of tomato (*Solanum lycopersicum*, 'Scotia') seedlings treated with pyroligneous acid (PA) under aluminum (Al) stress.

Treatment	CBC (nmol g^{-1} FW)	Glycolysis (μmol g^{-1} FW)	PPP (nmol g^{-1} FW)	TCA (μmol g^{-1} FW)	ETC (nmol g^{-1} FW)
Control	23.79	21.72	30.01	6.97	5.63
1 mM Al	15.54	15.18	20.70	6.08	5.38
4 mM Al	6.10	20.63	5.97	7.77	4.88
0.25% PA	16.15	20.72	22.90	8.66	6.51
0.25% PA + 1 mM Al	11.10	15.81	13.40	7.99	6.05
0.25% PA + 4 mM Al	5.30	15.60	5.23	8.03	5.75
1% PA	17.18	38.16	20.97	10.18	8.17
1% PA + 1 mM Al	6.50	32.36	6.33	11.97	7.07
1% PA + 4 mM Al	5.33	56.65	5.13	10.92	3.53
CV (%)	55.70	52.56	64.72	21.97	22.43

CBC, Calvin–Benson cycle; PPP, pentose phosphate pathway; TCA, tricarboxylic acid cycle; ETC, electron transport chain.

3.2.1. Calvin–Benson Cycle

The Calvin–Benson cycle is the main biochemical pathway for carbon fixation in the chloroplast stroma of C3 plants [30]. During this process, plants utilize light energy to synthesize sugars and other carbon intermediates [31]. The regulation of CBC under stressful environments has been identified as a survival strategy for plants [31]. In the present study, both PA application and Al stress greatly affected the total CBC metabolites in the leaves of tomato seedlings (Table 1). PA treatments slightly reduced the total CBC metabolites compared to the control (Table 1). Similarly, Al stress exhibited a phytotoxic effect on CBC metabolites, especially under 4 mM Al, and this effect was further pronounced in PA-treated plants, suggesting that PA treatment could lead to decreased photosynthetic capacity under Al stress (Table 1). Moreover, the reduction in CBC metabolites under 4 mM Al stress can be attributed to low levels of all CBC metabolites except for sedoheptulose-7P (Figure 2A). This result agrees with previous studies that demonstrated a decrease in carbon fixation in tomato [32], spinach (*Spinacia oleracea*) [33], citrus [5,34,35] and rice (*Oryza sativa*) seedlings [36] under Al stress. Similarly, the decrease in total CBC metabolites caused by 0.25% PA, irrespective of Al exposure, may be attributed to low contents of all CBC metabolites except for DHAP, ribulose-5P and sedoheptulose-7P. While 3-phosphoglyceric acid, DAP, fructose 1,6-bisP and ribulose-bisP were increased in the 1% PA condition under 4 mM Al, the reduction in total CBC metabolites in 1% PA-treated plants under Al stress could be due to decreased glyceraldehyde-3P, fructose 6-phosphate, sedoheptulose-bisP, sedoheptulose-7P, ribose-5P and ribulose-5P (Figure 2A). Although PA treatment led to a reduction in total CBC metabolites, the impact of PA under Al stress could be considered slightly lower compared to 4 mM Al alone.

Previous research has suggested that the Al-induced downregulation of CBC results from both stomatal and non-stomatal factors [31,34]. Stomatal closure is thought to reduce leaf CO_2 influx, which further impairs CBC processes [31]. However, some studies have indicated that stomatal closure alone cannot fully explain the Al-induced reduction in CBC. This is because intracellular CO_2 levels were higher in Al-stressed leaves or comparable to those of control leaves, irrespective of stomatal conductance [5,6,32,34,37]. Pereira et al. [38] also indicated that Al-induced reduction in CBC was associated with thylakoid structural damage in lemon (*Citrus limon*) seedlings. Although the activities of CBC enzymes were not examined in this study, several studies have reported that heavy metal stress inhibits the activities of CBC enzymes, leading to a reduction in the overall CBC process [39–42].

Hence, the decrease in total CBC metabolites in both control and PA-treated plants under Al stress could be due to Al-induced structural damage and reduction in CBC enzyme activities [38,40].

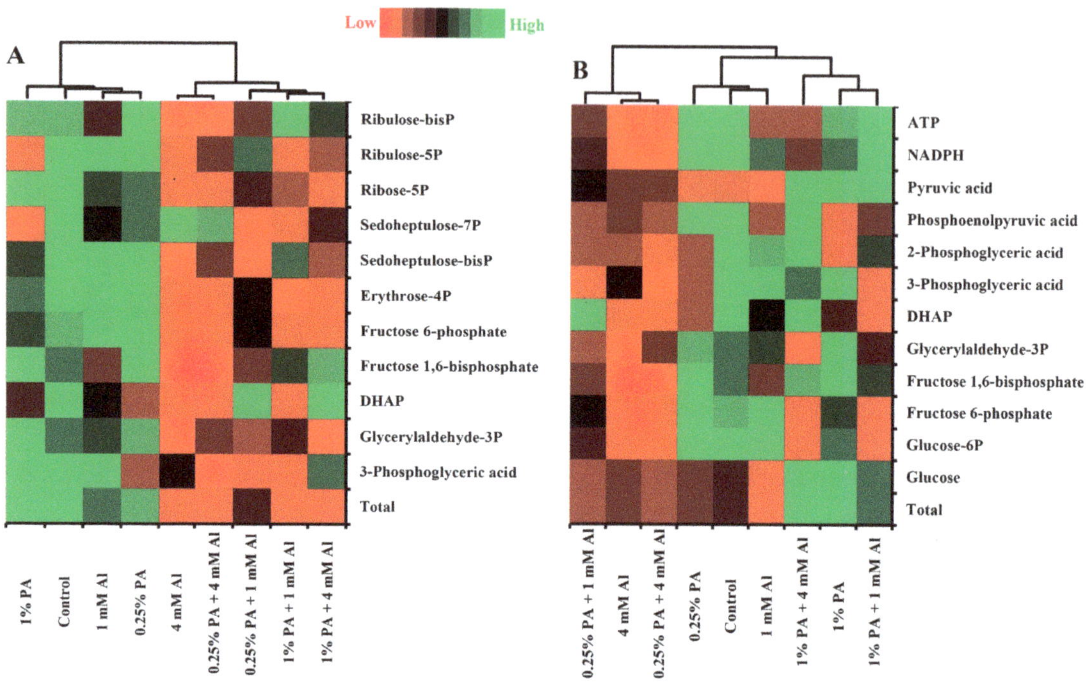

Figure 2. Heat map of metabolites involved in (**A**) the Calvin–Benson cycle and (**B**) the glycolysis pathway in leaves of Scotia tomato seedlings treated with pyroligneous acid (PA) under aluminum (Al) stress. Metabolite concentrations in each compartment are normalized across all data for an individual compound such that similar colour intensities between compounds can represent widely differing concentrations. The red colour represents a lower concentration, and the green colour represents a higher concentration of a particular metabolite.

3.2.2. Glycolysis

Balanced energy flow under environmental stress plays a crucial role in plant growth, development and stress tolerance. Glycolysis is a significant catabolic pathway that breaks down carbohydrates to provide energy for physiological and cellular operations in plants [9]. In this study, carbohydrate catabolism via the glycolysis pathway was higher in 1% PA-treated plants, irrespective of Al treatment (Table 1). In the absence of Al exposure, the total glycolysis metabolites were markedly increased in the 1% PA-treated plants by approximately 0.75-fold compared to the control. When exposed to Al stress, glycolysis metabolites were sightly downregulated in the control plants, while the 1% PA treatment increased the total glycolysis metabolites by approximately 2-fold and 1.75-fold compared to the 1 mM Al and 4 mM Al treatments, respectively (Table 1). Moreover, the increase in total glycolysis metabolites with 1% PA treatment in the absence of Al exposure was due to high glucose and pyruvate acid contents (Figure 2B). On the other hand, the increase in total glycolysis metabolites in 1% PA-treated plants under Al stress can be ascribed to high levels of glucose, fructose 1,6-phosphate, DHAP, 3- and 2-phosphoglyceric acid, PEP, pyruvic acid, NADPH and ATP (Figure 2B). These findings are consistent with the results reported by Niedziela et al. [43] who observed that key enzymes of glycolysis were enhanced in Al-treated plants, leading to accelerated production of pyruvate and acetyl CoA.

Studies have shown that Al stress leads to energy deprivation in plant growth and development. As a result, tolerant plants accumulate high levels of glucose to produce adequate ATP through glycolysis [12,13,44]. For instance, in Al-tolerant rice, Wang et al. [14] showed that glycolytic pathway proteins are enhanced, which could maintain basic respiration needs and/or generate ATP for Al stress tolerance. A similar increase in glycolytic flux was noted in Al-tolerant citrus leaves [12]. The magnitude of glucose production in plants through the CBC is crucial in confirming Al stress tolerance [36]. Accumulation of glucose under Al stress also functions as an osmoprotectant, which decreases osmotic potential, regulates turgor dynamics and promotes the maintenance of membrane integrity and overall cellular homeostasis [7,45]. This osmoregulation enables plants to maintain sufficient carbohydrate reserves for sustaining essential metabolism under Al stress [36]. These findings indicate that the 1% PA treatment can trigger a metabolic switch from growth to survival mode via glucose metabolism for energy generation. Furthermore, biostimulants have been reported to enhance plant tolerance to various stresses [46]. Similar to many other biostimulants, PA has been reported to enhance the production of proteins and metabolites associated with carbohydrate metabolism and energy production [3,23]. Consistent with the study of Wang et al. [23], PA treatment increased the abundance of glycolytic metabolic pathway proteins, increasing plant tolerance to drought. Additionally, the bioactive compounds in PA may act as signaling molecules to regulate key metabolic pathways, including CCM, and/or stimulate differential gene expression to promote plant growth and resilience to Al stress [3,21,23]. Hence, increasing glycolytic metabolites with 1% PA treatment can be considered as an effective strategy to meet the energy demand for Al tolerance and adaptation.

3.2.3. Pentose Phosphate Pathway

The PPP is a critical metabolic pathway for glucose degradation and is important for providing reducing power and intermediate metabolites for other pathways [47]. The present study showed that the total PPP metabolites were affected by both PA and Al treatment (Table 1; Figure 3A). The total PPP metabolites were reduced by approximately 0.23-fold and 0.3-fold with 0.25% PA and 1% PA, respectively, compared to the control. The decrease in total PPP metabolites in 1% PA-treated plants can be attributed to low levels of ribulose-5P and sedoheptulose-7P, although 6-phosphogluconate levels were higher compared to the control (Figure 3A). On the other hand, Al stress markedly reduced the total PPP metabolites, and PA treatment could not alleviate these effects compared to the control (Table 1). For instance, under 1 mM Al, the total PPP metabolites were substantially reduced by approximately 0.35-fold and 0.69-fold with 0.25% PA and 1% PA treatments, respectively, compared to 1 mM Al treatment alone. The reduction in total PPP metabolites in 0.25% PA-treated plants can be attributed to low glucose-6P, ribose-5P, glyceraldehyde-3P, fructose-6P, erythrose-4P and sedoheptulose-7P contents. Similarly, the reduction in total PPP metabolites following the application of 1% PA can be attributed to low glucose-6P, ribulose-5P, ribose 5-P, glyceraldehyde-3P, fructose-6P, erythrose-4P and sedoheptulose-7P contents, although 6-phosphogluconate levels were higher compared to the 1 mM Al treatment alone (Figure 3A).

The 4 mM Al significantly reduced the total PPP metabolites, which were not different from those of 0.25% and 1% PA-treated plants under the same conditions of Al stress (Table 1). Furthermore, the substantial reduction in total PPP metabolites in both control and PA-treated plants under 4 mM Al can be attributed to low levels of all PPP metabolites except 6-phosphogluconate and sedoheptulose-7P contents (Figure 3A). Nicotinamide adenine dinucleotide phosphate (NADPH) is the primary reducing power in the PPP pathway and is used to maintain cellular redox homeostasis via antioxidant production [47]. However, the levels of NADPH were not affected by PA treatment but were reduced under 4 mM Al stress (Figure 3A). It has been established that Al stress leads to ROS accumulation, which causes cellular damage and programmed cell death [48]. The results of this study indicate that a reduction in NADPH levels under Al stress could increase ROS generation and compromise the antioxidant defense system in plants.

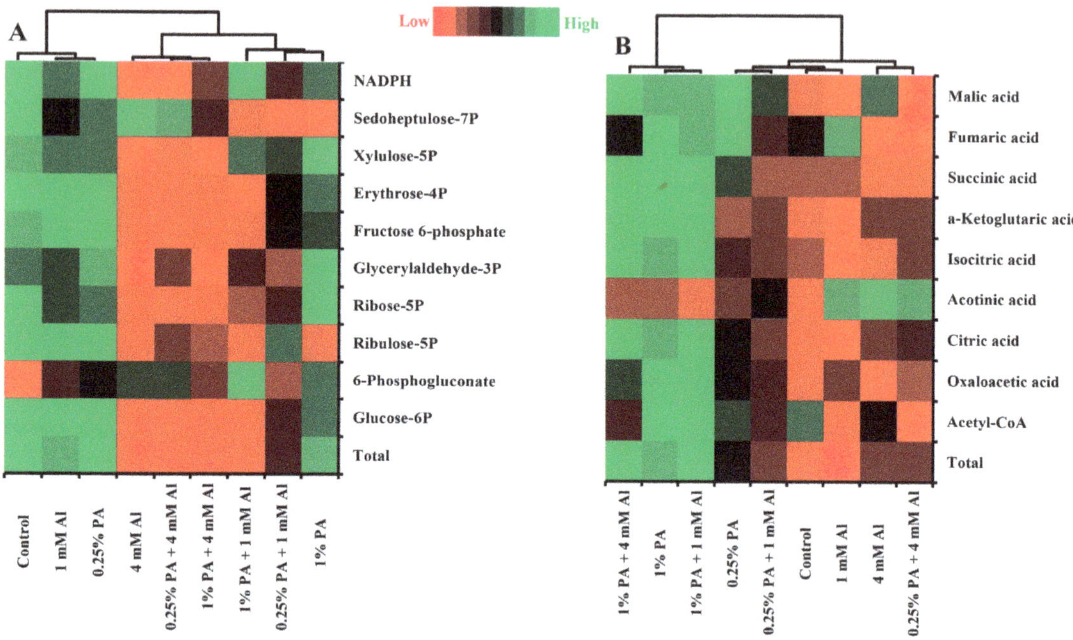

Figure 3. Heat map of metabolites profile involved in (**A**) pentose phosphate pathway and (**B**) tricarboxylic acid (TCA) cycle in leaves of Scotia tomato seedlings treated with pyroligneous acid (PA) under aluminum (Al) stress. Metabolite concentrations in every compartment are normalized across all data for an individual compound such that similar colour intensities between compounds can represent widely differing concentrations. The red colour represents a lower concentration, and the green colour represents a higher concentration of a particular metabolite.

3.2.4. Tricarboxylic Acid Cycle

The TCA cycle is a central biochemical pathways for respiratory substrate oxidation and the production of ATP for all cellular functions [49]. In response to Al stress, the TCA cycle serves as a reservoir for organic acids (OA) production, which has been strongly linked to Al tolerance in most plants [2,4]. Our study revealed that the total TCA cycle intermediate metabolites were affected by PA and Al treatments (Table 1; Figure 3B). In the absence of Al stress, the total TCA cycle metabolites increased by approximately 0.24-fold and 0.46-fold with PA treatment compared to the control. The increased total TCA cycle metabolites with 0.25% PA could be due to increased levels of oxaloacetic acid, citric acid, succinic acid, fumaric acid and malic acid (Figure 3B). Likewise, the increased total TCA cycle metabolites with 1% PA treatment can be ascribed to high contents of all TCA cycle metabolites except for acetyl-CoA and aconitic acid, which were not altered compared to the control (Figure 3B). Similarly, in response to Al stress, PA-treated plants accumulated high levels of TCA cycle metabolites compared to plants treated with Al alone. Exposing the plants to 4 mM Al without PA increased the total TCA cycle metabolites by approximately 0.11-fold due to high levels of citric acid, aconitic acid, α-ketoglutaric acid and malic acid (Figure 3B). It has been established that primary metabolic constituents associated with Al tolerance are strongly linked to mitochondrial metabolism and OA production [2,4,8]. Organic acid production in response to Al stress is the most-characterized mechanism for Al tolerance, with citrate and malate being the common TCA cycle metabolites identified in several plants [4,8,13]. For instance, tomatoes produce and secrets malate, which chelate

Al ions into non-toxic forms [50]. The increased OA production observed under Al stress in this study is consistent with previous studies in which mitochondrial citrate and malate levels were enhanced and shown to improve Al tolerance in tomato [50], maize (*Zea mays*) [8,13,15], buckwheat (*Fagopyrum esculentum*) [51], cabbage (*Brassica oleracea*) [52] and wheat (*Triticum aestivum*) [53].

Furthermore, both 0.25% PA and 1% PA treatments resulted in the accumulation of approximately 0.31-fold and 0.97-fold of the total TCA cycle metabolites, respectively, under 1 mM Al compared to plants treated with 1 mM Al alone. Under 4 mM Al, the total TCA cycle metabolites increased by approximately 0.03-fold and 0.41-fold in 0.25% PA and 1% PA-treated plants, respectively, compared to plants exposed to 4 mM Al alone (Table 1). The increase in total TCA cycle metabolites with 1% PA treatment under Al stress could be due to high levels of all TCA metabolites except for aconitic acid, which was reduced (Figure 3B). Similarly, Wang et al. [23] reported that PA treatment enhanced the TCA cycle in wheat seedlings under drought stress due to increased activities of malate dehydrogenase. Additionally, Sweetlove et al. [49] suggested that the activities of the different TCA cycle enzymes are independent of each other, and the TCA metabolites vary in their flux levels. Although the activities of these enzymes were not examined in this study, the increased TCA cycle metabolites following PA treatment could be associated with an increase in enzyme activities. This indicates that PA can promote the TCA cycle to produce sufficient ATP to accommodate the energy demand of plants under Al stress [23]. Evidence from numerous studies revealed that some plants accumulate high levels of TCA cycle metabolites internally to detoxify Al in roots and leaf cells and compartmentalize Al–OA complexes into the vacuole [2,51]. This internal Al complexation could prevent Al from interacting with macromolecules, including nucleic acids and proteins. Although this Al tolerance mechanism is mostly associated with hyperaccumulators [4,51], the increased TCA cycle metabolites in the leaves of both PA and control-treated plants under Al stress could suggest the stimulation of internal detoxification of Al ions, thereby promoting Al tolerance. Moreover, the TCA cycle metabolites are not only known to chelate Al ions. For instance, citrate plays an important role in antioxidant production and is involved in respiratory assimilation to produce energy for stress defense [54]. Additionally, α-ketoglutaric acid is critical in respiration and nitrogen metabolism for amino acid biosynthesis, which regulate osmotic potential and mediate stress tolerance in plants [55]. Hence, the increased TCA cycle metabolites in PA and control plants under Al stress can also be linked to amino acid and nucleic acid biosynthesis for Al tolerance.

3.2.5. Electron Transport Chain (ETC)

The electron transport chain is one of the most critical pathways for both cellular respiration and photosynthesis [49]. It consists of an array of electron transporters that oxidize reducing equivalents for energy generation *via* ATP biosynthesis [11]. The results of this study revealed that PA and Al treatment altered the total ETC metabolites in tomato leaves (Table 1; Figure 4). The total ETC metabolites were increased by approximately 0.16-fold and 0.45-fold in 0.25% PA and 1% PA-treated plants, respectively, compared to the control in the absence of Al stress (Table 1). This increase in ETC metabolites following PA treatments could be due to high levels of FMN, FAD, NAD and NADH contents (Figure 4). Nonetheless, the levels of ADP and ATP remained unchanged with PA treatments (Figure 4). When exposed to Al stress, the tomato plants slightly reduced their total ETC metabolites by approximately 0.13-fold under 4 mM Al, while those under 1 mM Al exhibited no considerable change. The reduction in total ETC metabolites with 4 mM Al can be attributed to low ADP and ATP contents compared to the control (Figure 4). Accumulating evidence has revealed that Al stress impairs the activities of ETC complexes and could result in decreased metabolite production [38,56–58].

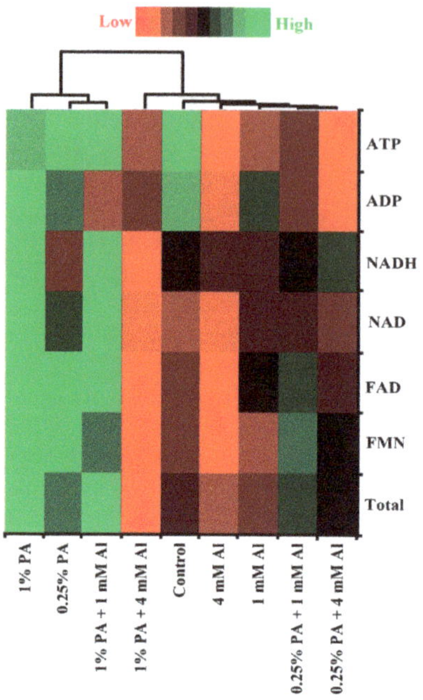

Figure 4. Heat map of metabolites profile involved in electron transport chain in the leaves of Scotia tomato seedlings treated with pyroligneous acid (PA) under aluminum (Al) stress. Metabolite concentrations in each compartment are normalized across all data for an individual compound, such that similar colour intensities between compounds can represent widely differing concentrations. The red colour represents a lower concentration, and the green colour represents a higher concentration of a particular metabolite.

Moreover, both 0.25% and 1% PA treatments under 1 mM Al increased the total ETC metabolites by approximately 0.12-fold and 0.31-fold, respectively, compared to plants exposed to 1 mM Al. The increased total ETC metabolites in the 0.25% PA-treated plants can be associated with increased FMN, FAD, NAD and ATP contents, while that of 1% PA-treated plants is due to high FMN, FAD, NAD, NADH and ATP contents (Figure 4). However, the total ETC metabolites were slightly increased in plants treated with 0.25% PA under 4 mM Al, while 1% PA decreased the total ETC metabolites compared to 4 mM Al alone (Table 1).

Additionally, the increase in ETC metabolites with 0.25% PA can be attributed to a slight increase in FMN, FAD, NAD and NADH contents, while the reduction in ETC metabolites with 1% PA treatments was due to low NADH contents (Figure 4). ATP synthase is critical for ATP production in ETC [11]. According to Su et al. [58], Al stress targets and damage subunits of ATP synthase, thereby reducing the amount of ATP generated. With a limitation in ATP production and electron transport, electrons leakage is enhanced, which could result in uncontrollable ROS generation and oxidative stress-induced cell death [11,57]. Interestingly, Wang et al. [23] reported that the relative abundance of proteins related to ATP synthesis was increased in PA-treated wheat seedlings under drought stress. although ATP-related proteins were not examined in this study, the increase in ATP production in PA-treated plants alone and with 1 mM Al stress partially aligns with the findings of Wang et al. [23]. In plant stress adaptation, the energy cost is high, and the enhancement of ETC processes with PA treatments suggests that restoring energy supply via ATP production is critical to protect several basal metabolic processes under Al stress.

3.3. *Associations between Central Carbon Metabolites*

To further examine the association among the intermediate metabolites of the CCM routes in the PA-treated tomato seedlings under Al stress, a 2D PCA and Pearson correlation were used (Table 2; Figures 5 and 6). The Pearson correlation analysis showed that the total CBC metabolites had a significantly ($p < 0.001$) strong positive association with the total PPP metabolites and a weak negative association with glycolysis and TCA cycle metabolites (Table 2). Additionally, the total glycolytic metabolites had a significantly ($p < 0.05$) moderate positive correlation with total TCA cycle metabolites and a weak negative association with total PPP and ETC metabolites (Table 2). These findings are consistent with the results reported by Niedziela et al. [43], who demonstrated that key enzymes of glycolysis were enhanced in Al-treated plants, leading to accelerated pyruvate and acetyl CoA production for OA synthesis. These results suggest that energy production plays a crucial role in supporting basic cellular function under Al stress, and that plants channel their photosynthetic carbon for stress adaption [14]. On the other hand, total PPP metabolites exhibited non-significant ($p > 0.05$) weak negative and positive associations with the total TCA cycle and ETC metabolites, respectively (Table 2). Similarly, the total TCA cycle metabolites had a weak positive correlation with the total ETC metabolites. The increase in glycolytic flux under Al stress proved that carbon skeletons produced via glycolysis are used primarily for OA production to detoxify Al ions, thereby promoting Al tolerance [13,50,51].

Table 2. Pearson correlation coefficients (r) amongst the specific central carbon metabolic pathways in Scotia tomato seedlings treated with pyroligneous acid (PA) under aluminum (Al) stress and their corresponding significance levels at $p \leq 0.05$.

Variables	CBC	Glycolysis	PPP	TCA
Glycolysis	r = −0.240 p = 0.533			
PPP	r = 0.993 p = 0.000	r = −0.274 p = 0.475		
TCA	r = −0.436 p = 0.240	r = 0.762 p = 0.017	r = −0.461 p = 0.211	
ETC	r = 0.319 p = 0.403	r = −0.303 p = 0.429	r = 0.302 p = 0.430	r = 0.147 p = 0.706

CBC, Calvin–Benson cycle; PPP, pentose phosphate pathway; TCA, tricarboxylic acid cycle.

Although some of the total metabolites showed a non-significant ($p > 0.05$) weak correlation with other metabolites, the CBC, glycolysis and PPP shared common metabolites which showed significant ($p < 0.05$) associations (Figure 5; Table S2). For example, RuBP had a significantly ($p < 0.01$) strong positive association with fructose 1,6BP and ATP, and a moderate positive association with glucose-6P and xylulose-5P (Figure 5; Table S2). Similarly, sedoheptulose-7BP showed a significantly ($p < 0.01$) moderate positive correlation with fructose-6P, erythrose-4P and glucose-6P, and a non-significant ($p > 0.05$) moderate positive association with ATP and xylulose-5P. Additionally, both ribose-5P and ribulose-5P had a significantly ($p < 0.01$) strong positive association with erythrose-4P and glucose-6P, and a moderate association with fructose-6P. However, both sedoheptulose-7P and ribulose-5P exhibited a significantly ($p < 0.01$) strong negative association with pyruvic acid, while ribulose-5P had a moderate association with xylulose-5P (Figure 5; Table S2). Moreover, the correlation analysis among the individual metabolites confirmed that the glycolysis intermediates had a significant ($p < 0.05$) association with the TCA cycle intermediate metabolites (Figure 5; Table S2). Fructose-1,6BP had a strong positive association with succinic acid and fumaric acid and a negative association with aconitic acid. Glucose had a strong association with isocitric acid, α-ketoglutaric acid and succinic acid, and a moderate association with oxaloacetic acid and citric acid (Figure 5; Table S2). Similarly, pyruvic acid had a strong positive association with oxaloacetic acid, citric acid, isocitric acid and α-ketoglutaric acid, and a moderate positive association with succinic acid and acetyl-CoA (Figure 5; Table S2).

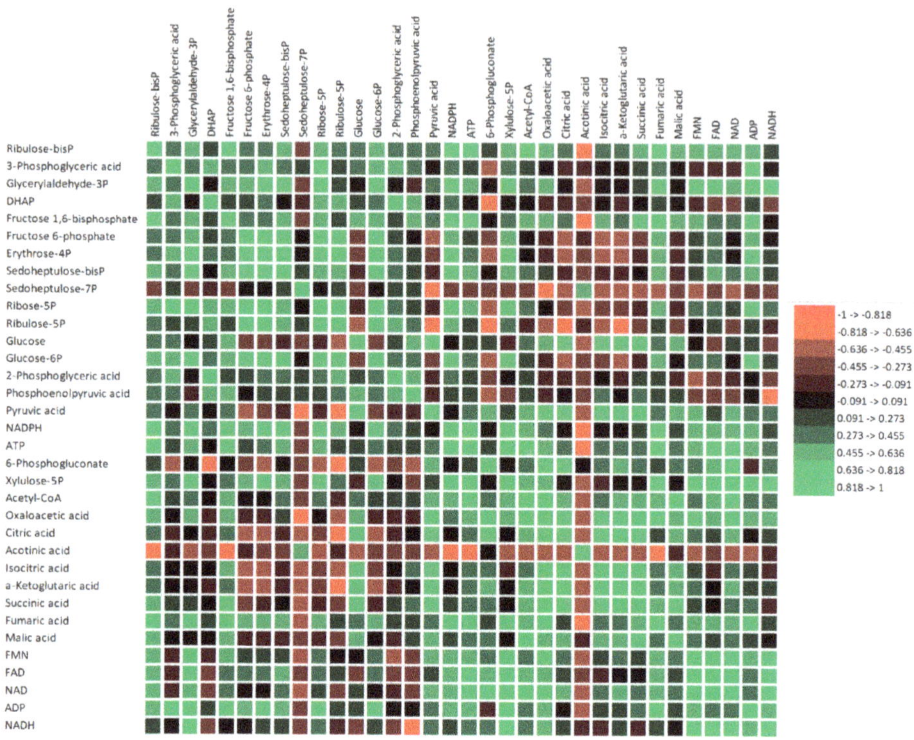

Figure 5. Pearson correlation matrix among individual metabolites of the central carbon metabolic pathway in Scotia tomato seedlings treated with pyroligneous acid (PA) under aluminum (Al) stress. The red colour represents a strong negative association, and the green colour represents a strong positive association. DHAP, dihydroxyacetone phosphate; NADP, nicotinamide adenine dinucleotide phosphate; FMN, flavin mononucleotides; FAD, flavin adenine dinucleotide; ADP, adenosine diphosphate; ATP, adenosine triphosphate.

A 2D PCA biplot revealed a projection of response variables in the factor spaces (F1 and F2) and explained approximately 82% of the total disparity in the dataset (Figure 6). The control treatment exhibited a strong influence on both the CBC and PPP, and a moderate influence on ETC. Similarly, both 0.25% PA and 1% PA showed a strong influence on ETC and a moderate influence on the CBC and PPP (Figure 6). However, under Al stress, 0.25% PA and 1 mM Al moderately influenced the CBC and PPP, while 1% PA showed a strong influence on TCA and glycolysis. Additionally, 1% PA under 4 mM Al showed a moderate influence on glycolysis and TCA, while both 0.25% PA under 4 mM Al and 4 mM alone had no influence on the CCM metabolites (Figure 6). These findings are consistent with the findings of Wang et al. [23], who reported that PA promotes the glycolysis pathway, which fuels the TCA cycle to produce sufficient energy for stress tolerance. In summary, the coordinated association between the determined metabolites confirmed that PA stimulates alterations in plant metabolism to increase energy production and organic acids biosynthesis for Al stress tolerance (Figure 7).

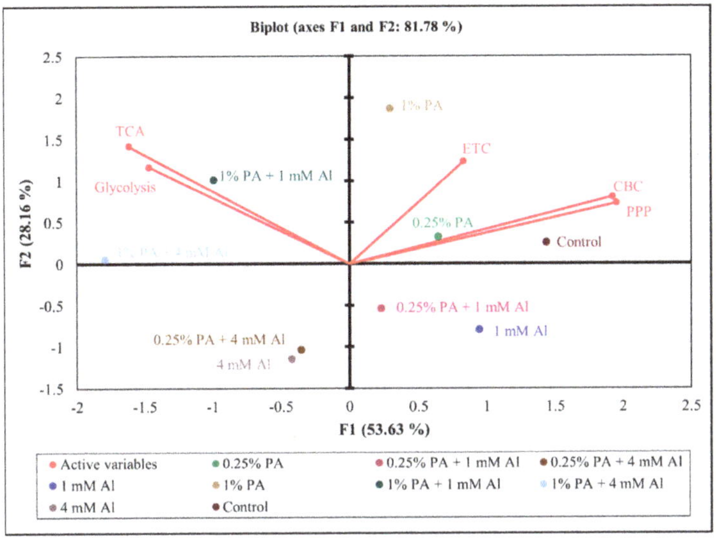

Figure 6. A two-dimensional principal component analysis (2-D PCA) biplot showing relationships amongst the explanatory variables (total metabolites involved in specific central carbon metabolic pathways) Calvin–Benson cycle (CBC), glycolysis, pentose phosphate pathway (PPP), tricarboxylic acid (TCA) cycle and electron transport chain (ETC) of Scotia tomato seedlings treated with pyroligneous acid (PA) under aluminum (Al) stress. The projection of the variables in the 2-D factor space (F1 and F2) explained a total of 81.78% of the variations in the dataset. Variables that are closely located are not different compared to variables located at a distance within a quadrant or between quadrants.

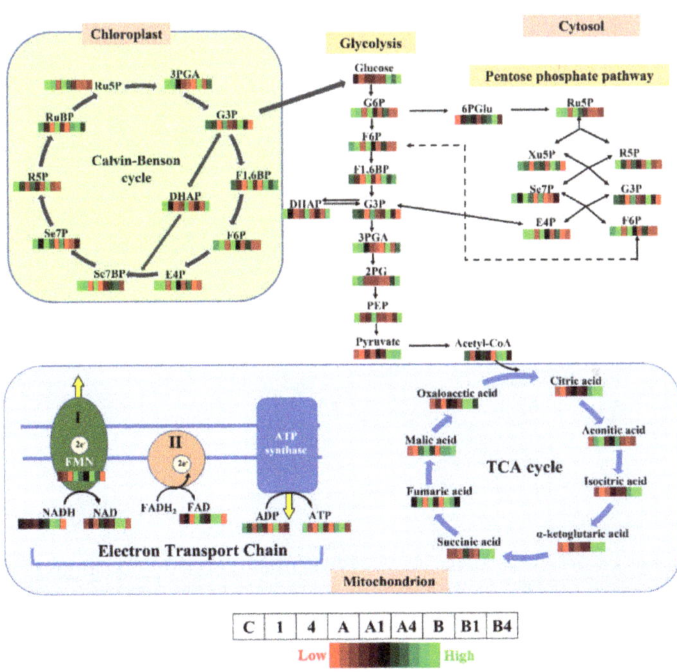

Figure 7. Overview of metabolic changes of key central carbon pathways in the leaves of Scotia tomato seedlings treated with pyroligneous acid (PA) under aluminum (Al) stress. Ru5P, ribulose-5-phosphate;

3PGA, 3-phosphoglyceric acid; G3P, glyceraldehyde-3-phosphate; F1,6BP, fructose-1,6-bisphosphate; F6P, fructose-6-phosphate; E4P, erythrose-4-phosphate; Se7BP, sedoheptulose-1,7-bisphosphate; Se7P, sedoheptulose-7-phosphate; R5P, ribose-5-phosphate; RuBP, ribulose-1,5-bisphosphate; G6P, glucose-6-phosphate; DHAP, dihydroxyacetone phosphate; 2PG, 2-phosphoglyceric acid; PEP, phosphoenolpyruvic acid; 6PGlu, 6-phosphogluconate; Xu5P, xylulose-5-phosphate; NADP, nicotinamide adenine dinucleotide phosphate; FMN, flavin mono-nucleotides; FAD, flavin adenine dinucleotide; ADP, adenosine diphosphate; ATP, adenosine triphosphate. The red colour of the metabolic pattern represents a lower concentration, and the green colour represents a higher concentration of a particular metabolite. The treatments that match the corresponding colour patterns were arranged from left to right as control (C), 1 mM Al (1), 4 mM Al (4), 0.25% PA (A), 0.25% PA + 1 mM Al (A1), 0.25% PA + 4 mM Al (A4), 1% PA (B), 1% PA + 1 mM Al (B1) and 1% PA + 4 mM Al (B4).

4. Conclusions

Acidic soils are widespread, and Al stress is a major limiting factor for plant growth and productivity in acidic soils. In the present study, we comprehensively investigated the metabolic response of key intermediate metabolites in central carbon metabolism routes in PA-treated tomato seedlings under Al stress. A total of 48 differentially expressed metabolites involved in the Calvin–Benson cycle, glycolysis, pentose phosphate pathway, tricarboxylic acid cycle and electron transport chain were identified in the leaves of both control and PA-treated tomato plants under Al stress. Aluminum stress considerably reduced the levels of these metabolites, while PA treatment triggered dynamic metabolic alterations and played an important role in Al stress adaptation. The coordinated association among these identified metabolites revealed that PA stimulates changes in plant metabolism to modulate energy production and the biosynthesis of organic acids for Al stress tolerance. However, further investigation is required to examine how PA treatment influences the genes and enzymes involved in these CCM routes under Al stress.

Supplementary Materials: The following supporting information can be downloaded at: https://www.mdpi.com/article/10.3390/metabo13060770/s1, Figure S1: Chromatograms for all identified metabolites of CCM routes. The sections (1–3) show detection of CCM metabolites. The compound names and the ion transition of each compound are shown on the extracted ion chromatograms (XICs); Figure S2: Classification of metabolites involved in CCM routes (Calvin–Benson cycle, glycolysis, pentose phosphate pathway (PPP), tricarboxylic acid (TCA) cycle and electron transport chain (ETC)) in leaves of tomato 'Scotia' seedlings treated with pyroligneous acid (PA) under aluminum (Al) stress.; Table S1: Metabolite content from tomato leaf tissues analyzed by LC-MRM-MS; Table S2: Pearson correlation and p-values between individual metabolites of the CCM routes in leaves of tomato 'Scotia' seedlings treated with pyroligneous acid (PA) under aluminum (Al) stress.

Author Contributions: Conceptualization, L.A. and R.O.; formal analysis, R.O.; investigation, R.O.; resources, L.A.; writing—original draft preparation, R.O.; writing—review and editing, R.O., R.H.T. and L.A.; visualization, R.O.; supervision, L.A. and R.H.T.; project administration, L.A.; funding acquisition, L.A. and R.H.T. All authors have read and agreed to the published version of the manuscript.

Funding: This work was financially supported by the Natural Sciences and Engineering Research Council of Canada (NSERC). Grant #CRDPJ532183-18.

Institutional Review Board Statement: Not applicable.

Informed Consent Statement: Not applicable.

Data Availability Statement: The data presented in this study are available from the corresponding author upon request. Data is not publicly available due to privacy or ethical restrictions.

Acknowledgments: The lead author wishes to thank Samuel K. Asiedu and all the lab members for their suggestions during the study, and the University of Victoria Genome BC (UVic GBC)—Proteomics Centre of The Metabolomics Innovation Centre, Canada, for conducting the metabolomic analysis.

Conflicts of Interest: The authors declare no conflict of interest.

References

1. Slessarev, E.W.; Lin, Y.; Bingham, N.L.; Johnson, J.E.; Dai, Y.; Schimel, J.P.; Chadwick, O.A. Water balance creates a threshold in soil pH at the global scale. *Nature* **2016**, *540*, 567–569. [CrossRef]
2. Kochian, L.V.; Piñeros, M.A.; Liu, J.; Magalhaes, J.V. Plant adaptation to acid soils: The molecular basis for crop aluminum resistance. *Annu. Rev. Plant Biol.* **2015**, *66*, 571–598. [CrossRef]
3. Ofoe, R.; Gunupuru, L.R.; Wang-Pruski, G.; Fofana, B.; Thomas, R.H.; Abbey, L. Seed priming with pyroligneous acid mitigates aluminum stress, and promotes tomato seed germination and seedling growth. *Plant Stress* **2022**, *4*, 100083. [CrossRef]
4. Ofoe, R.; Thomas, R.H.; Asiedu, S.K.; Wang-Pruski, G.; Fofana, B.; Abbey, L. Aluminum in plant: Benefits, toxicity and tolerance mechanisms. *Front. Plant Sci.* **2023**, *13*, 1085998. [CrossRef] [PubMed]
5. Chen, L.-S.; Qi, Y.-P.; Liu, X.-H. Effects of aluminum on light energy utilization and photoprotective systems in citrus leaves. *Ann. Bot.* **2005**, *96*, 35–41. [CrossRef] [PubMed]
6. Peixoto, H.P.; Da Matta, F.M.; Da Matta, J.C. Responses of the photosynthetic apparatus to aluminum stress in two sorghum cultivars. *J. Plant Nutr.* **2002**, *25*, 821–832. [CrossRef]
7. Pandey, P.; Srivastava, R.K.; Rajpoot, R.; Rani, A.; Pandey, A.K.; Dubey, R.S. Water deficit and aluminum interactive effects on generation of reactive oxygen species and responses of antioxidative enzymes in the seedlings of two rice cultivars differing in stress tolerance. *Environ. Sci. Pollut. Res.* **2016**, *23*, 1516–1528. [CrossRef]
8. Siqueira, J.A.; Barros, J.A.S.; Dal-Bianco, M.; Martins, S.C.V.; Magalhães, P.C.; Ribeiro, D.M.; DaMatta, F.M.; Araújo, W.L.; Ribeiro, C. Metabolic and physiological adjustments of maize leaves in response to aluminum stress. *Theor. Exp. Plant Physiol.* **2020**, *32*, 133–145. [CrossRef]
9. Timm, S.; Arrivault, S. Regulation of Central Carbon and Amino Acid Metabolism in Plants. *Plants* **2021**, *10*, 430. [CrossRef]
10. Fuchs, G.; Berg, I.A. Unfamiliar metabolic links in the central carbon metabolism. *J. Biotechnol.* **2014**, *192 Pt B*, 314–322. [CrossRef]
11. Fernie, A.R.; Carrari, F.; Sweetlove, L.J. Respiratory metabolism: Glycolysis, the TCA cycle and mitochondrial electron transport. *Curr. Opin. Plant Biol.* **2004**, *7*, 254–261. [CrossRef]
12. Li, H.; Yang, L.-T.; Qi, Y.-P.; Guo, P.; Lu, Y.-B.; Chen, L.-S. Aluminum Toxicity-Induced Alterations of Leaf Proteome in Two Citrus Species Differing in Aluminum Tolerance. *Int. J. Mol. Sci.* **2016**, *17*, 1180. [CrossRef] [PubMed]
13. Pinto, V.B.; Almeida, V.C.; Pereira-Lima, Í.A.; Vale, E.M.; Araújo, W.L.; Silveira, V.; Viana, J.M.S. Deciphering the major metabolic pathways associated with aluminum tolerance in popcorn roots using label-free quantitative proteomics. *Planta* **2021**, *254*, 132. [CrossRef]
14. Wang, Z.Q.; Xu, X.Y.; Gong, Q.Q.; Xie, C.; Fan, W.; Yang, J.L.; Lin, Q.S.; Zheng, S.J. Root proteome of rice studied by iTRAQ provides integrated insight into aluminum stress tolerance mechanisms in plants. *J. Proteom.* **2014**, *98*, 189–205. [CrossRef]
15. Wang, L.; Fan, X.-W.; Pan, J.-L.; Huang, Z.-B.; Li, Y.-Z. Physiological characterization of maize tolerance to low dose of aluminum, highlighted by promoted leaf growth. *Planta* **2015**, *242*, 1391–1403. [CrossRef]
16. Pareek, A.; Dhankher, O.P.; Foyer, C.H. Mitigating the impact of climate change on plant productivity and ecosystem sustainability. *J. Exp. Bot.* **2020**, *71*, 451–456. [CrossRef]
17. Hasanuzzaman, M.; Parvin, K.; Bardhan, K.; Nahar, K.; Anee, T.I.; Masud, A.A.C.; Fotopoulos, V. Biostimulants for the Regulation of Reactive Oxygen Species Metabolism in Plants under Abiotic Stress. *Cell* **2021**, *10*, 2537. [CrossRef]
18. Francesca, S.; Raimondi, G.; Cirillo, V.; Maggio, A.; Barone, A.; Rigano, M.M. A Novel Protein Hydrolysate-Based Biostimulant Improves Tomato Performances under Drought Stress. *Plants* **2021**, *10*, 783. [CrossRef]
19. Grewal, A.; Abbey, L.; Gunupuru, L.R. Production, prospects and potential application of pyroligneous acid in agriculture. *J. Anal. Appl. Pyrolysis* **2018**, *135*, 152–159. [CrossRef]
20. Ofoe, R.; Qin, D.; Gunupuru, L.R.; Thomas, R.H.; Abbey, L. Effect of Pyroligneous Acid on the Productivity and Nutritional Quality of Greenhouse Tomato. *Plants* **2022**, *11*, 1650. [CrossRef] [PubMed]
21. Ofoe, R.; Gunupuru, L.R.; Abbey, L. Metabolites, elemental profile and chemical activities of Pinus strobus high temperature-derived pyroligneous acid. *Chem. Biol. Technol. Agric.* **2022**, *9*, 85. [CrossRef]
22. Wei, Q.; Ma, X.; Zhao, Z.; Zhang, S.; Liu, S. Antioxidant activities and chemical profiles of pyroligneous acids from walnut shell. *J. Anal. Appl. Pyrolysis* **2010**, *88*, 149–154. [CrossRef]
23. Wang, Y.; Qiu, L.; Song, Q.; Wang, S.; Wang, Y.; Ge, Y. Root proteomics reveals the effects of wood vinegar on wheat growth and subsequent tolerance to drought stress. *Int. J. Mol. Sci.* **2019**, *20*, 943. [CrossRef]
24. Han, J.; Gagnon, S.; Eckle, T.; Borchers, C.H. Metabolomic analysis of key central carbon metabolism carboxylic acids as their 3-nitrophenylhydrazones by UPLC/ESI-MS. *Electrophoresis* **2013**, *34*, 2891–2900. [CrossRef] [PubMed]
25. Han, J.; Tschernutter, V.; Yang, J.; Eckle, T.; Borchers, C.H. Analysis of selected sugars and sugar phosphates in mouse heart tissue by reductive amination and liquid chromatography-electrospray ionization mass spectrometry. *Anal. Chem.* **2013**, *85*, 5965–5973. [CrossRef]
26. Das, A.; Rushton, P.J.; Rohila, J.S. Metabolomic profiling of soybeans (*Glycine max* L.) reveals the importance of sugar and nitrogen metabolism under drought and heat stress. *Plants* **2017**, *6*, 21. [CrossRef]
27. Muhammad, N.; Zvobgo, G.; Zhang, G.-p. A review: The beneficial effects and possible mechanisms of aluminum on plant growth in acidic soil. *J. Integr. Agric.* **2019**, *18*, 1518–1528. [CrossRef]

28. Liu, X.; Zhan, Y.; Li, X.; Li, Y.; Feng, X.; Bagavathiannan, M.; Zhang, C.; Qu, M.; Yu, J. The use of wood vinegar as a non-synthetic herbicide for control of broadleaf weeds. *Ind. Crops Prod.* **2021**, *173*, 114105. [CrossRef]
29. Liu, Y.J.; Tao, J.Y.; Cao, J.; Zeng, Y.P.; Li, X.; Ma, J.; Huang, Z.; Jiang, M.Y.; Sun, L.X. The Beneficial Effects of Aluminum on the Plant Growth in *Camellia japonica*. *J. Soil Sci. Plant Nutr.* **2020**, *20*, 1799–1809. [CrossRef]
30. Michelet, L.; Zaffagnini, M.; Morisse, S.; Sparla, F.; Pérez-Pérez, M.E.; Francia, F.; Danon, A.; Marchand, C.H.; Fermani, S.; Trost, P. Redox regulation of the Calvin–Benson cycle: Something old, something new. *Front. Plant Sci.* **2013**, *4*, 470. [CrossRef]
31. Sharma, S.; Joshi, J.; Kataria, S.; Verma, S.K.; Chatterjee, S.; Jain, M.; Pathak, K.; Rastogi, A.; Brestic, M. Regulation of the Calvin cycle under abiotic stresses: An overview. In *Plant Life under Changing Environment*; Academic Press: Cambridge, MA, USA, 2020; pp. 681–717. [CrossRef]
32. Simon, L.; Kieger, M.; Sung, S.S.; Smalley, T.J. Aluminum toxicity in tomato. Part 2. Leaf gas exchange, chlorophyll content, and invertase activity. *J. Plant Nutr.* **1994**, *17*, 307–317. [CrossRef]
33. Hampp, R.; Schnabl, H.J.Z.f.P. Effect of aluminium ions on 14CO_2-fixation and membrane system of isolated spinach chloroplasts. *Z. Für Pflanzenphysiol.* **1975**, *76*, 300–306. [CrossRef]
34. Chen, L.-S.; Qi, Y.-P.; Smith, B.R.; Liu, X.-H. Aluminum-induced decrease in CO_2 assimilation in citrus seedlings is unaccompanied by decreased activities of key enzymes involved in CO_2 assimilation. *Tree Physiol.* **2005**, *25*, 317–324. [CrossRef]
35. Jiang, H.-X.; Tang, N.; Zheng, J.-G.; Chen, L.-S. Antagonistic actions of boron against inhibitory effects of aluminum toxicity on growth, CO_2 assimilation, ribulose-1, 5-bisphosphate carboxylase/oxygenase, and photosynthetic electron transport probed by the JIP-test, of Citrus grandis seedlings. *BMC Plant Biol.* **2009**, *9*, 102. [CrossRef]
36. Mishra, P.; Dubey, R.S. Effect of aluminium on metabolism of starch and sugars in growing rice seedlings. *Acta Physiol. Plant.* **2008**, *30*, 265–275. [CrossRef]
37. Jiang, H.X.; Tang, N.; Zheng, J.G.; Li, Y.; Chen, L.S. Phosphorus alleviates aluminum-induced inhibition of growth and photosynthesis in Citrus grandis seedlings. *Physiol. Plant.* **2009**, *137*, 298–311. [CrossRef] [PubMed]
38. Pereira, W.E.; de Siqueira, D.L.; Martínez, C.A.; Puiatti, M. Gas exchange and chlorophyll fluorescence in four citrus rootstocks under aluminium stress. *J. Plant Physiol.* **2000**, *157*, 513–520. [CrossRef]
39. Sharma, A.; Kapoor, D.; Gautam, S.; Landi, M.; Kandhol, N.; Araniti, F.; Ramakrishnan, M.; Satish, L.; Singh, V.P.; Sharma, P. Heavy metal induced regulation of plant biology: Recent insights. *Physiol. Plant.* **2022**, *174*, e13688. [CrossRef] [PubMed]
40. Manzoor, Z.; Hassan, Z.; Ul-Allah, S.; Khan, A.A.; Sattar, A.; Shahzad, U.; Amin, H.; Hussain, M. Transcription factors involved in plant responses to heavy metal stress adaptation. In *Plant Perspectives to Global Climate Changes*; Elsevier: Amsterdam, The Netherlands, 2022; pp. 221–231.
41. Ahsan, N.; Renaut, J.; Komatsu, S. Recent developments in the application of proteomics to the analysis of plant responses to heavy metals. *Proteomics* **2009**, *9*, 2602–2621. [CrossRef]
42. D'Alessandro, A.; Taamalli, M.; Gevi, F.; Timperio, A.M.; Zolla, L.; Ghnaya, T. Cadmium Stress Responses in *Brassica juncea*: Hints from Proteomics and Metabolomics. *J. Proteome Res.* **2013**, *12*, 4979–4997. [CrossRef]
43. Niedziela, A.; Domżalska, L.; Dynkowska, W.M.; Pernisová, M.; Rybka, K. Aluminum Stress Induces Irreversible Proteomic Changes in the Roots of the Sensitive but Not the Tolerant Genotype of Triticale Seedlings. *Plants* **2022**, *11*, 165. [CrossRef]
44. Koch, K. Sucrose metabolism: Regulatory mechanisms and pivotal roles in sugar sensing and plant development. *Curr. Opin. Plant Biol.* **2004**, *7*, 235–246. [CrossRef] [PubMed]
45. Ejaz, S.; Fahad, S.; Anjum, M.A.; Nawaz, A.; Naz, S.; Hussain, S.; Ahmad, S. *Role of Osmolytes in the Mechanisms of Antioxidant Defense of Plants*; Springer: Cham, Switzerland, 2020; Volume 39, pp. 95–117.
46. Conselvan, G.B.; Fuentes, D.; Merchant, A.; Peggion, C.; Francioso, O.; Carletti, P. Effects of humic substances and indole-3-acetic acid on Arabidopsis sugar and amino acid metabolic profile. *Plant Soil* **2018**, *426*, 17–32. [CrossRef]
47. Lehmann, M.; Laxa, M.; Sweetlove, L.J.; Fernie, A.R.; Obata, T. Metabolic recovery of *Arabidopsis thaliana* roots following cessation of oxidative stress. *Metabolomics* **2012**, *8*, 143–153. [CrossRef] [PubMed]
48. Huang, W.; Yang, X.; Yao, S.; LwinOo, T.; He, H.; Wang, A.; Li, C.; He, L. Reactive oxygen species burst induced by aluminum stress triggers mitochondria-dependent programmed cell death in peanut root tip cells. *Plant Physiol. Biochem.* **2014**, *82*, 76–84. [CrossRef]
49. Sweetlove, L.J.; Beard, K.F.M.; Nunes-Nesi, A.; Fernie, A.R.; Ratcliffe, R.G. Not just a circle: Flux modes in the plant TCA cycle. *Trends Plant Sci.* **2010**, *15*, 462–470. [CrossRef]
50. Ye, J.; Wang, X.; Hu, T.; Zhang, F.; Wang, B.; Li, C.; Yang, T.; Li, H.; Lu, Y.; Giovannoni, J.J.; et al. An InDel in the promoter of AI-ACTIVATED MALATE TRANSPORTER9 selected during tomato domestication determines fruit malate contents and aluminum tolerance. *Plant Cell* **2017**, *29*, 2249–2268. [CrossRef]
51. Wang, H.; Chen, R.F.; Iwashita, T.; Shen, R.F.; Ma, J.F. Physiological characterization of aluminum tolerance and accumulation in tartary and wild buckwheat. *New Phytol.* **2015**, *205*, 273–279. [CrossRef]
52. Zhang, L.; Wu, X.X.; Wang, J.F.; Qi, C.D.; Wang, X.Y.; Wang, G.L.; Li, M.Y.; Li, X.S.; Guo, Y.D. BoALMT1, an Al-Induced Malate Transporter in Cabbage, Enhances Aluminum Tolerance in *Arabidopsis thaliana*. *Front. Plant Sci.* **2018**, *8*, 9. [CrossRef]
53. Sasaki, T.; Yamamoto, Y.; Ezaki, B.; Katsuhara, M.; Ahn, S.J.; Ryan, P.R.; Delhaize, E.; Matsumoto, H. A wheat gene encoding an aluminum-activated malate transporter. *Plant J.* **2004**, *37*, 645–653. [CrossRef] [PubMed]
54. Zhao, Z.; Hu, L.; Hu, T.; Fu, J. Differential metabolic responses of two tall fescue genotypes to heat stress. *Acta Prataculturae Sin.* **2015**, *24*, 58–69. [CrossRef]

55. Xu, Y.; Fu, X. Reprogramming of plant central metabolism in response to abiotic stresses: A metabolomics view. *Int. J. Mol. Sci.* **2022**, *23*, 5716. [CrossRef] [PubMed]
56. Li, Z.; Xing, F.; Xing, D. Characterization of target site of aluminum phytotoxicity in photosynthetic electron transport by fluorescence techniques in tobacco leaves. *Plant Cell Physiol.* **2012**, *53*, 1295–1309. [CrossRef] [PubMed]
57. Li, Z.; Xing, D. Mechanistic study of mitochondria-dependent programmed cell death induced by aluminium phytotoxicity using fluorescence techniques. *J. Exp. Bot.* **2011**, *62*, 331–343. [CrossRef]
58. Su, C.; Jiang, Y.; Yang, Y.; Zhang, W.; Xu, Q. Responses of duckweed (*Lemna minor* L.) to aluminum stress: Physiological and proteomics analyses. *Ecotoxicol. Environ. Saf.* **2019**, *170*, 127–140. [CrossRef]

metabolites

Article

How Central Carbon Metabolites of Mexican Mint (*Plectranthus amboinicus*) Plants Are Impacted under Different Watering Regimes

Lord Abbey *, Raphael Ofoe, Zijing Wang and Sparsha Chada

Department of Plant, Food, and Environmental Sciences, Faculty of Agriculture, Dalhousie University, Truro, NS B2N 5E3, Canada
* Correspondence: loab07@gmail.com

Abstract: Plants are sessile, and their ability to reprogram their metabolism to adapt to fluctuations in soil water level is crucial but not clearly understood. A study was performed to determine alterations in intermediate metabolites involved in central carbon metabolism (CCM) following exposure of Mexican mint (*Plectranthus amboinicus*) to varying watering regimes. The water treatments were regular watering (RW), drought (DR), flooding (FL), and resumption of regular watering after flooding (DHFL) or after drought (RH). Leaf cluster formation and leaf greening were swift following the resumption of regular watering. A total of 68 key metabolites from the CCM routes were found to be significantly ($p < 0.01$) impacted by water stress. Calvin cycle metabolites in FL plants, glycolytic metabolites in DR plants, total tricarboxylic acid (TCA) cycle metabolites in DR and DHFL plants, and nucleotide biosynthetic molecules in FL and RH plants were significantly ($p < 0.05$) increased. Pentose phosphate pathway (PPP) metabolites were equally high in all the plants except DR plants. Total Calvin cycle metabolites had a significantly ($p < 0.001$) strong positive association with TCA cycle (r = 0.81) and PPP (r = 0.75) metabolites. Total PPP metabolites had a moderately positive association with total TCA cycle metabolites (r = 0.68; $p < 0.01$) and a negative correlation with total glycolytic metabolites (r = −0.70; $p < 0.005$). In conclusion, the metabolic alterations of Mexican mint plants under different watering regimes were revealed. Future studies will use transcriptomic and proteomic approaches to identify genes and proteins that regulate the CCM route.

Keywords: edaphic stress; Jamaican thyme; Crassulacean acid metabolism; TCA cycle; Calvin cycle; nucleotide pathway; glycolysis; pentose phosphate pathway

1. Introduction

A universal shift in weather patterns due to sustained global climate change has resulted in fluctuations in edaphic stress conditions, particularly drought, flooding, soil salinity, and soil impedance [1]. Plants are subject to such harsh and variable edaphic stress conditions because they are sessile. Typically, plant responses to the onset of edaphic stress include inhibition of plant root hydraulic conductance; reductions in stomatal conductance and photosynthesis; reprogramming of affected metabolic pathways; and, ultimately, cessation of plant growth and cell death [2–6]. The initial biochemical response to soil water stress is the production of free radicals and reactive oxygen species such as hydrogen peroxide and superoxide, which adversely affect plant morpho-physiological functions and interrupt metabolic processes to the detriment of the plant [7,8]. According to Stoychev et al. [9], such stress conditions lead to alterations in energy metabolism for survival and adaptation.

The Calvin cycle occurs in the stroma of the chloroplast after photolysis (light reaction), and so it is referred to as the dark reaction phase of photosynthesis. From the light reaction, energy in the form of ATP and NADPH is generated and later used to produce glucose and downstream carbohydrate molecules in the Calvin cycle [10]. The carbohydrates from the Calvin cycle enter the central carbon metabolism (CCM) route via the glycolytic pathway [11,12]. The CCM is a complex series of enzymatic steps for the conversion of sugars into metabolic precursors in cells [7,8,13]. The CCM route, which consists of the glycolytic pathways, the pentose phosphate pathway (PPP) for sugar interconversion, and the tricarboxylic acid (TCA) cycle for final complete oxidation [7,14], provides 12 metabolites that are the basic carbon precursors of all biosynthetic pathways [15]. PPP is divided into an oxidative phase that produces ribose-5-phosphate for the synthesis of nucleic acids and a non-oxidative phase that produces NADPH [16]. The latter is an important reducing power for the synthesis of fatty acids, nucleotides, and non-essential amino acids for numerous metabolic pathways. So far, the impact of drought, flooding, or water stress recovery on CCM is understudied, but there are few reports on the individual pathways.

Cellular respiratory metabolism in plants via the TCA cycle occurs in the matrix of the mitochondrion, and it is essential for energy supply to different organelles for the maintenance of various physiological functions. A study conducted by Moradi et al. [6] revealed that drought disrupted the thyme (*Thymus* sp.) TCA cycle and reduced the synthesis of amino acids. In contrast, drought-induced metabolic reprogramming in Arabidopsis resulted in an increase in TCA cycle intermediate metabolites [17,18]. Kumar et al. [5] examined inbred maize (*Zea mays*) plants under water stress conditions and revealed that TCA cycle metabolites like succinate, α-ketoglutarate, and fumarate were reduced in drought-tolerant varieties while citrate and isocitrate were increased in drought-sensitive varieties. In addition, drought altered NADP+ concentration and electron receptor potential for the electron transport chain due to a reduction in NADPH+ and H+ availability for the TCA cycle. However, none of these studies reported on flooding or drought-flooding cycles. The ability of a plant to tolerate water stress is dependent on cellular concentrations of metabolic solutes like proline, betaine, fructose, and sucrose [19]. Research conducted using drought-tolerant wheat (*Triticum aestivum*) varieties showed that under severe water stress conditions, pyruvic acid, phenylpyruvate, fructose-6-phosphate, glucose, sucrose, and fructose were remarkably increased via the glycolytic pathway [3,5], suggesting an increase in sugar synthesis under water stress conditions. Nucleotide metabolism is the most critical cellular component for plant growth and affects many metabolic processes [20]. Purine and pyrimidine nucleotides are the building blocks for nucleic acid biosynthesis, which provides the required energy for the biosynthesis of carbohydrates, proteins, lipids, and secondary metabolites known to be central to all cellular metabolisms [21,22].

Water is crucial for plant growth and development, and globally, increasing numbers of farm operations are impacted by drought or flooding conditions because of changes in precipitation patterns caused by global climate change and competition with the rising global population and the manufacturing industries requirements for water [2,23]. Therefore, a fundamental understanding of plant response to fluctuations in water stress will be highly valuable for the development of stress-tolerant crops and the management of crop water requirements. Mexican mint (*Plectranthus amboinicus*) is an herbaceous perennial plant that belongs to the Lamiaceae family with a diverse array of ethnobotanical characteristics, culinary properties, and aroma-medicinal compounds [6,24]. Like most other plants, these properties can be affected by drought or flooding to varying degrees (Moradi et al. 2014 [6]). Although Mexican mint can survive drought conditions to some extent, its growth and chemical composition may be remarkably altered in prolonged stress conditions.

Mexican mint is physiologically a Crassulacean acid metabolism (CAM) plant that fixes CO_2 into C4 acids at night [25]. Because CAM plants absorb CO_2 at night, their stomata are closed much of the day to conserve water and can tolerate limited drought conditions. However, the biochemical and physiological mechanisms underpinning the Mexican mint plant's response to drought or flooding are not reported. Based on current knowledge, it was hypothesized that exposure of Mexican mint plants to prolonged drought or flooding will cause extensive disruption to CCM, but the plant will recover as soon as regular watering resumes. Therefore, the present study determined variations in metabolites of the different CCM routes following prolonged exposure of Mexican mint plants to drought or flooding and reversals to regular watering. This knowledge may be extended to other plants under the current dispensation of climate change and global warming.

2. Materials and Methods

2.1. Location

This research was conducted in the Plant Physiology and Stress Laboratory of the Department of Plant, Food, and Environmental Sciences, Faculty of Agriculture, Dalhousie University, and targeted metabolite quantitation was performed at the University of Victoria Genome BC (UVic GBC)—Proteomics Centre of The Metabolomics Innovation Centre, Canada, between December 2021 and April 2022.

2.2. Preparation and Rooting of Cuttings

Cuttings of soft tissue branches from the youngest second and third nodes on the main stem of healthy and well-watered Mexican mint mother plants were collected from the PFES greenhouse plant stock. Each cutting was trimmed to 5 cm in height with four pairs of corresponding leaves. Mexican mint is an easy rooting plant, so rooting hormone was not used. Four nodes on the stem part of the cuttings were embedded in moist perlite (Perlite Canada Inc., Montreal, QC, Canada) contained in a plastic flat tray of dimensions 50 cm length × 28 cm width × 6.5 cm depth. The planted trays were covered with a dome-shaped transparent plastic cover to maintain ≥95% relative humidity environment to induce rooting. The trays were placed on a planting shelf with 24 h fluorescence light at 22 °C. The cuttings were watered every other day, but the leaves were finely sprayed with water twice a day with no addition of fertilizer. The rooted cuttings were ready for transplanting after three weeks.

2.3. Planting of Rooted Cuttings and Growing Condition

Uniformly rooted cuttings were transplanted into 15 cm diameter plastic pots with saucers underneath, and each pot contained 200 g of Promix-BX mixed with 150 g of vermicast. The Promix-BX potting medium (Premier Horticulture Inc., Quakertown, PA, USA) contained 75–85% sphagnum peat moss, horticultural grade perlite, vermiculite, and dolomitic and calcitic limestone, and the vermicast was produced by Red Wiggler (*Eisenia fetida*) worms purchased from Growing Green Earthworm Castings (Lower Wedgeport, NS, Canada). The potted plants were arranged in a completely randomized design with four replications in a Biotronette Mark II Environmental Chamber (Lab-Line Instruments Inc., Melrose Park, IL, USA). The growth chamber was set at a 24°/20 °C day/night temperature cycle and a 12/12 h day/night light cycle. The plants were grown for 120 days, during which they were regularly watered to field capacity every three days prior to the imposition of water stress treatment. Pots were rearranged weekly to offset unpredictable occurrences due to variations in the environment.

2.4. Water Stress Treatment

To understand how the central carbon metabolism of plants recovering from water stress after reversal of drought (i.e., ≤10% field capacity) or flooding (i.e., oversaturation) is altered, the treatments used were regular watering (RW) at field capacity; continuous drought (DR); continuous flooding (FL); dehydration and resumption of regular watering after continuous flooding for 8 weeks (DHFL); and rehydration and resumption of regular watering after continuous drought for 8 weeks (RH). Regular watering was done every other day to maintain the field capacity of the growing medium. Flooding was simulated by complete submergence of the 15 cm diameter pot with the Mexican mint plant in water contained in a 20 cm diameter plastic pot. The experiment was arranged in a completely randomized design with four replications in the growth chamber.

2.5. Central Carbon Metabolites

2.5.1. Sample Preparation

The green leaves of the potted Mexican mint plants were harvested after 8 weeks of water stress treatment and immediately dipped in liquid nitrogen (N), ground into a fine powder, and stored at −80 °C until analysis. The ground leaf tissue samples on ice were shipped by overnight courier to UVic GBC for targeted metabolite quantitation. Triplicate samples (50 mg) of ground leaf tissue per treatment were separately added to 500 L of 80% methanol and homogenized using a MM400 mill mixer (Retsch, Haan, Germany). The mixture was then sonicated for 5 min in an ice-water bath, followed by centrifugation at 21,000× *g* for 20 min at 5 °C. A quantity of 250 µL of the supernatants was removed and combined with 150 µL of dichloromethane and 150 µL of water. After centrifugal clarification of the mixture and vortex mixing for 30 s, triplicate samples of 80 µL aliquots of the supernatant were dried in a N gas flow. There were three biological replicates and two technical replicates. The obtained residues were used for the assays described below.

2.5.2. Tricarboxylic Acid Cycle Assay

Standard stock solutions were prepared based on the method description provided by Han et al. (2013) [26]. The standard stock solutions of all targeted carboxylic acids were prepared in 80% methanol at concentrations of 200–1000 µM. A total of 50 µL of each standard solution or the supernatant of each sample was mixed and reacted with 50 µL of a 200 mM 3-nitrophenylhydrazines (NPH) solution at 30 °C. Following the reaction, 450 µL of water was added to each solution, and 10 µL of the resulting solution was injected into a C18 liquid chromatography (LC) column (2.1 × 100 mm, 1.8 µm) for quantification of the carboxylic acid by ultrahigh LC-multiple reaction monitoring/mass spectroscopy (UPLC-MRM/MS), and ion detection was performed on an Agilent 1290 UHPLC coupled to a Sciex 4000 QTRAP MS instrument (AB Sciex, Concord, ON, Canada) [26].

2.5.3. Glucose and Selected Sugar Phosphates

An 80 L aliquot of the dried residue of each sample was mixed with 50 L of 50% methanol. After that, 50 mL of standard solutions of glucose, ribose, ribose-5-phosphate, glucose-6-phosphate, and mannose-6-phosphate were serially diluted before being combined and reacted with 100 mL of 25 mM 3-amino-9-ethylcarbazole (AEC) solution, 50 mL of 50 mM sodium cyanoborohydride ($NaCBH_3$) solution, and 20 mL of LC/MS grade acetic acid for 70 min at 60 °C. A quantity of 300 µL of chloroform and 200 µL of water were added during the reaction period, after which each supernatant was centrifuged at 12,500× *g* for 5 min and vortexed for 15 s before adding 50 mL of water. Finally, 10 µL of the mixture was injected into a pentafluorophenylpropyl (PFP) LC column (2.1 × 150 mm, 1.7 µm) to run UPLC-MRM/MS on an Agilent 1290 UHPLC coupled to an Agilent 6495B QQQ instrument (Conquer Scientific, Poway, CA, USA) with positive-ion detection as previously described by [26].

2.5.4. Other Metabolites

In 50% methanol, a solution known as an internal standard (IS) was created that contained 25 isotope-labelled metabolites including adenosine monophosphate (AMP), adenosine-5-triphosphate (ATP), uridine monophosphate (UMP), uridine triphosphate (UTP), uridine diphosphate (UDP)-glucose, fructose-6-phosphate (fructose-6P), fructose-bisphosphate, glycerol-3-phosphate, nicotinamide adenine dinucleotide (NAD), NADH, glucose-1-phosphate, ribose-5-phosphate, and others. Moreover, all the targeted metabolites were produced as serially diluted standard solutions in the IS solution at concentrations ranging from 0.00002 to 10 M. Eighty liter aliquots of dried residue were dissolved in 100 L of the IS solution. Ten microliters of each sample solution or each standard solution was injected onto a C18 LC column (2.1 × 100 mm, 1.9 µm) to run UPLC-MRM/MS and negative ion detection on a Waters Acquity UPLC system coupled to a Sciex QTRAP 6500 Plus

MS instrument using tributylamine acetate buffer-acetonitrile/methanol (1:1, v/v) as the mobile phase with a gradient of 0.25 mL/min (from 10% to 50% B over 25 min) and 60 °C.

2.6. Calculation and Statistical Analysis

Concentrations of the detected analytes in the above assays were calculated with IS calibration by interpolating the constructed linear regression curves of individual compounds using the analyte-to-internal standard peak area ratios measured from injections of the sample solution. All data obtained were subjected to a one-way analysis of variance (ANOVA) using Minitab version 21 (Minitab, Inc., State College, PA, USA). Tukey's honestly significant difference post-test was used to separate treatment means at $p \leq 0.05$. A multivariate statistical analysis of grouped compounds, two-dimensional principal component analysis (PCA), and hierarchical clustering for differential metabolism per group were constructed with Euclidean distance using XLSTAT version 2022.3 (Addinsoft, New York, NY, USA).

3. Result and Discussion

3.1. Plant Growth

The effects of prolonged drought and flooding stresses on plant growth and productivity are well documented [13,27], but not much has been reported on stress reversal. The results of the present study revealed that although morphological recuperation of Mexican mint through rehydration of drought-stressed (RH) plants and dehydration of flooded (DHFL) plants can be fast and occur within 4 h, the overall recovery of total plant size and leaf greening similar to that of regularly watered (RW) plants can be delayed (Figure 1A). This was proven by the resumption of leaf cluster formation on RH and DHFL plants compared to the stressed, i.e., drought (DR) and flooded (FL) plants, respectively (Figure 1A).

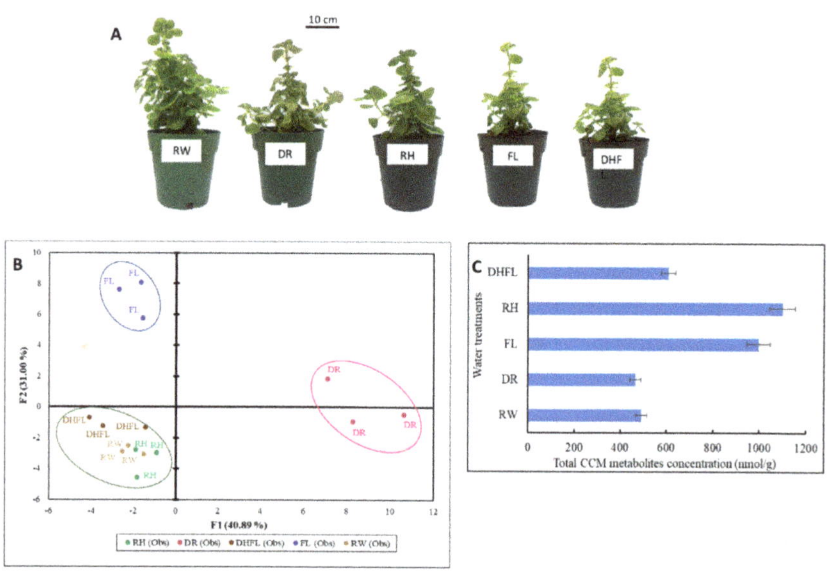

Figure 1. Response of Mexican mint (*Plectranthus amboinicus*) plants to varying watering regimes. (**A**) Plant growth, (**B**) two-dimensional principal component analysis of metabolic profiles, and (**C**) a bar chart of total concentrations of intermediate metabolites along the central carbon metabolic (CCM) routes with error bars. Regular watering (RW), drought (DR), flooding (FL), resumption of regular watering after flooding (DHFL) or after drought (RH).

3.2. Total CCM Metabolites

Plants produce large quantities of specialized metabolites, which are end-products of cellular regulatory activities [28], but the compositions of these metabolites can be highly modulated by stress conditions [5,6,29], as demonstrated in the present study (Figure 1B,C). Anecdotally, Mexican mint as a CAM plant is known to tolerate some extent of water deficit stress [25], but its tolerance level and the mechanistic response to prolonged and continuous water stress conditions are unknown up to now. To understand this mechanism, changes in metabolites involved in CCM under different watering regimes were assessed. A two-dimensional PCA biplot presented in Figure 1B explained ca. 72% of the overall variations in total metabolites. The first (F1) and second (F2) factors represent ca. 41% and 31%, respectively. The PCA revealed three distinct clusters that clearly discriminated between the different water treatments with respect to the total metabolites of the CCM routes. Based on the cluster formation, it seemed the metabolism of RH and DHFL plants was adjusted after 8 weeks of stress recovery and resumed normal growth and metabolism that was similar to the RW plants. These results suggested that, irrespective of the prolonged drought and flooding, the Mexican mint plants underwent metabolic reprogramming when regular watering was resumed, as previously reported for intermediate metabolites in flood-stressed clover (*Trifolium* spp.) [9], drought-stressed Arabidopsis [17], and corn [5]. Drought or flooding causes remarkable degradation of chlorophyll and a significant impairment of photosynthesis, in addition to increased osmotic adjustment and MDA content [7,27]. These may explain the yellowing and stunting of the DR and FL plants compared to the RW plants.

3.3. Metabolites Profile of the Different CCM Routes

The results showed that the abundance of intermediate metabolites involved in the CCM routes was significantly ($p < 0.05$) altered due to water stress (Table 1). During water stress, plants go through rapid metabolic adjustment to maintain proper metabolism as an adaptation mechanism [12]. The process of metabolic adjustment requires several regulatory mechanisms to mediate signaling between multiple metabolic pathways and to initiate alterations in the composition of metabolites throughout the various CCM routes [7]. Metabolic analysis in the present study revealed the presence of 68 putative metabolites in the Mexican mint plants, which were categorized into five carbon-mediated metabolic groups for Calvin cycle, glycolysis, TCA cycle, PPP, and nucleotide biosynthesis [14].

Table 1. Total metabolites involved in specific central carbon metabolic routes in Mexican mint (*Plectranthus amboinicus*) plants under varying watering regimes.

Treatment	Calvin Cycle (nmol/g)	Glycolysis (mmol/g)	TCA (mmol/g)	PPP (nmol/g)	Nucleotide Biosynthesis (nmol/g)
DR	31.33b	29.42a	59.86a	56.67b	288.65b
FL	67.75a	9.04b	36.62bc	182.62a	701.27a
DHFL	47.61ab	2.57c	51.62ab	196.32a	310.62b
RH	41.06b	2.07c	27.59c	149.38a	879.76a
RW	36.95b	1.90c	37.83bc	147.73a	267.60b
p-value	0.002	0.00	0.002	0.000	0.000

PPP, pentose phosphate pathway; TCA, tricarboxylic acid. Regular watering (RW), drought (DR), flooding (FL), resumption of regular watering after flooding (DHFL) or after drought (RH). Values are means of three replications, and different alphabetical letters denote significant differences according to Tukey's honestly significant difference post-test analyses at a significant level of $p < 0.05$.

Remarkable shifts in total concentrations of intermediate metabolites and energy generation were noticed with variations in watering regimes (Table 1). It is reported that plant response to water stress begins with rapid production of reactive oxygen species that act as an alarm signal that triggers acclimatory-defense responses by specific signal transduction pathways [7,8]. Consequently, these reactions result in a series of metabolic reprogramming in the plant. This can explain the significant ($p < 0.05$) alterations in total metabolites of the different CCM routes and energy generation in the Mexican mint plants following water stress imposition and water stress recovery, as shown in Table 1 and explained further below.

3.3.1. Calvin Cycle Intermediates

The production of downstream carbohydrate molecules, including glucose in the Calvin cycle, was significantly ($p < 0.05$) influenced by the different watering regimes (Table 1, Figure 2). Interestingly, the concentrations of total metabolites involved in the Calvin cycle were reduced non-significantly ($p > 0.05$) by ca. 15% in Mexican mint plants exposed to DR compared to those under RW (Table 1). On the contrary, the concentrations of total Calvin cycle metabolites were significantly ($p < 0.001$) increased by ca. 83% in FL plants compared to RW plants (Table 1). Plants that recovered from water stress, i.e., RH and DHFL plants, had their total Calvin cycle metabolites slightly increased by ca. 11% and 29%, respectively, although not significantly ($p > 0.05$) different from that of the RW plants (Table 1). This means that the water-stressed Mexican mint plants recovered their photosynthetic capabilities upon rehydration after a period of drought or upon dehydration after a period of flooding, and their photosynthetic capabilities approached those of the RW plants.

The ANOVA shows that the individual intermediate metabolites involved in the Calvin cycle varied significantly ($p < 0.05$) with variation in watering regime (Supplementary Table S1). Specifically, fructose-1,6-bisphosphate (F1,6BP) and ribulose-1,5-bisphosphate (RuBP) were significantly ($p < 0.05$) highest in DR plants. However, the overall reduction in total Calvin cycle metabolites in DR plants (Table 1) can be attributed to low concentrations of sedoheptulose-7-phosphate (Se7P), erythrose-4-phosphate (E4P), F1,6BP, dihydroxyacetone phosphate (DHAP), 3-phosphoglyceric acid (3PG), and ribulose-5-phosphate (Ru5P), as shown by the heatmap in Figure 2. The result suggested a possible reduction in the rate of photosynthesis in the DR plants, which was consistent with previous reports for other plants where drought reduced the abundance of numerous Calvin cycle proteins and other associated enzymes involved in photosynthesis [30–32]. Although enzymatic activities were not investigated, Wingler et al. [33] suggested that the reduction in Calvin cycle metabolites can be ascribed to impairment in photorespiratory enzyme activities, which can affect photosynthesis under drought stress. In soybean (*Glycine max*), Chen et al. [34] revealed that drought stress repressed the expression of genes involved in the Calvin cycle and resulted in a substantial reduction in photosynthetic abilities.

On the other hand, the increase in Calvin cycle metabolites in FL plants can be ascribed to increased concentrations of Se7P, E4P, F6P, ribose-5-phosphate (R5P), and Ru5P (Figure 2). Previous studies demonstrated that the ability to maintain photosynthesis was strongly associated with flood tolerance in several plant species [35,36]. It was found that under flooding conditions, most flood-tolerant plants switch from aerobic to anaerobic respiration to increase photosynthetic carbon production. Such a flood adaptation mechanism could have been one of the reasons for increased Calvin cycle metabolites in FL plants, as reported for *Kandelia candel* by Pan et al. [37]. Furthermore, Se7P produced from sedoheptulose-1,7-bisphosphate (Se1,7BP) through the action of the enzyme sedoheptulose-bisphosphatase is critical for RuBP regeneration. Therefore, the increase in Se7P concentration in FL plants probably indicated an increase in photosynthetic rate and carbohydrate synthesis, as previously noticed by [38]. RH and DHFL plants had high concentrations of DHAP and Glyceraldehyde-3-phosphate (G3P), similar to concentrations found in RW plants (Figure 2). Collectively, G3P and DHAP are termed triosephosphates and can interconvert via the catalyzation of isomerase [39]. G3P and DHAP are important intermediates for photosynthesis and glycolysis.

Studies have shown that during glycolysis, ca. 0.05–0.3% of glucose is non-enzymatically transformed into methylglyoxal, which can be increased following plant exposure to stressful conditions [40–42]. The present finding suggested that G3P, a key end-product of photosynthesis, was increased in plants under regular watering (i.e., RW) or plants that resumed regular watering after stress (i.e., RH and DHFL) compared to DR or FL plants. Sugar accumulation in response to drought and flooding has been reported by many authors [43–45]. Glucose is produced at the end of the Calvin cycle, and therefore, the increase in concentrations of glucose in both DR (high) and FL (from moderate to high) plants (Figure 3) can be considered a biochemical strategy to supply substrates for high-energy generation through glycolysis [11] as the stressed plants switched to a survival mode [9].

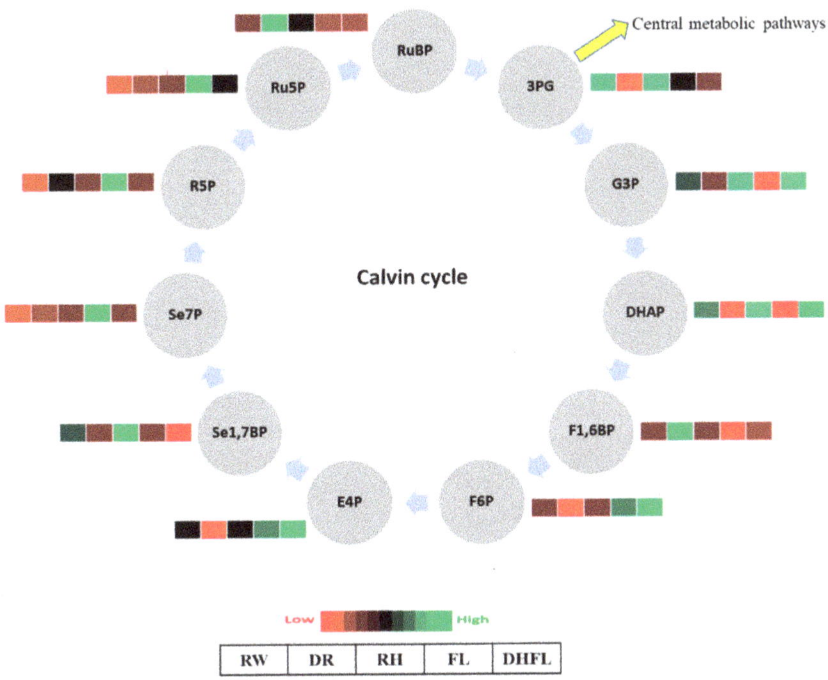

Figure 2. A heat map depicting the individual concentrations of intermediate metabolites of the Calvin cycle in Mexican mint (*Plectranthus amboinicus*) plants under varying watering regimes (n = 3). 3-phosphoglyceric acid (3PG), glyceraldehyde-3-phosphate (G3P), dihydroxyacetone phosphate (DHAP), fructose-1,6-bisphosphate (F1,6BP), fructose-6-phosphate (F6P), erythrose-4-phosphate (E4P), sedoheptulose-1,7-bisphosphate (Se1,7BP), sedoheptulose-7-phosphate (Se7P), ribose-5-phosphate (R5P), ribulose-5-phosphate (Ru5P), and ribulose-1,5-bisphosphate (RuBP). Metabolite concentrations in each block compartment are normalized across all data for an individual compound such that similar color intensities between compounds can represent widely differing concentrations. The red color represents a lower concentration, and the green color represents a higher concentration of a particular metabolite. The compartments were arranged from left to right as follows: regular watering (RW), drought (DR), resumption of regular watering after drought (RH), flooding (FL), and resumption of regular watering after flooding (DHFL).

3.3.2. Glycolytic Pathway Intermediates

The 3C sugar from the Calvin cycle enters the glycolytic pathway. In this study, we observed that the enzymatic breakdown of glucose via glycolysis was significantly ($p < 0.001$) higher in DR plants followed by FL plants when compared to RW plants (Supplementary Table S2). Total glycolytic metabolites were increased by ca. 1448% in DR plants and ca. 376% in FL plants compared to those of the RW plants (Table 1). Specifically, the increase in glycolytic metabolites in DR plants can be attributed to high concentrations of glucose, F1,6BP, phosphoenolpyruvic acid (PEP), pyruvic acid (PA), NADH, and adenosine triphosphate (ATP) (Figure 3; Supplementary Table S2). Additionally, the increase in glycolytic metabolites in FL plants can be attributed to high concentrations of F6P, PEP, and PA. These observations suggested that PEP and PA levels were increased in Mexican mint plants under both drought and flood conditions. Because total glycolytic pathway metabolites were not significantly ($p > 0.05$) altered in RH and DHFL plants, we suggested that glucose metabolism was altered by DR and FL conditions but was normalized after the plants were exposed to regular watering conditions, as portrayed by the RH and DHFL plants, respectively.

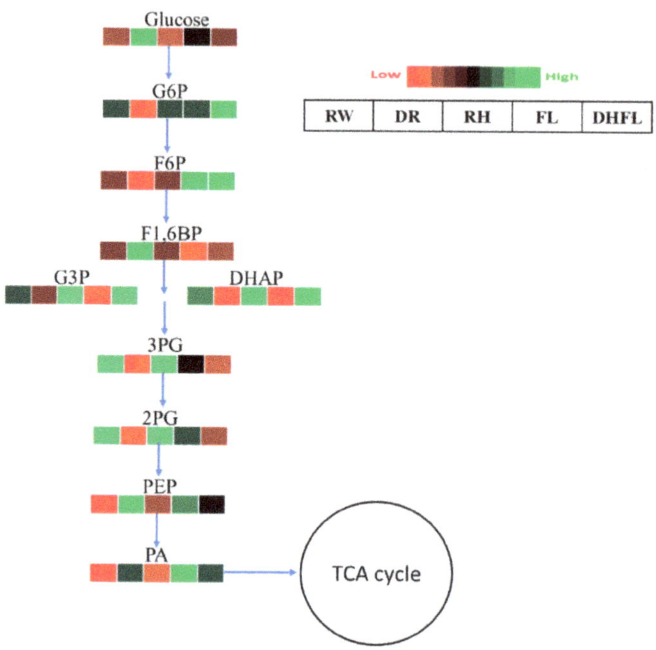

Figure 3. A heat map depicting the individual concentrations of intermediate metabolites of the glycolysis pathway in Mexican mint (*Plectranthus amboinicus*) plants under varying watering regimes (n = 3). Glucose-6-phosphate (G6P), fructose-6-phosphate (F6P), fructose-1,6-bisphosphate (F1,6BP), glyceraldehyde-3-phosphate (G3P), dihydroxyacetone phosphate (DHAP), 3-phosphoglyceric acid (3PG), 2-phosphoglyceric acid (2PG), phosphoenolpyruvic acid (PEP), and pyruvic acid (PA). Metabolite concentrations in each block compartment are normalized across all data for an individual compound such that similar color intensities between compounds can represent widely differing concentrations. The red color represents a lower concentration, and the green color represents a higher concentration of a particular metabolite. The compartment was arranged from left to right as follows: regular watering (RW), drought (DR), resumption of regular watering after drought (RH), flooding (FL), and resumption of regular watering after flooding (DHFL).

Numerous studies have consistently reported that stress imposed by drought and flooding conditions accelerates glycolysis as a strategy for providing energy for stress defense activation and adaptation. For instance, Guo et al. [3] reported that glucose, PA, and PEP concentrations in wheat were considerably increased in drought-tolerant genotypes. Similarly, glucose and other glycolytic intermediate metabolites accumulated in *Pinus painaster* [46], *Thymus vulgaris*, *T. Kotschyanus* [23], and *Lotus japonicus* [47] in response to drought. Moreover, water stress instigated "energy crises" through the impairment of oxidative phosphorylation of the mitochondrion, resulting in a drastic reduction of ATP production [48,49]. Researchers have shown that sugar accumulation plays a crucial role in carbon resource allocation and plant growth [50]. To survive the "energy crises" under such stressful conditions, tolerant plants increase their glycolytic influx by accumulating more glucose to produce sufficient ATP via glycolysis to maintain basic cellular functions and regenerate NAD^+ to maintain the glycolytic flux [48]. Furthermore, it has been reported that pyruvate metabolism plays a critical role in plant water stress tolerance [3,37,51]. A previous study by Pan et al. [37] also showed that *Kandelia candel* plants accumulated high concentrations of PA under flooding due to increased activities of 6-phosphofructokinase (PFK) and pyruvate kinase (PK). Additionally, PA production from glycolysis can be directed to a recycled pool of NAD+ through the fermentation pathway and/or serve as a hub for the biosynthesis of amino acids, fats, and sugars [52,53]. In poplar (*Populus* sp.) plants, PA accumulation under flood conditions was strongly linked to amino acid biosynthesis as a flood tolerance mechanism [54]. Although the activities of these enzymes were not examined in the present study, the increased PA concentrations in DR and FL plants (Figure 3) were consistent with previous findings. This suggested that the glycolytic pathway was crucial for modulating energy metabolism, which can lead to increased amino acid production in *Mexican mint* plants under water stress. Nevertheless, further studies are required to examine the detailed mechanisms involved.

3.3.3. Pentose Phosphate Pathway Intermediates

Changes in the cellular abundance of PPP intermediate metabolites have been reported to strongly affect several other metabolic pathways [55]. In the present study, it was found that the total concentration of PPP metabolites was not significantly ($p > 0.05$) different amongst FL, DHFL, RH, and RW plants, and their average of 168.93 nmol/g was ca. 189% more than that of the DR plants (Table 1). Comparatively, the concentration of total PPP metabolites in DHFL plants was non-significantly ($p > 0.05$) the highest, followed by FL plants, and moderate in RW and RH plants. The low concentration of total PPP metabolite in DR plants can be associated with significant ($p < 0.001$) reductions in G6P and E4P, although the concentrations of R5P and xylulose-5-phosphate (Xu5P) were significantly ($p < 0.001$) enhanced compared to the RW plants (Figure 4; Supplementary Table S3). On the contrary, there was an increase in total PPP metabolite concentration in the FL plants (Table 1) that can be ascribed to a significant ($p < 0.001$) increase in the concentrations of R5P, Ru5P, Se7P, E4P, and F6P (Figure 4).

Similarly, increased concentrations of G6P, E4P, G3P, and F6P accounted for the high concentrations of total PPP metabolites in DHFL plants (Figure 4; Supplementary Table S3). It was established that PPP is primarily responsible for the major supply of NADPH for several biosynthetic pathways in cells and contributes to antioxidant production [56]. The increased NADPH in DR plants possibly indicated an enhancement of the antioxidant system, as reported for drought-stressed soybean plants by [57]. This suggested that maintaining redox potential could be a necessity for Mexican mint plant protection against drought-induced oxidative stress [58]. Overall, the RH plants had a similar PPP metabolite profile to the RW plants, and the abundance of G6P, 6PG, R5P, Xu5P, G3P, and Se7P in RW, RH, and DHFL plants were similar (Figure 4). Consequently, the concentration of the resultant product, i.e., glucose, in RW, RH, and DHFL plants entering the glycolytic pathway was similar (Figure 3), since PPP runs parallel to glycolysis in the cytosol. Interestingly, the concentration of NADPH in FL and DHFL plants was similarly low as that found in RW plants (Supplementary Table S3). The increase in R5P, E4P, and F6P concentrations in FL plants as a result of the flooding, suggests stress tolerance mechanisms can lead to enhanced production of downstream compounds, including nucleotides, aromatic amino acids, and fatty acids [16,59,60]. Overall, Mexican mint plants increased PPP metabolic flux under flooding stress to achieve energy homeostasis—a mechanism reported for *K. candel* by Pan et al. [37].

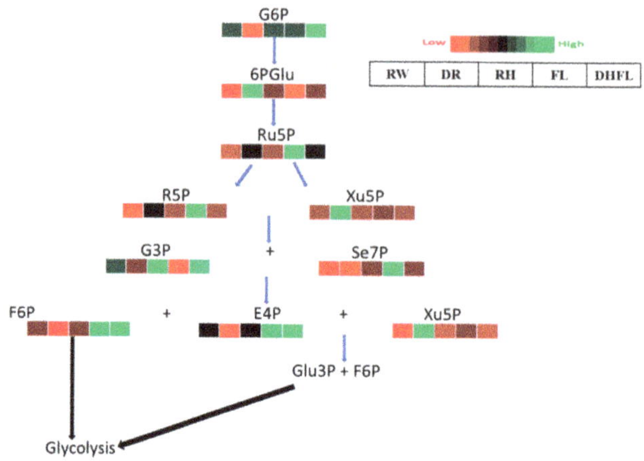

Figure 4. A heat map depicting the individual concentrations of intermediate metabolites of the pentose phosphate pathway in Mexican mint (*Plectranthus amboinicus*) plants under varying watering regimes (n = 3). Glucose-6-phosphate (G6P), 6-phosphogluconate (6PGlu), ribulose-5-phosphate (Ru5P), ribose-5-phosphate (R5P), xylulose-5-phosphate (Xu5P), glyceraldehyde-3-phosphate (G3P), sedoheptulose-7-phosphate (Se7P), fructose-6-phosphate (F6P), erythrose-4-phosphate (E4P), and glucose-3-phosphate (Glu3P). Metabolite concentrations in each block compartment are normalized across all data for an individual compound such that similar color intensities between compounds can represent widely differing concentrations. The red color represents a lower concentration, and the green color represents a higher concentration of a particular metabolite. The compartment was arranged from left to right as follows: regular watering (RW), drought (DR), resumption of regular watering after drought (RH), flooding (FL), and resumption of regular watering after flooding (DHFL).

3.3.4. Tricarboxylic Acid Cycle Intermediates

It was found that both total and individual TCA cycle intermediate metabolites were significantly ($p < 0.002$) affected by the different watering regimes (Table 1; Figure 5). Total TCA cycle metabolites were significantly ($p < 0.001$) increased by ca. 58% in DR plants and ca. 36% in DHFL plants compared to their RW counterparts (Table 1). In contrast, drought stress reduced TCA cycle metabolites in maize, rice, and sesame (*Sesamum indicum*) plants [32,61,62]. This contrasting report could be attributed to differences in plant genotype, i.e., Mexican mint versus maize, rice, and sesame. The fewest total TCA cycle metabolites were found in RH plants and were not significantly ($p > 0.05$) different from those of the RW and FL plants. When compared to RW plants, flooding stress did not alter the total concentration of the TCA metabolites in the Mexican mint plants. This conformed with previous reports for flood-tolerant soybean [63,64], *Acanthus ilicifolius* [65], *Lotus japonicus* [66], and rice [67] plants. According to Menezes-Silva et al. [68], plants can intensify their defense against stress by maintaining information from previous stress events. This "stress memory" in plants involves physiochemical processes such as photosynthesis, maintenance of water status, and osmotic adjustment [69]. These can explain the results of the present study, which showed that the removal of flooding (i.e., DHFL plants) increased total TCA cycle metabolites compared to FL or RW plants (Table 1). It therefore seemed obvious that the ability of DHFL plants to withstand future stress was enhanced compared to FL and RW plants. Although the reductions in TCA cycle metabolites did not significantly ($p > 0.05$) impact ATP production in this study, such reductions could likely alter TCA-mediated amino acid production [70], which must be investigated in future studies.

More specifically, there was a remarkable increase in eight out of nine of the individual TCA cycle intermediate metabolites in the FL plants (Supplementary Table S4) compared to the other treatments. The concentrations of incoming PA from glycolysis and the TCA cycle intermediate metabolites—isocitric acid, α-ketoglutaric acid, succinic acid, fumaric acid, and malic acid—were particularly high in FL plants compared to all the other treatments (Figure 5). This suggests greater mitochondrial activity in FL plants to generate carbon skeletons for amino acid biosynthesis [70]. The highest concentrations of isocitric acid, aconitic acid, and citric acid were noticed in DR plants, which had the least concentrations of oxaloacetic acid, acetyl-CoA, malic acid, and α-ketoglutaric acid.

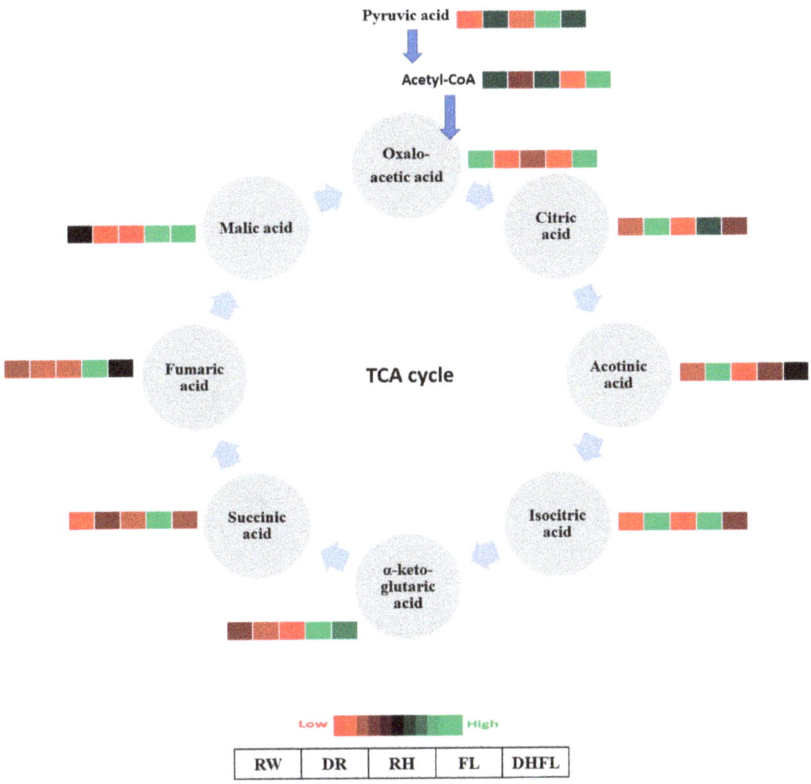

Figure 5. A heat map depicting the individual concentrations of intermediate metabolites of the tricarboxylic acid cycle in Mexican mint (*Plectranthus amboinicus*) plants under varying watering regimes (n = 3). Metabolite concentrations in each block compartment are normalized across all data for an individual compound such that similar color intensities between compounds can represent widely differing concentrations. The red color represents a lower concentration, and the green color represents a higher concentration of a particular metabolite. The compartments were arranged from left to right as follows: regular watering (RW), drought (DR), resumption of regular watering after drought (RH), flooding (FL), and resumption of regular watering after flooding (DHFL).

Apart from oxaloacetic acid, which was comparatively high in RW plants, all the other metabolites were low. The low concentrations of succinic acid, isocitric acid, acotinic acid, and citric acid in RW plants were similar to those in RH and DHFL plants (Figure 5). Particularly, the RH plants had low concentrations of all the individual TCA cycle intermediate metabolites. For the DHFL plants, the increase in TCA cycle metabolites can be ascribed to high levels of acetyl-CoA, oxaloacetic acid, and malic acid. According to Sweetlove et al. (2010) [70], TCA cycle intermediate metabolites do not have similar flux, and the activities of the different enzymes in the different steps are independent of each other. As such, the suppression of one enzyme within the cycle does not alter the activity of the other enzymes. Therefore, the high concentrations of TCA cycle metabolites in FL and DHFL plants can be attributed to the link between the TCA cycle and other associated pathways, including ammonium assimilation and the biosynthesis of amino acids, nucleotides, and secondary metabolites that contribute to plant stress tolerance [69,70].

Studies showed that changes in malate, citrate, α-ketoglutaric acid, and fumarate during flood stress occurred in many different plants [63,64,66]. The accumulation of succinate is obvious during flood-induced hypoxia conditions since succinate dehydrogenase requires oxygen [52]. Additionally, both drought and flood stresses stimulate the accumulation of citrate in plants [23,64]. In addition, citrate is not only involved in amino acid metabolism but also serves as an antioxidant and intermediate in respiratory metabolisms to generate energy for defense pathways in stress adaptation mechanisms [71]. Furthermore, α-ketoglutaric acid plays an important role in respiration and N assimilation for the biosynthesis of proline, glutamate, glutamine, and arginine. These amino acids function in regulating osmotic potential and act as osmolytes to maintain protein integrity and to mediate water stress tolerance in plants [3,52]. Therefore, like many other plants, the accumulation of these metabolites can be essential for water stress tolerance in Mexican mint plants.

3.3.5. Nucleotide Biosynthetic Pathway Intermediates

Apart from guanine monophosphate (GMP), all the determined nucleotide pathway intermediate metabolites were significantly ($p < 0.05$) influenced by variation in watering regime (Table 1; Supplementary Table S5). Nucleotide metabolism is the most critical cellular component for plant growth and affects several metabolic processes [20]. These nucleotides are essential for information storage and recovery in dividing and expanding tissues. The results revealed that total nucleotide intermediate concentrations were significantly ($p < 0.001$) highest in RH (ca. 229%) and FL (ca. 162%) plants compared to RW plants (Table 1). Total concentrations of nucleotide intermediates in DR, DHFL, and RW plants were not significantly ($p > 0.05$) different and ranged from 267 to 311 nmol/g. Evidence from several studies revealed that nucleotide biosynthesis increased drought tolerance in plants such as Arabidopsis [72], orchids (*Dendrobium* spp.) [73,74], pitaya (*Hylocereus undatus*) [75], drought-tolerant wild wheat (*Triticum boeoticum*) [76], and soybeans [55]. Based on these previous reports, we surmised that nucleotide metabolism would be a critical mechanism for water stress tolerance in Mexican mint. These nucleotides may also allow for the repair and maintenance of water stress-induced cellular damages, which, as a result, might have promoted water stress tolerance in RH and DHFL plants.

Furthermore, the *de novo* biosynthesis of purine is characterized by the formation of adenosine monophosphate (AMP) and GMP [20]. AMPs are synthesized from activated ribose (5-phosphoribosyl-1-pyrophosphate), key amino acids such as glutamine and glycine, and 10-formyl tetrahydrofolate [20]. GMPs are obtained from deamination of AMP or direct transport of inosine 5′-monophosphate (IMP) in the chloroplast [20]. The results of the present study showed that the relative concentrations of the individual nucleotide intermediates, i.e., guanine diphosphate (GDP), GTP, AMP, and adenosine diphosphate (ADP), known to be involved in purine biosynthesis, were significantly ($p < 0.001$) increased in DR plants, except for GMP (Figure 6). Conversely, these metabolites were significantly ($p < 0.05$) reduced in the FL plants, except for GMP, which was highest in both FL and RW plants.

During the *de novo* biosynthesis of pyrimidine, uridine monophosphate (UMP) is typically formed from phosphoribosyl diphosphate (PRPP), aspartate, and carbamoylphosphate to play a critical role in cytosine monophosphate (CMP) biosynthesis [20,22]. The results of the present study indicated that concentrations of UMP, uridine triphosphate (UTP), uridine diphosphate (UDP), and CMP metabolites were significantly ($p < 0.001$) increased in DR plants compared to RW plants (Figure 6). Ultimately, UMP, UDP, and CMP concentrations were increased in FL plants, while UTP concentrations were slightly reduced. Intriguingly, purine and pyrimidine nucleotides may also function as co-substrates for carbohydrate metabolism. For example, ADP-glucose is an activated precursor for starch synthesis, while UDP and UTP are directly involved in the synthesis, degradation, and transportation (i.e., in the form of UDP-glucose or UTP-glucose) of various forms of carbohydrate, including sucrose, sugar components of glycoproteins, and cell wall matrix polysaccharides [22,77]. UDP-glucose acts as a glucosyl donor for the biosynthesis of hormones and secondary metabolites [77]. Furthermore, ADP-glucose and UDP-glucose/UDP-mannose concentrations in the Mexican mint plants were significantly ($p < 0.05$) elevated in FL plants but were not altered in DR plants compared to the RW plants (Figure 6). These findings suggested that carbohydrate metabolism and secondary metabolite syntheses can enhance flood tolerance. This provides a new insight into drought and flooding tolerance in Mexican mint plants.

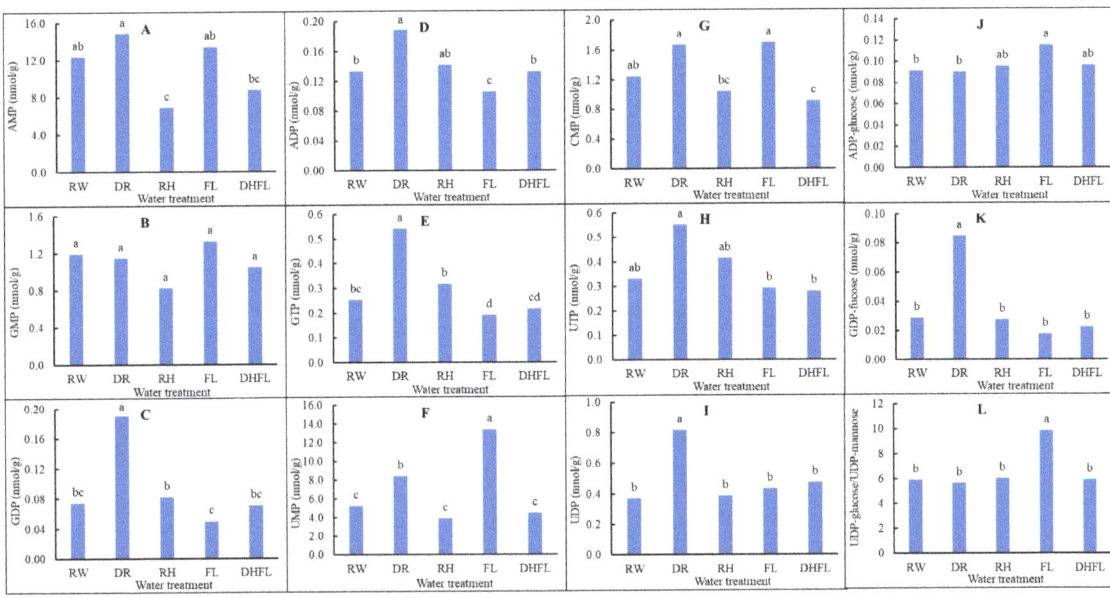

Figure 6. Changes in nucleotide biosynthetic pathway metabolites in Mexican mint (*Plectranthus amboinicus*) plants in response to varying watering regimes (n = 3). (**A**) Adenosine monophosphate (AMP), (**B**) guanine monophosphate (GMP), (**C**) guanine diphosphate (GDP), (**D**) adenosine diphosphate (ADP), (**E**) guanine triphosphate (GTP), (**F**) uridine monophosphate (UMP), (**G**) cytosine monophosphate (CMP), (**H**) uridine triphosphate (UTP), (**I**) uridine diphosphate (UDP), (**J**) ADP-glucose, (**K**) GDP-fucose, and (**L**) UDP-glucose/UDP-mannose. Different alphabetical letters on the bars denote significant differences according to Tukey's honestly significant difference post-test analyses at a significant level of $p < 0.05$.

3.4. Association between the Central Carbon Metabolism Routes

A multivariate analysis using 2-D PCA biplots and the Pearson correlation coefficient (r) was used to further demonstrate the nature of the association amongst the different routes involved in CCM as influenced by variation in watering regime. The PCA biplot in Figure 7 showed a projection of the response variables in the factor space, which explained ca. 96% of the total variations in the dataset. It was found that total metabolites in the Calvin cycle had a significantly ($p < 0.001$) strong positive association with those of the TCA cycle (r = 0.81) and PPP (r = 0.75) intermediate metabolites. Total PPP intermediate metabolites had a significantly ($p < 0.01$) moderately positive association with the TCA cycle intermediate metabolites (r = 0.68), and a negative correlation with total glycolysis intermediate metabolites (r = −0.70; $p < 0.005$; Table 2).

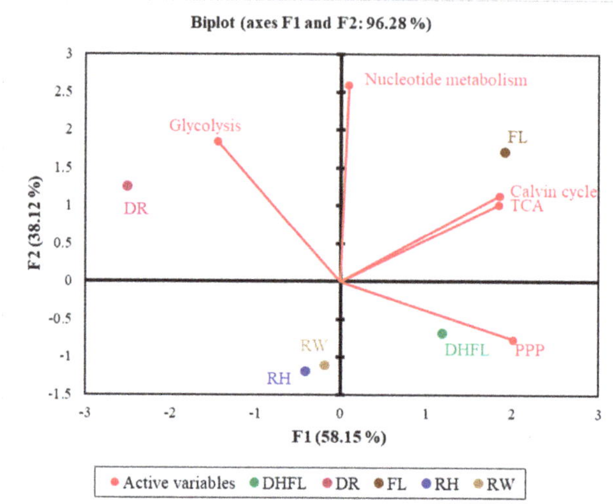

Figure 7. A two-dimensional principal component analysis biplot showing relationships amongst the total metabolites involved in the Calvin cycle, glycolysis, pentose phosphate pathway (PPP), tricarboxylic acid (TCA) cycle, and nucleotide biosynthetic pathway of Mexican mint (*Plectranthus amboinicus*) plants under varying watering regimes. The projection of the variables in the 2D factor space (F1 and F2) explained a total of ca. 96% of the variations in the dataset. Variables that are closely located are not different compared to variables located at a distance within a quadrant or between quadrants (n = 3). Regular watering (RW), drought (DR), resumption of regular watering after drought (RH), flooding (FL), and resumption of regular watering after flooding (DHFL).

Table 2. Pearson correlation coefficients (r) amongst the specific central carbon metabolic pathways in Mexican mint (*Plectranthus amboinicus*) under varying watering regimes at a significance level of $p \leq 0.05$.

	TCA	Glycolysis	Calvin Cycle	PPP
Glycolysis (nmol/g)	r = −0.268 p = 0.335			
Calvin cycle (nmol/g)	r = 0.814 p = 0.000	r = −0.215 p = 0.442		
PPP (nmol/g)	r = 0.684 p = 0.005	r = −0.702 p = 0.004	r = 0.749 p = 0.001	
Nucleotide metabolism (nmol/g)	r = 0.380 p = 0.162	r = 0.645 p = 0.009	r = 0.501 p = 0.057	r = −0.168 p = 0.549

PPP, pentose phosphate pathway; TCA, tricarboxylic acid cycle.

Furthermore, total nucleotide biosynthetic metabolites had a significant ($p < 0.01$) moderately positive correlation with total glycolysis metabolites (r = 0.65) and a moderate non-significant ($p > 0.05$) association (r = 0.50) with total Calvin cycle metabolites (Table 2). Moreover, since the Calvin cycle, PPP, and glycolytic pathways have some metabolites in common, the association amongst the various individual pathway metabolites involved in the CCM routes showed a distinct and significant ($p < 0.05$) positive association (Supplementary Table S6).

The results in Figure 8 confirmed that most of the specific Calvin cycle intermediate metabolites had a significant ($p < 0.05$) positive association with major TCA cycle intermediate metabolites. For instance, Se7P, R5P, and Ru5P concentrations positively correlated with fumaric acid and succinic acid concentrations, whereas RuBP and xylose-5-phosphate (X5P) exhibited a moderate and strong positive association with aconitic acid and citric acid, respectively. Additionally, PPP intermediate metabolites exhibited a significant ($p < 0.05$) positive correlation with most of the TCA cycle metabolites (Figure 8; Supplementary Table S6). In addition to Se7P, R5P, and Ru5P, 6-phosphogluconate exhibited a strong positive association with aconitic acid and citric acid. Glucose had a strong positive association with isocitric acid, aconitic acid, and citric acid, while PA—a key product of glycolysis—had a strong positive association with fumaric acid, succinic acid, α-ketoglutaric acid, isocitric acid, and citric acid (Figure 8; Supplementary Table S6). These associations with respect to Mexican mint plants agree with the report by Rocha et al. [66]. They explained that the products of glucose metabolism feed the TCA cycle for energy generation and the synthesis of other metabolic compounds for other pathways. Additionally, ATP and NADPH, which are essential for energy production in plants during water stress adaptation, had a moderately negative correlation with malic acid, fumaric acid, and succinic acid and a moderately positive association with aconitic acid and citric acid (Figure 8; Supplementary Table S6).

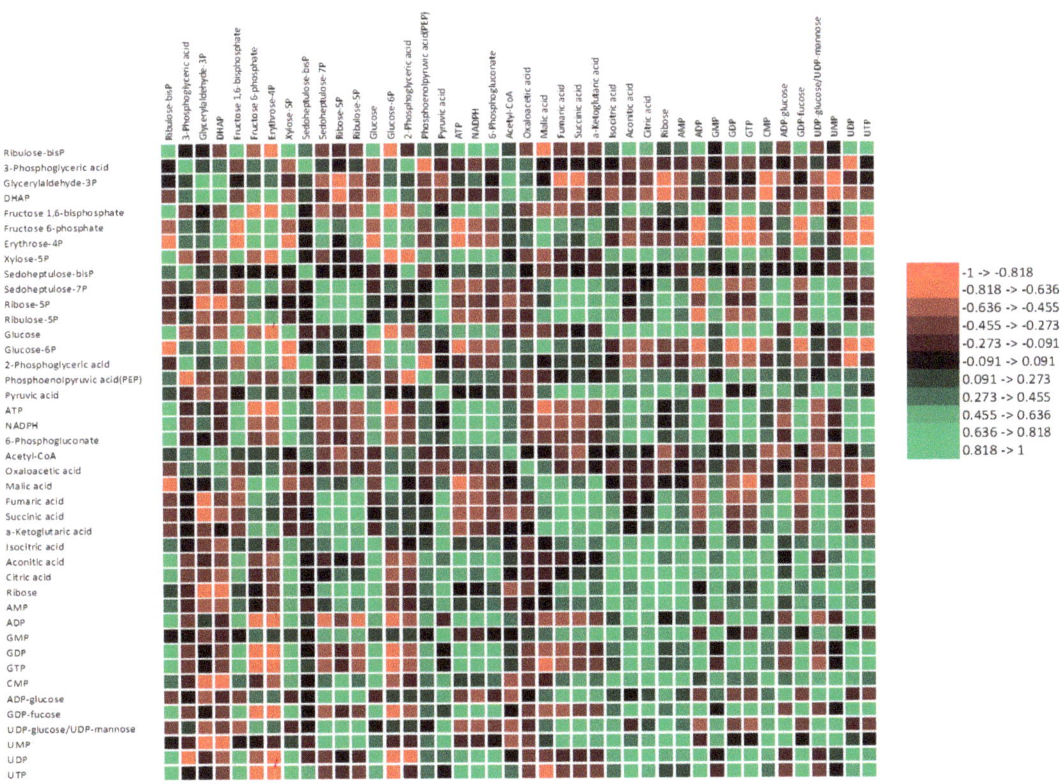

Figure 8. A heat map of the composite correlation matrix between individual metabolites of the central carbon metabolic routes in Mexican mint (*Plectranthus amboinicus*) plants in response to varying watering regimes (n = 3). The red color represents a strong negative association, and the green color represents a strong positive association. Adenosine triphosphate (ATP), adenosine monophosphate (AMP), guanine monophosphate (GMP), guanine diphosphate (GDP), adenosine diphosphate (ADP), guanine triphosphate (GTP), uridine monophosphate (UMP), cytosine monophosphate (CMP), uridine triphosphate (UTP), uridine diphosphate (UDP), ADP-glucose, GDP-fucose, and UDP-glucose/UDP-mannose.

The present study showed moderately positive associations between glucose and AMP, ADP, GDP, GTP, CMP, and UTP, and strong positive associations between glucose and UDP and UDP-glucose (Figure 8). Additionally, PA had moderate associations with ribose, AMP, GMP, CMP, ADP-glucose, and UDP-glucose. Similarly, key individual Calvin cycle intermediate metabolites, i.e., RuBP, Ru5P, and R5P, showed strong positive associations with the different nucleotide biosynthetic metabolites.

Overall, the coordinated association between the various pathway metabolites confirmed that water stress instigated metabolic reprogramming of Mexican mint plants to regulate energy production and the biosynthesis of essential compounds necessary for water stress tolerance and adaptation. Additionally, the PCA biplot showed that flooding had a strong influence on the Calvin cycle and TCA cycle, which suggested the possibility of modulating nucleotide biosynthesis and the PPP pathway. On the contrary, drought stress tends to strongly affect glycolysis and moderately influence nucleotide metabolism, with less influence on the Calvin cycle and TCA cycle. This is consistent with previous studies that revealed stimulation of glucose metabolism in plants during drought as a key strategy for energy production to maintain basic cellular functions and thereby mediate drought tolerance [3,23,46]. Additionally, flood-tolerant plants undergo anaerobic respiration and drive an increase in photosynthetic carbon production via the Calvin cycle [37]. In addition, the function of the TCA cycle is to support ATP synthesis for stress tolerance. Furthermore, dehydration after flooding showed a strong influence on the PPP pathway and a moderate effect on the Calvin cycle and the TCA cycle. Nevertheless, rehydration after drought had only a moderate influence on the PPP pathway and no influence on the other CCM routes.

4. Conclusions

Global climate change and a shift in weather patterns have made water stress one of the major limiting factors for crop growth and productivity worldwide. In this study, we demonstrated that water stress affects CCM routes associated with plant resilience and growth. The determined 68 key metabolites involved in the CCM routes, i.e., the Calvin cycle, glycolysis, tricarboxylic acid cycle, pentose phosphate pathway, and nucleotide biosynthesis, were remarkably influenced by both flooding and drought and their respective reversals. This indicates that the accumulation of intermediate metabolites associated with these CCM routes is crucial for energy production and provides carbon skeletons for further macromolecule biosynthesis during internal metabolic reprogramming for water stress tolerance and adaptation. However, plants undergo a fast recovery process following rehydration of drought plants or dehydration of flooded plants to normal growth conditions with no obvious effect on the CCM routes, although it is impacted. For the first time, we revealed the metabolic alteration of Mexican mint plants under different watering regimes and elucidated the mechanism of its water stress tolerance, which can be extended to other members of the Lamiaceae family. We recommend further analysis using transcriptomic and proteomic approaches to identify genes and proteins that regulate these CCM routes in response to water stress.

Supplementary Materials: The following supporting information can be downloaded at: https://www.mdpi.com/article/10.3390/metabo13040539/s1, Table S1: Calvin cycle metabolite profile of Mexican mint (*Plectranthus amboinicus*) in response to varying water regimes. Regular watering (RW), drought (DR), flooding (FL), resumption of regular watering after flooding (DHFL) or after drought (RH). Values are means of three replications, and different alphabetical letters denote significant differences according to Tukey's honestly significant difference post-test analyses at a significant level of $p < 0.05$. Table S2: Glycolysis metabolite profile of Mexican mint (*Plectranthus amboinicus*) in response to varying water regimes. Regular watering (RW), drought (DR), flooding (FL), resumption of regular watering after flooding (DHFL) or after drought (RH). Values are means of three replications, and different alphabetical letters denote significant differences according to Tukey's honestly significant difference post-test analyses at a significant level of $p < 0.05$. Table S3: Pentose phosphate pathway metabolites profile of Mexican mint (*Plectranthus amboinicus*) in response to varying water regimes. Regular watering (RW), drought (DR), flooding (FL), resumption of regular watering after flooding (DHFL) or after drought (RH). Values are means of three replications, and different alphabetical letters denote significant differences according to Tukey's honestly significant difference post-test analyses at a significant level of $p < 0.05$. Table S4: Tricarboxylic acid cycle metabolites profile of Mexican mint (*Plectranthus amboinicus*) in response to varying water regimes. Regular watering (RW), drought (DR), flooding (FL), resumption of regular watering after flooding (DHFL) or after drought (RH). Values are means of three replications, and different alphabetical letters denote significant differences according to Tukey's honestly significant difference post-test analyses at a significant level of $p < 0.05$. Table S5: Nucleotide biosynthetic molecules of Mexican mint (*Plectranthus amboinicus*) in response to varying water regimes. Regular watering (RW), drought (DR), flooding (FL), resumption of regular watering after flooding (DHFL) or after drought (RH). Values are means of three replications; and different alphabetical letters denote significant differences according to Tukey's honestly significant difference post-test analyses at a significant level of $p < 0.05$. Table S6: Correlation analysis between individual metabolites of the central carbon metabolic routes in Mexican mint (*Plectranthus aboinicus*) plants in response to varying watering regimes.

Author Contributions: Conceptualization, L.A.; formal analysis, R.O. and S.C.; funding acquisition, L.A.; investigation, L.A. and Z.W.; methodology, L.A.; project administration, L.A.; resources, L.A.; supervision, L.A.; validation, R.O., S.C. and L.A.; writing—original draft, R.O., S.C., Z.W. and L.A.; writing—review and editing, L.A. and R.O. All authors have read and agreed to the published version of the manuscript.

Funding: This work was financially supported by the Natural Sciences and Engineering Research Council of Canada (NSERC). Grant #CRDPJ523129–2017.

Institutional Review Board Statement: Not applicable.

Informed Consent Statement: Not applicable.

Data Availability Statement: The data presented in this study are available on request from the corresponding author. The data are not publicly available due to privacy.

Acknowledgments: The lead author wishes to thank Samuel Kwaku Asiedu for his suggestions during the study and the University of Victoria Genome BC (UVic GBC)—Proteomics Centre of The Metabolomics Innovation Centre, Canada, for the metabolomic analysis.

Conflicts of Interest: The authors declare no conflict of interest.

References

1. Vaccaro, S.; Muscolo, A.; Pizzeghello, D.; Spaccini, R.; Piccolo, A.; Nardi, S. Effect of a compost and its water-soluble fractions on key enzymes of nitrogen metabolism in maize seedlings. *J. Agric. Food Chem.* **2009**, *57*, 11267–11276. [CrossRef] [PubMed]
2. Reddy, A.R.; Chaitanya, K.V.; Vivekanandan, M. Drought-induced responses of photosynthesis and antioxidant metabolism in higher plants. *J. Plant Physiol.* **2004**, *161*, 1189–1202. [CrossRef] [PubMed]
3. Guo, R.; Shi, L.; Jiao, Y.; Li, M.; Zhong, X.; Gu, F.; Liu, Q.; Xia, X.; Li, H. Metabolic responses to drought stress in the tissues of drought-tolerant and drought-sensitive wheat genotype seedlings. *AoB Plants* **2018**, *10*, ply016. [CrossRef] [PubMed]
4. Klassen, A.; Faccio, A.T.; Canuto, G.A.; da Cruz, P.L.; Ribeiro, H.C.; Tavares, M.F.; Sussulini, A. Metabolomics: Definitions and significance in systems biology. *Adv. Exp. Med. Biol.* **2017**, *965*, 3–17. [CrossRef]
5. Kumar, M.; Kumar Patel, M.; Kumar, N.; Bajpai, A.B.; Siddique, K.H.M. Metabolomics and molecular approaches reveal drought stress tolerance in plants. *Int. J. Mol. Sci.* **2021**, *22*, 9108. [CrossRef]
6. Moradi, P.; Ford-Lloyd, B.; Pritchard, J. Plant-water responses of different medicinal plant thyme (*Thymus* spp.) species to drought stress condition. *Aust. J. Crop Sci.* **2014**, *8*, 666–673.
7. Cramer, G.R.; Urano, K.; Delrot, S.; Pezzotti, M.; Shinozaki, K. Effects of abiotic stress on plants: A systems biology perspective. *BMC Plant Biol.* **2011**, *11*, 163. [CrossRef]
8. Kleinwächter, M.; Selmar, D. Influencing the Product Quality by Applying Drought Stress During the Cultivation of Medicinal Plants. In *Physiological Mechanisms and Adaptation Strategies in Plants under Changing Environment: Volume 1*; Ahmad, P., Wani, M.R., Eds.; Springer: New York, NY, USA, 2014; pp. 57–73.
9. Stoychev, V.; Simova-Stoilova, L.; Vaseva, I.; Kostadinova, A.; Nenkova, R.; Feller, U.; Demirevska, K. Protein changes and proteolytic degradation in red and white clover plants subjected to waterlogging. *Acta Physiol. Plant.* **2013**, *35*, 1925–1932. [CrossRef]
10. Sharma, S.; Joshi, J.; Kataria, S.; Verma, S.K.; Chatterjee, S.; Jain, M.; Pathak, K.; Rastogi, A.; Brestic, M. Regulation of the Calvin cycle under abiotic stresses: An overview. *Plant Life Chang. Environ.* **2020**, 681–717. [CrossRef]

11. Fernie, A.R.; Carrari, F.; Sweetlove, L.J. Respiratory metabolism: Glycolysis, the TCA cycle and mitochondrial electron transport. *Curr. Opin. Plant Biol.* **2004**, *7*, 254–261. [CrossRef]
12. Timm, S.; Arrivault, S. Regulation of Central Carbon and Amino Acid Metabolism in Plants. *Plants* **2021**, *10*, 430. [CrossRef] [PubMed]
13. Yadav, B.; Jogawat, A.; Rahman, M.S.; Narayan, O.P. Secondary metabolites in the drought stress tolerance of crop plants: A review. *Gene Rep.* **2021**, *23*, 101040. [CrossRef]
14. Fuchs, G.; Berg, I.A. Unfamiliar metabolic links in the central carbon metabolism. *J. Biotechnol.* **2014**, *192 Pt B*, 314–322. [CrossRef]
15. Ataman, M.; Hatzimanikatis, V. lumpGEM: Systematic generation of subnetworks and elementally balanced lumped reactions for the biosynthesis of target metabolites. *PLoS Comput. Biol.* **2017**, *13*, e1005513. [CrossRef] [PubMed]
16. Ge, T.; Yang, J.; Zhou, S.; Wang, Y.; Li, Y.; Tong, X. The Role of the Pentose Phosphate Pathway in Diabetes and Cancer. *Front. Endocrinol.* **2020**, *11*, 365. [CrossRef]
17. Pires, M.V.; Pereira Júnior, A.A.; Medeiros, D.B.; Daloso, D.M.; Pham, P.A.; Barros, K.A.; Engqvist, M.K.M.; Florian, A.; Krahnert, I.; Maurino, V.G. The influence of alternative pathways of respiration that utilize branched–chain amino acids following water shortage in Arabidopsis. *Plant Cell Environ.* **2016**, *39*, 1304–1319. [CrossRef]
18. Fàbregas, N.; Lozano-Elena, F.; Blasco-Escámez, D.; Tohge, T.; Martínez-Andújar, C.; Albacete, A.; Osorio, S.; Bustamante, M.; Riechmann, J.L.; Nomura, T.; et al. Overexpression of the vascular brassinosteroid receptor BRL3 confers drought resistance without penalizing plant growth. *Nat. Commun.* **2018**, *9*, 4680. [CrossRef]
19. Wang, H.; Zhang, M.; Guo, R.; Shi, D.; Liu, B.; Lin, X.; Yang, C. Effects of salt stress on ion balance and nitrogen metabolism of old and young leaves in rice (*Oryza sativa* L.). *BMC Plant Biol.* **2012**, *12*, 1–11. [CrossRef]
20. Witte, C.-P.; Herde, M. Nucleotide Metabolism in Plants. *Plant Physiol.* **2019**, *182*, 63–78. [CrossRef]
21. Stasolla, C.; Katahira, R.; Thorpe, T.A.; Ashihara, H. Purine and pyrimidine nucleotide metabolism in higher plants. *J. Plant Physiol.* **2003**, *160*, 1271–1295. [CrossRef]
22. Zrenner, R.; Stitt, M.; Sonnewald, U.; Boldt, R. Pyrimidine and purine biosynthesis and degradation in plants. *Annu. Rev. Plant Biol.* **2006**, *57*, 805–836. [CrossRef] [PubMed]
23. Ashrafi, M.; Azimi-Moqadam, M.-R.; Moradi, P.; MohseniFard, E.; Shekari, F.; Kompany-Zareh, M. Effect of drought stress on metabolite adjustments in drought tolerant and sensitive thyme. *Plant Physiol. Biochem.* **2018**, *132*, 391–399. [CrossRef] [PubMed]
24. Lukhoba, C.W.; Simmonds, M.S.J.; Paton, A.J. Plectranthus: A review of ethnobotanical uses. *J. Ethnopharmacol.* **2006**, *103*, 1–24. [CrossRef] [PubMed]
25. Paul, M.J. Photosynthetic Carbon Dioxide Fixation. In *Encyclopedia of Biological Chemistry*; Lennarz, W.J., Lane, M.D., Eds.; Elsevier: New York, NY, USA, 2004; pp. 336–341.
26. Han, J.; Gagnon, S.; Eckle, T.; Borchers, C.H. Metabolomic analysis of key central carbon metabolism carboxylic acids as their 3-nitrophenylhydrazones by UPLC/ESI-MS. *Electrophoresis* **2013**, *34*, 2891–2900. [CrossRef]
27. Xiong, Q.; Deng, Y.; Zhong, L.; He, H.; Chen, X. Effects of drought-flood abrupt alternation on yield and physiological characteristics of rice. *Int. J. Agric. Biol.* **2018**, *20*, 1107–1116.
28. Wang, S.; Alseekh, S.; Fernie, A.R.; Luo, J. The structure and function of major plant metabolite modifications. *Mol. Plant* **2019**, *12*, 899–919. [CrossRef]
29. Zhang, T.; Zhang, A.; Qiu, S.; Yang, S.; Wang, X. Current trends and innovations in bioanalytical techniques of metabolomics. *Crit. Rev. Anal. Chem.* **2016**, *346*, 342–351. [CrossRef]
30. Brito, C.; Dinis, L.-T.; Moutinho-Pereira, J.; Correia, C.M. Drought stress effects and olive tree acclimation under a changing climate. *Plants* **2019**, *8*, 232. [CrossRef]
31. Abdallah, M.B.; Trupiano, D.; Polzella, A.; De Zio, E.; Sassi, M.; Scaloni, A.; Zarrouk, M.; Youssef, N.B.; Scippa, G.S. Unraveling physiological, biochemical and molecular mechanisms involved in olive (*Olea europaea* L. cv. Chétoui) tolerance to drought and salt stresses. *J. Plant Physiol.* **2018**, *220*, 83–95. [CrossRef]
32. Todaka, D.; Zhao, Y.; Yoshida, T.; Kudo, M.; Kidokoro, S.; Mizoi, J.; Kodaira, K.S.; Takebayashi, Y.; Kojima, M.; Sakakibara, H. Temporal and spatial changes in gene expression, metabolite accumulation and phytohormone content in rice seedlings grown under drought stress conditions. *Plant J.* **2017**, *90*, 61–78. [CrossRef]
33. Wingler, A.; Lea, P.J.; Quick, W.P.; Leegood, R.C. Photorespiration: Metabolic pathways and their role in stress protection. *Philos. Trans. R. Soc. London. Ser. B Biol. Sci.* **2000**, *355*, 1517–1529. [CrossRef] [PubMed]
34. Chen, W.; Yao, Q.; Patil, G.B.; Agarwal, G.; Deshmukh, R.K.; Lin, L.; Wang, B.; Wang, Y.; Prince, S.J.; Song, L. Identification and comparative analysis of differential gene expression in soybean leaf tissue under drought and flooding stress revealed by RNA-Seq. *Front. Plant Sci.* **2016**, *7*, 1044. [CrossRef]
35. Irfan, M.; Hayat, S.; Hayat, Q.; Afroz, S.; Ahmad, A. Physiological and biochemical changes in plants under waterlogging. *Protoplasma* **2010**, *241*, 3–17. [CrossRef] [PubMed]
36. Ren, B.; Zhang, J.; Dong, S.; Liu, P.; Zhao, B. Effects of waterlogging on leaf mesophyll cell ultrastructure and photosynthetic characteristics of summer maize. *PLoS ONE* **2016**, *11*, e0161424. [CrossRef] [PubMed]
37. Pan, D.; Wang, L.; Tan, F.; Lu, S.; Lv, X.; Zaynab, M.; Cheng, C.-L.; Abubakar, Y.S.; Chen, S.; Chen, W. Phosphoproteomics unveils stable energy supply as key to flooding tolerance in *Kandelia candel*. *J. Proteom.* **2018**, *176*, 1–12. [CrossRef]
38. Parry, M.A.J.; Andralojc, P.J.; Scales, J.C.; Salvucci, M.E.; Carmo-Silva, A.E.; Alonso, H.; Whitney, S.M. Rubisco activity and regulation as targets for crop improvement. *J. Exp. Bot.* **2013**, *64*, 717–730. [CrossRef]

39. Li, Z.-G. Methylglyoxal: A Novel Signaling Molecule in Plant Responses to Abiotic Stresses. In *Plant Signaling Molecules*; Khan, M.I.R., Reddy, P.S., Ferrante, A., Khan, N.A., Eds.; Woodhead Publishing Ltd.: Sawston, UK, 2019; pp. 219–233.
40. Hossain, M.A.; Burritt, D.J.; Fujita, M. Cross-Stress Tolerance in Plants: Molecular Mechanisms and Possible Involvement of Reactive Oxygen Species and Methylglyoxal Detoxification Systems. In *Abiotic Stress Response in Plants*; Tuteja, N., Gill, S.S., Eds.; John Wiley & Sons, Inc.: Hoboken, NJ, USA, 2016; pp. 327–380.
41. Li, Z.-G. Methylglyoxal and Glyoxalase System in Plants: Old Players, New Concepts. *Bot. Rev.* **2016**, *82*, 183–203. [CrossRef]
42. Singh, P.; Dhaka, N. Glyoxalase system and salinity stress in plants. In *Managing Salt Tolerance in Plants: Molecular and Genomic Perspectives*; CRC Press: London, UK, 2016; pp. 173–185.
43. Polacik, K.A.; Maricle, B.R. Effects of flooding on photosynthesis and root respiration in saltcedar (*Tamarix ramosissima*), an invasive riparian shrub. *Environ. Exp. Bot.* **2013**, *89*, 19–27. [CrossRef]
44. Hossain, Z.; Komatsu, S. Potentiality of Soybean Proteomics in Untying the Mechanism of Flood and Drought Stress Tolerance. *Proteomes* **2014**, *2*, 107–127. [CrossRef]
45. Tamang, B.G.; Li, S.; Rajasundaram, D.; Lamichhane, S.; Fukao, T. Overlapping and stress-specific transcriptomic and hormonal responses to flooding and drought in soybean. *Plant J.* **2021**, *107*, 100–117. [CrossRef]
46. de Miguel, M.; Guevara, M.Á.; Sánchez-Gómez, D.; De María, N.; Díaz, L.M.; Mancha, J.A.; De Simón, B.F.; Cadahía, E.; Desai, N.; Aranda, I. Organ-specific metabolic responses to drought in *Pinus pinaster* Ait. *Plant Physiol. Biochem.* **2016**, *102*, 17–26. [CrossRef] [PubMed]
47. Sanchez, D.H.; Schwabe, F.; Erban, A.; Udvardi, M.K.; Kopka, J. Comparative metabolomics of drought acclimation in model and forage legumes. *Plant Cell Environ.* **2012**, *35*, 136–149. [CrossRef] [PubMed]
48. Bailey-Serres, J.; Voesenek, L. Flooding stress: Acclimations and genetic diversity. *Annu. Rev. Plant Biol.* **2008**, *59*, 313. [CrossRef] [PubMed]
49. Bailey-Serres, J.; Fukao, T.; Gibbs, D.J.; Holdsworth, M.J.; Lee, S.C.; Licausi, F.; Perata, P.; Voesenek, L.A.C.J.; van Dongen, J.T. Making sense of low oxygen sensing. *Trends Plant Sci.* **2012**, *17*, 129–138. [CrossRef] [PubMed]
50. Koch, K. Sucrose metabolism: Regulatory mechanisms and pivotal roles in sugar sensing and plant development. *Curr. Opin. Plant Biol.* **2004**, *7*, 235–246. [CrossRef]
51. Good, A.G.; Zaplachinski, S.T. The effects of drought stress on free amino acid accumulation and protein synthesis in *Brassica napus*. *Physiol. Plant.* **1994**, *90*, 9–14. [CrossRef]
52. Xu, Y.; Fu, X. Reprogramming of Plant Central Metabolism in Response to Abiotic Stresses: A Metabolomics View. *Int. J. Mol. Sci.* **2022**, *23*, 5716. [CrossRef]
53. Ricoult, C.; Echeverria, L.O.; Cliquet, J.-B.; Limami, A.M. Characterization of alanine aminotransferase (AlaAT) multigene family and hypoxic response in young seedlings of the model legume *Medicago truncatula*. *J. Exp. Bot.* **2006**, *57*, 3079–3089. [CrossRef]
54. Peng, Y.; Zhou, Z.; Zhang, Z.; Yu, X.; Zhang, X.; Du, K. Molecular and physiological responses in roots of two full-sib poplars uncover mechanisms that contribute to differences in partial submergence tolerance. *Sci. Rep.* **2018**, *8*, 1–15. [CrossRef]
55. Das, A.; Rushton, P.J.; Rohila, J.S. Metabolomic Profiling of Soybeans (*Glycine max* L.) Reveals the Importance of Sugar and Nitrogen Metabolism under Drought and Heat Stress. *Plants* **2017**, *6*, 21. [CrossRef]
56. Siddhartha, S.; Asha, A.; Pramod, K.S. Regulation and properties of glucose-6-phosphate dehydrogenase: A review. *Int. J. Plant Physiol. Biochem.* **2012**, *4*, 1–19. [CrossRef]
57. Liu, J.; Wang, X.; Hu, Y.; Hu, W.; Bi, Y. Glucose-6-phosphate dehydrogenase plays a pivotal role in tolerance to drought stress in soybean roots. *Plant Cell Rep.* **2013**, *32*, 415–429. [CrossRef]
58. Juhnke, H.; Krems, B.; Kötter, P.; Entian, K.D. Mutants that show increased sensitivity to hydrogen peroxide reveal an important role for the pentose phosphate pathway in protection of yeast against oxidative stress. *Mol. Gen. Genet. MGG* **1996**, *252*, 456–464. [CrossRef] [PubMed]
59. Huang, J.; Zhang, H.; Wang, J.; Yang, J. Molecular cloning and characterization of rice 6-phosphogluconate dehydrogenase gene that is up-regulated by salt stressa. *Mol. Biol. Rep.* **2003**, *30*, 223–227. [CrossRef] [PubMed]
60. Kruger, N.J.; von Schaewen, A. The oxidative pentose phosphate pathway: Structure and organisation. *Curr. Opin. Plant Biol.* **2003**, *6*, 236–246. [CrossRef]
61. Yang, L.; Fountain, J.C.; Ji, P.; Ni, X.; Chen, S.; Lee, R.D.; Kemerait, R.C.; Guo, B. Deciphering drought-induced metabolic responses and regulation in developing maize kernels. *Plant Biotechnol. J.* **2018**, *16*, 1616–1628. [CrossRef]
62. You, J.; Zhang, Y.; Liu, A.; Li, D.; Wang, X.; Dossa, K.; Zhou, R.; Yu, J.; Zhang, Y.; Wang, L.; et al. Transcriptomic and metabolomic profiling of drought-tolerant and susceptible sesame genotypes in response to drought stress. *BMC Plant Biol.* **2019**, *19*, 267. [CrossRef]
63. Komatsu, S.; Yamamoto, A.; Nakamura, T.; Nouri, M.-Z.; Nanjo, Y.; Nishizawa, K.; Furukawa, K. Comprehensive analysis of mitochondria in roots and hypocotyls of soybean under flooding stress using proteomics and metabolomics techniques. *J. Proteome Res.* **2011**, *10*, 3993–4004. [CrossRef]
64. Wang, X.; Zhu, W.; Hashiguchi, A.; Nishimura, M.; Tian, J.; Komatsu, S. Metabolic profiles of flooding-tolerant mechanism in early-stage soybean responding to initial stress. *Plant Mol. Biol.* **2017**, *94*, 669–685. [CrossRef]
65. Liu, Y.-l.; Zheng, H.-l. Physiological and Proteomic Analyses of Two Acanthus Species to Tidal Flooding Stress. *Int. J. Mol. Sci.* **2021**, *22*, 1055. [CrossRef]

66. Rocha, M.; Licausi, F.; Araújo, W.L.; Nunes-Nesi, A.; Sodek, L.; Fernie, A.R.; van Dongen, J.T. Glycolysis and the Tricarboxylic Acid Cycle Are Linked by Alanine Aminotransferase during Hypoxia Induced by Waterlogging of *Lotus japonicus*. *Plant Physiol.* **2010**, *152*, 1501–1513. [CrossRef]
67. Locke, A.M.; Barding, G.A., Jr.; Sathnur, S.; Larive, C.K.; Bailey-Serres, J. Rice SUB1A constrains remodelling of the transcriptome and metabolome during submergence to facilitate post-submergence recovery. *Plant Cell Environ.* **2018**, *41*, 721–736. [CrossRef]
68. Menezes-Silva, P.E.; Sanglard, L.M.V.P.; Ávila, R.T.; Morais, L.E.; Martins, S.C.V.; Nobres, P.; Patreze, C.M.; Ferreira, M.A.; Araújo, W.L.; Fernie, A.R. Photosynthetic and metabolic acclimation to repeated drought events play key roles in drought tolerance in coffee. *J. Exp. Bot.* **2017**, *68*, 4309–4322. [CrossRef]
69. Jacques, C.; Salon, C.; Barnard, R.L.; Vernoud, V.; Prudent, M. Drought Stress Memory at the Plant Cycle Level: A Review. *Plants* **2021**, *10*, 1873. [CrossRef]
70. Sweetlove, L.J.; Beard, K.F.M.; Nunes-Nesi, A.; Fernie, A.R.; Ratcliffe, R.G. Not just a circle: Flux modes in the plant TCA cycle. *Trends Plant Sci.* **2010**, *15*, 462–470. [CrossRef]
71. Zhao, Z.; Hu, L.; Hu, T.; Fu, J. Differential metabolic responses of two tall fescue genotypes to heat stress. *Acta Prataculturae Sin.* **2015**, *24*, 58–69.
72. Watanabe, S.; Kounosu, Y.; Shimada, H.; Sakamoto, A. Arabidopsis xanthine dehydrogenase mutants defective in purine degradation show a compromised protective response to drought and oxidative stress. *Plant Biotechnol.* **2014**, *31*, 173–178. [CrossRef]
73. Zhao, D.; Shi, Y.; Senthilkumar, H.A.; Qiao, Q.; Wang, Q.; Shen, Y.; Hu, G. Enriched networks 'nucleoside/nucleotide and ribonucleoside/ribonucleotide metabolic processes' and 'response to stimulus' potentially conferred to drought adaptation of the epiphytic orchid *Dendrobium wangliangii*. *Physiol. Mol. Biol. Plants* **2019**, *25*, 31–45. [CrossRef] [PubMed]
74. Zhang, C.; Chen, J.; Huang, W.; Song, X.; Niu, J. Transcriptomics and Metabolomics Reveal Purine and Phenylpropanoid Metabolism Response to Drought Stress in *Dendrobium sinense*, an Endemic Orchid Species in Hainan Island. *Front. Genet.* **2021**, *12*, 692702. [CrossRef]
75. Fan, Q.-J.; Yan, F.-X.; Qiao, G.; Zhang, B.-X.; Wen, X.-P. Identification of differentially-expressed genes potentially implicated in drought response in pitaya (*Hylocereus undatus*) by suppression subtractive hybridization and cDNA microarray analysis. *Gene* **2014**, *533*, 322–331. [CrossRef]
76. Liu, H.; Sultan, M.A.R.F.; Liu, X.L.; Zhang, J.; Yu, F.; Zhao, H.X. Physiological and comparative proteomic analysis reveals different drought responses in roots and leaves of drought-tolerant wild wheat (*Triticum boeoticum*). *PLoS ONE* **2015**, *10*, e0121852. [CrossRef] [PubMed]
77. Lim, E.K.; Bowles, D.J. A class of plant glycosyltransferases involved in cellular homeostasis. *EMBO J.* **2004**, *23*, 2915–2922. [CrossRef] [PubMed]

metabolites

MDPI

Article

Data-Driven Characterization of Metabolome Reprogramming during Early Development of Sorghum Seedlings

Ian A. Dubery *, Lerato P. Nephali, Fidele Tugizimana and Paul A. Steenkamp

Research Centre for Plant Metabolomics, Department of Biochemistry, University of Johannesburg, P.O. Box 524, Auckland Park 2006, South Africa; 201495571@student.uj.ac.za (L.P.N.); ftugizimana@uj.ac.za (F.T.); psteenkamp@uj.ac.za (P.A.S.)
* Correspondence: idubery@uj.ac.za

Abstract: Specialized metabolites are produced via discrete metabolic pathways. These small molecules play significant roles in plant growth and development, as well as defense against environmental stresses. These include damping off or seedling blight at a post-emergence stage. Targeted metabolomics was followed to gain insights into metabolome changes characteristic of different developmental stages of sorghum seedlings. Metabolites were extracted from leaves at seven time points post-germination and analyzed using ultra-high performance liquid chromatography coupled to mass spectrometry. Multivariate statistical analysis combined with chemometric tools, such as principal component analysis, hierarchical clustering analysis, and orthogonal partial least squares–discriminant analysis, were applied for data exploration and to reduce data dimensionality as well as for the selection of potential discriminant biomarkers. Changes in metabolome patterns of the seedlings were analyzed in the early, middle, and late stages of growth (7, 14, and 29 days post-germination). The metabolite classes were amino acids, organic acids, lipids, cyanogenic glycosides, hormones, hydroxycinnamic acid derivatives, and flavonoids, with the latter representing the largest class of metabolites. In general, the metabolite content showed an increase with the progression of the plant growth stages. Most of the differential metabolites were derived from tryptophan and phenylalanine, which contribute to innate immune defenses as well as growth. Quantitative analysis identified a correlation of apigenin flavone derivatives with growth stage. Data-driven investigations of these metabolomes provided new insights into the developmental dynamics that occur in seedlings to limit post-germination mortality.

Keywords: developmental stage; metabolome; multivariate data analysis; specialized metabolite; *Sorghum bicolor*

1. Introduction

Sorghum (*Sorghum bicolor* (L.) Moench) is a food and staple crop that is indigenous to the African continent, of particular use as a grain crop in arid areas. Sorghum is cultivated for either the production of bioenergy, animal feed, and/or human consumption in over 30 countries; therefore, the production of this crop plays a significant role in the global economy, as well as in alleviating food insecurity and unemployment [1]. Sorghum seedlings are very delicate during the emergence period and extremely vulnerable to soil-borne pathogens under suboptimal growing conditions. Many fungal pathogens that are common soil inhabitants (e.g., *Rhizoctonia*, *Fusarium*, *Sclerotinia*, *Verticillium*, and *Pythium* spp.) can cause pre- and post-emergence damping-off or seedling blight. Affected seedlings may show yellowing, wilting, and death of leaves, with the roots of diseased plants discolored and rotten (https://infonet-biovision.org/plant_pests/damping-diseases, accessed on 15 December 2023). Some soil-borne diseases have been a serious problem for many decades and are responsible for restrictions on agricultural yield. Thus far, no biological control strategies have been developed [2].

The ability of a plant to overcome attempted pathogen attacks determines the level of resistance thereof. Small-molecule metabolites are closely linked to the phenotypic characteristics of a plant or cultivar. The phenotypically observable changes in the developmental trajectory of sorghum seedlings reflect the underlying metabolomic reconfigurations. Generally, the type and concentration(s) of specialized metabolites are determined by the species, genotype, physiology, developmental stage, and environmental factors during growth [3,4]. Sorghum contains various specialized phytochemicals that contribute to the nutritional properties and the development of the crop. The most predominant metabolites in sorghum plants are phenolic compounds, which include flavonoids, tannins, anthocyanins, and cinnamic acids [5]. Flavonoids and phenolic acids support crop growth in extreme environmental conditions and provide the plant with adaptive coping mechanisms to deal with both abiotic and biotic stresses. These phenolic compounds have associated antioxidant properties, known to reduce oxidative stress [6], as well as antimicrobial properties [7–9].

To descriptively understand developmental changes of sorghum, 'omics' sciences, such as metabolomics, have evolved to be indispensable in interrogating cellular biochemistry. As such, it contributes to a comprehensive characterization of the metabolome and cellular dynamics of the biological system under consideration [10,11]. Such studies contribute towards improving breeding strategies (correlating agronomical traits to a metabolic phenotype), creating stress-resilient crops, and increasing crop quality and yield [9,11]. Thus, reported herein is a liquid chromatography–mass spectrometry (LC-MS)-based metabolomics study to elucidate metabolic changes of sorghum seedlings over the early growth period following germination. Chemometrics methods, such as principal component analysis (PCA) and orthogonal partial least squares–discriminant analysis (OPLS-DA), were applied to mine and interpret the generated metabolomics datasets, elucidating differential metabolic profiles at different stages, i.e., time points post-germination. Such insights would point to the dynamics of seedling metabolism, also revealing possible biochemical events that are involved in protection and adaptation at this early growth stage of the plants.

2. Materials and Methods

2.1. Sorghum Seedling Cultivation

Seeds from the *Sorghum bicolor* cv. NS 5511, a red/bitter seed variety [9], were obtained from a commercial seed supplier, (Agricol, Pretoria, South Africa), and cultivated in germination mix soil (Culterra, Muldersdrift, South Africa). The seeds were sown in trays (23 × 36 × 6 cm) under controlled greenhouse conditions: a light/dark cycle of 12 h/12 h, an average light intensity of 85 μmol/m^2/s, and the temperature regulated to between 22 and 24 °C. The seedlings were harvested at 7, 11, 14, 18, 22, 25, and 29 days post-germination (d.p.g.). For data analysis, days 7, 14, and 29 were designated as corresponding to 'early', 'middle', and 'late' developmental stages, respectively. The experimental design included three independent biological replicates of each time point. The harvested leaves were weighed, snap-frozen to quench metabolic activity, and stored at −80 °C until extraction (Supplementary Figure S1A,B).

2.2. Metabolite Extraction and Pre-Analytical Sample Preparation

Metabolites were extracted as previously described [9]. Briefly, frozen leaf tissue was mixed with a cold extraction solvent (80% aqueous methanol) in a ratio of 1:15 (*m*/*v*). The mixture was homogenized using an Ultra-Turrax homogenizer (CAT Scientific, Berlin, Germany), followed by sonication for 15 s with a probe sonicator (Bandelin Sonopuls, Berlin, Germany) set at 55% power. Homogenates were centrifuged at 5000× *g* and 4 °C for 25 min. The supernatants of each sample were then concentrated by evaporation under vacuum to 1 mL using a rotary evaporator set at 55 °C. The 1 mL extracts from each sample were further evaporated to complete dryness with a speed vacuum concentrator (Eppendorf, Merck, Johannesburg, South Africa) set at 45 °C. The final step of sample preparation consisted of resuspending the dried extracts in 50% LC-grade methanol (Romil,

Cambridge, UK) in a 1:10 m/v ratio. This was followed by filtering samples through 0.22 µm nylon syringe filters into glass chromatography vials fitted with 500 µL inserts. The filtered extracts were capped and kept at −20 °C until analysis.

2.3. Ultra-High Performance Liquid Chromatography (UHPLC) Coupled to High-Definition Mass Spectrometry (MS) and Data Processing

Analyses were performed on a Waters Acquity UHPLC connected in tandem to a SYNAPT G1 Q-TOF mass spectrometer via an electrospray ionization (ESI) interface and operated with MassLynxTM software (ver. 4.1, Waters Corporation, Manchester, UK). Sample extracts were chromatographically separated on a reverse-phase C18 column (150 mm × 2.1 mm × 1.8 µm—HSS T3, Waters Corporation, Milford, MA, USA) at 60 °C. The mobile phase consisted of 0.1% formic acid in MilliQ water (solvent A) and 0.1% formic acid in acetonitrile (Romil, Cambridge, UK) (solvent B) with a flow rate of 0.4 mL/min. Gradient elution was used, and the initial conditions were 2% B and maintained for 1 min. The gradient was ramped up to 95% B at 15 min and maintained for 2 min, and then changed to the initial conditions at 18 min, followed by a 2 min equilibration time of the column. The total chromatographic run time was 20 min, and the injection volume was 2 µL. Each sample, originating from three independent biological replicates, was analyzed in triplicate (n = 9) in both positive and negative ESI modes. Sample acquisition was randomized, and the quality control (QC) sample used to monitor the performance and stability of the UHPLC-MS system was repeatedly injected to evaluate any analytical variability. The conditions were set as follows: capillary voltage of 2.5 kV, sampling cone at 30 V, extraction cone at 4 V, cone gas flow of 50 L/h, desolvation gas flow of 550 L/h, source temperature at 120 °C, desolvation temperature at 450 °C, scan time of 0.1 s, and mass range of 100–1000 Da. Leucine enkephalin (50 pg/mL, $[M+H]^+$ = 556.2771 Da and $[M–H]^-$ = 554.2615 Da) was used as a reference calibrant at a flow rate of 0.1 mL/min. It was sampled every 15 s and produced an average intensity of 350 counts/scan in centroid mode. The mass accuracy window was 0.5 Da, with a typical mass accuracy ranging from 1 to 3 mDa.

In addition, a data-independent acquisition (DIA) method, namely MS^E, was applied; the MS analyses were set to carry out non-fragmented as well as five fragmenting experiments simultaneously by applying alternating collision energy of 0 eV (unfragmented) and from 10 to 50 eV (fragmented). This was performed to generate molecular fragmentation data for downstream structural elucidation required for compound annotation or identification.

Pre-processing of raw MS data was performed using the MarkerLynx^{TM} application manager for MassLynx^{TM} XS software version 4.1 (Waters Corporation, Manchester, UK), for detection and alignment of peaks, as well as cleaning of data matrices for reduced noise and redundancy.

2.4. Metabolite Annotation

Annotation of spectral features was based upon physicochemical properties and/or spectral similarity with public/commercial spectral libraries and according to level 2 of the Metabolomics Standards Initiative [12]. Based on accurate mass determinations from the UHPLC-MS analysis, a potential empirical formula was calculated for each peak of interest using the m/z values and searched against the databases such as PubChem (https://pubchem.ncbi.nlm.nih.gov/, accessed on 30 June 2023), the Dictionary of Natural Products (http://dnp.chemnetbase.com/faces/chemical/ChemicalSearch.xhtml, accessed on 31 May 2023), MS-DIAL (Mass Spectrometry-Data Independent Analysis software, http://prime.psc.riken.jp, accessed on 16 July 2023, version 4.9.221218) and ChemSpider (http://www.chemspider.com/, accessed on 16 July 2023), also taking possible adduct formation into account. The chemical structures were confirmed by inspecting the MSE information derived from the MS analyses at the five different fragmentation conditions.

In addition, the generated data matrices were also processed using the Taverna workbench (www.taverna.org.uk, accessed on 5 January 2019) for PUTMEDID_LCMS identification of metabolite workflows by correlation analysis, annotation of metabolic features, and putative identification of metabolites, as previously described [10,13]. For the Taverna workbench analysis, data matrices were formatted from MarkerLynx-based data processing. The Taverna Metabolite ID procedure consists of three key workflows: (i) Pearson-based correlation analysis (List CorrData), (ii) metabolic feature annotation (annotate Massmatch), permitting the grouping of ion peaks with comparable properties like Rt, and annotating features with the type of m/z ion (molecular ion, isotope, possible adducts, etc.) assumed to be derived from the same metabolite. The elemental composition/molecular formula (MF) of each m/z ion was then computed, and (iii) metabolite annotation (matchMF-MF) of the computed MF was automatically compared and matched to the MF from a pre-defined reference list of sorghum metabolites [9,10,14].

2.5. *Visualization and Comparison of Annotated Metabolite Trends*

A triangle/ternary plot, constructed with Microsoft Excel, was used for the overall comparison of the annotated compounds in the analyzed samples [15]. Triangle plots are graphical representations of variables that sum to a constant (100%), represented within a two-dimensional triangle. The original data were normalized to 100% and transformed into X and Y coordinates, which were then plotted on a scatter plot with coordinates for a triangle. Heatmaps support the visualization of multidimensional datasets and identify metabolic patterns under similar experimental conditions. In addition, heat maps can be used to locate hidden groups among identified metabolites and associations between experimental groups and metabolic changes [16]. Following annotation of the discriminant features, heat maps were constructed for the corresponding metabolites using the MetaboAnalyst bioinformatics tool suite (version 4.0; http://www.metaboanalyst.ca/, accessed on 3 December 2022) [17]. Average peak intensities (n = 9) were used to construct heat maps illustrating differences in the relative concentrations of the selected analytes from the different groups.

2.6. *Data Mining, Multivariate Data Analysis, and Statistical Modeling*

For data mining and multivariate data statistical analysis (MVDA), annotated metabolites (43) were further analyzed using SIMCA 15 (Soft Independent Modelling of Class Analogy, including the 'omics' skin) (Sartorius, Stedim Data Analytics AB, Umeå, Sweden) and MetaboAnalyst 4.0 (http://www.metaboanalyst.ca/, accessed on 23 March 2022). Such analyses included data exploration and clustering. Before MVDA and computation of chemometric models, log transformation and Pareto scaling were performed on the data for variable normalization. As unsupervised methods, principal component analysis (PCA) was applied to reduce the dimensionality of the data and to obtain an overview of the metabolic data, general clustering, and trends. Hierarchical cluster analysis (HiCA) was used to analyze the natural structure and patterns within the datasets. The information derived from these unsupervised methods was used to obtain more insights by applying a supervised method, orthogonal partial least squares–discriminant analysis (OPLS-DA), as a binary classification method within a reduced dimensional space. OPLS-DA also identifies discriminant molecules specific to the different sample group classifications. Here, the comparisons included the 7 vs. 14 d, 7 vs. 29 d, and 14 vs. 29 d groups. OPLS-DA models were validated using various multivariate statistical tools and included explained variation (R^2) and predictive ability (Q^2) metrics, the analysis of variance testing of cross-validated predictive residuals (CV-ANOVA, p-value < 0.05 as a cut-off), and response permutation tests (with n = 100) [18], as described in the legends to the figures. These MVDA models were constructed for comparison of the seedling sample datasets for all the time points post-germination. OPLS-DA S (loadings/scatter) plots were used to identify m/z features or variables with both high correlation and covariation, [p(corr) $\geq$ 0.5, ≤ -0.5 and (p1) $\geq$ 0.1, ≤ -0.1]. From the OPLS-DA analyses, variable importance in projection (VIP) plots were used to identify the most significantly altered metabolites extracted from the OPLS-DA models to explore their potential biological significance. VIP values > 1 were used as the cut-off for statistical significance and to avoid possible bias in feature selection.

2.7. Metabolomics Pathway Analysis and Network Correlation Analyses

The identified significant metabolites (with their respective Kyoto Encyclopedia of Genes and Genomes (KEGG; https://www.genome.jp/kegg/, accessed on 23 May 2022) identifiers) were uploaded into the MetPA tool for identification, analysis, and visualization of the affected metabolic pathways (MetaboAnalyst 4.0 (http://www.metaboanalyst.ca/, accessed on 10 October 2022). MetPA performs pathway topological analysis, and the possible biological roles can be inferred/evaluated through enrichment analysis [14,18]. The pathway analysis algorithms specified for over-representation analysis were the hypergeometric test, and for pathway topology analysis, it was the relative betweenness centrality [10]. The global significance of a pathway enrichment is estimated by ranking the p-value from real data among the p-values from permutation data to adjust for type I error [18]. MetPA-computed metabolic pathway analysis generates a visual representation of information showing all matched pathways according to the log p-values and impact scores, as shown in Table S2. As a complementary approach, the interconnectedness of the active pathways was modeled using KEGG MAPPER (https://www.genome.jp/kegg/mapper.html, accessed on 24 May 2022) by uploading KEGG identifiers of the annotated metabolites via a searcher pathway option, where compounds are searched against KEGG pathway maps. Statistical analyses are also used to describe these pathways by the p-values and false discovery rate (FDR) of the individual metabolites [19]. The p-value was set at <0.1 and the FDR cut-off was <0.5.

Network correlation analyses were developed to examine direct biochemical associations. Here, assigned KEGG identifiers of each annotated metabolite were uploaded on the KEGG mapping tool (https://www.genome.jp/kegg/tool/map_pathway1.html, accessed on 24 May 2022), using the organism-specific search mode for *Arabidopsis thaliana*. The network was visualized using the Cytoscape version 2.8.2 tool (https://cytoscape.org/, accessed on 28 May 2022), and network characteristic mapping reflected chemometric modeling information via network edge (or link) and node (or vertex) features [20]. The centrality parameter is a quantitative measure of the position of a node relative to the other nodes, commonly applied in the estimation of a node's relative significance in network organization. Considering that metabolic networks are directed graphs, the significance role played by a compound is determined using 'relative betweenness centrality' and 'out degree centrality' in MetPA. The pathway impact is measured as the collection of the significance measures of the corresponding metabolites normalized by the sum of the significance measures of the total metabolites in each pathway.

2.8. Multiple Reaction Monitoring (MRM) UHPLC-MS/MS Method for the Quantification of Targeted Defense-Related Flavonoids

Pure, authentic standards were obtained from Chengdu Biopurify Phytochemicals (Chengdu, Sichuan, China), while D-fluorophenylalanine was obtained from Sigma-Aldrich Merck (Johannesburg, South Africa). To determine the selectivity of the method in the separation of the targeted flavonoids, 100 ppm stock solutions of the pure standards (apigenin, apigetrin, luteolin, luteoloside, naringenin, vicenin-2, vicenin-3, vitexin, isovitexin, and internal standard D-fluorophenylalanine (Table S3)) were separated using reverse phase chromatography followed by multiple reaction monitoring (MRM) quantification [21]. The prepared stock standard solutions were analyzed using a C18 reverse phase chromatography column (Restek AQ, 100 mm × 2.1 mm × 3 μm) on a Shimadzu Nexera 20A UHPLC system connected to a Shimadzu 8050 triple quadrupole mass spectrometer with an ESI interface switching between both positive and negative ionization modes (Shimadzu, Kyoto, Japan). The standards were injected in triplicate and separated using a binary gradient (Solvent A: Pure MilliQ Water/ 0.1% (v/v) formic acid, Solvent B: UHPLC grade methanol/

0.1% (*v*/*v*) formic acid) at a 0.40 mL/min flow rate. In the binary gradient, the concentration of solvent B was increased in 5% (*v*/*v*) increments to 25% (*v*/*v*) at 2–18 min and 95% (*v*/*v*) at 25–30 min before reducing the concentration to 2% at 31 min.

The instrumental conditions for MS/MS detection performed in positive and negative ion modes with MRM scanning were set as follows: 3 L/min nebulizing gas flow, 15 L/min drying gas flow, 4.5 kV interface voltage, 400 °C heat block temperature, 250 °C desolvation temperature, 1.6×10^{-3} Pa ion gauge vacuum. The precursor ions were directly infused into the triple quadrupole MS in the multiple reaction monitoring mode to determine the optimum conditions for generating product ions to be used for quantification. Briefly, the individual precursor ions were selected in the first quadrupole (Q1), followed by collision-induced dissociation (Q2) with nitrogen gas at 230 kPa, and transitioned into product ions, which were detected in the third quadrupole (Q3). The built-in (vendor specific) LabSolutions optimization software (Shimadzu, Kyoto, Japan) optimized the individual collision energies (CEs) for the authentic standard compounds. The most favorable ESI mode was determined from the peaks with the highest intensity. The optimized CE, precursor-to-product ion transitions, and product ions of the individual standard compounds are summarized in Table S3.

For the preparation of calibration curves, standard stock solutions of 100 ppm were prepared from the pure standard compounds dissolved in a 50% (*v*/*v*) UHPLC Grade Methanol/MilliQ Water solvent. Serial dilutions were prepared to 5, 1, 0.5, 0.25, 0.1, and 0.05 parts per million (ppm, 1 ppm = 1 mg/L). The standards were prepared in triplicate and analyzed as mentioned above, with each sample injected in triplicate. The standard curves were constructed by plotting the acquired integrated peak area against the standard concentration. The standard curve equations and R^2 values are summarized in Table S4. The obtained values in ppm were converted to ng/g plant leaf material (wet weight).

The limit of detection (LOD), the lowest analyte concentration that can be detected with a 1:10 signal-to-noise (S/N) ratio, and the limit of quantification (LOQ), the lowest level at which the analyte can be quantified with a 1:3 signal-to-noise (S/N) ratio [21], were determined by the preparation of serial dilutions (0.00025–5 ppm) of the pure standard solutions to determine the lowest detectable concentrations, as analyzed on the UHPLC-3Q-MS system. The LOD and LOQ of all the standard compounds and internal standard were determined at 0.025 ppm and 0.05 ppm, respectively.

SPSS software (IBM SPSS Statistics, version 29 (IBM Corp., Armonk, NY, USA) was used for descriptive statistics. Here, one-way analysis of variance (ANOVA) was performed to compare the mean values of individual metabolites at different time points. ANOVA was followed by the Tukey post hoc test, where differences between the means were considered significant at $p < 0.05$, indicated in graphs with an asterisk.

3. Results and Discussion

Early development is a perilous stage in the life of seedlings, as the plant must establish itself in the environment during this time. As a strategy to contribute to survival, plants utilize the synthesis of specialized metabolites during this phase [22], which involves interactions with beneficial or pathogenic microorganisms and the deployment of possible defense mechanisms [23,24]. Thus, the focus was to interrogate the metabolomic reprogramming of sorghum seedlings over the early growth period following germination to identify key metabolite markers that define the early development of the plant.

3.1. UHPLC-MS Analyses of Sorghum Leaf Extracts and Initial Data Analysis

The methanolic extracts from different developmental stages displayed inherent multidimensionality due to the complex physicochemical characteristics (e.g., polar vs. nonpolar; aglycone vs. glycosides) of the sample constituents. The UHPLC-MS chromatograms revealed differential profiles, which included variation in peak intensities and presence/absence of peaks. This indicates differential metabolite composition and content and is indicative of time-related changes to the metabolomic architecture (Figure S2A,B).

Untargeted metabolomics generates very large datasets that can be approached using two methods: (i) the traditional route, whereby the initial steps entail the application of chemometrics methods, followed by the annotation of the selected discriminant variables, and (ii) targeted profiling, whereby metabolite annotation is initially performed, then followed by the application of chemometrics methods [25–27]. The latter route was followed in this study, with the focus on previously identified key metabolites in the defense response of sorghum seedlings [9,10,14].

The metabolites are of different chemical classes, including amino acids, cyanogenic glycosides, flavonoids, hydroxycinnamic acid (HCA) derivatives, hormones, lipids, organic acids, and other phenolics (Table S1). As shown in Figure 1, the flavonoid class of phytochemicals forms the largest group of metabolites, while the cyanogenic glucoside class forms the smallest group compared to all of the other classes of metabolites.

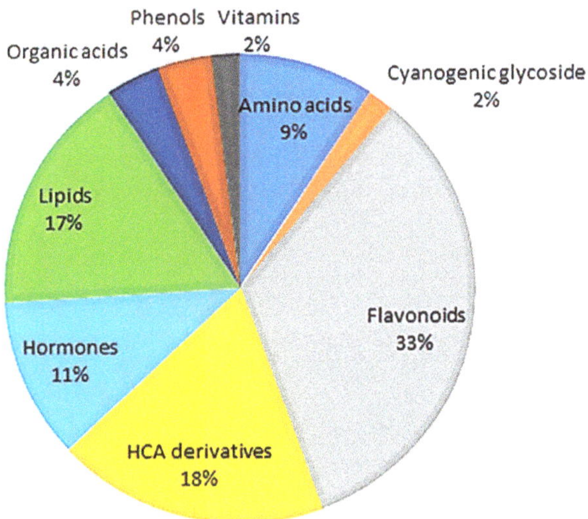

Figure 1. Classes of annotated metabolites present in hydromethanolic extracts of developing *Sorghum bicolor* seedlings. Of all the annotated metabolites (Table S1), the flavonoids, derivatives of hydroxycinnamic acids, and lipids were the major classes associated with early growth and development across all time points, days 7–29.

A ternary plot (Figure 2) was generated to depict the global relative quantification levels of different classes of metabolites across the early, middle, and late seedling developmental stages (corresponding to days 7, 14, and 29 post-germination, respectively). This plot shows the relative quantification of members of each metabolite class at each of the three stages, as represented on the ternary at each point. At each apex of the plot is the maximum relative quantification of the metabolites at that particular growth stage (i.e., 100%), which then decreases along the axes towards the next apex. It allows us to see how the points cover the space and to detect potential zones that should be explored in further experiments. From a global point of view, a larger content of metabolites is observed at the late stage as compared to the middle and early stages. It can also be deduced that the phenylpropanoid-derived phenolic compounds (HCA derivatives and flavonoids) are more prevalent in the late stage compared to early and middle stages, whereas lipids are more concentrated in the early and middle stages compared to the late stage. Other classes, such as amino acids and hormones, are clustered between the early and late stages.

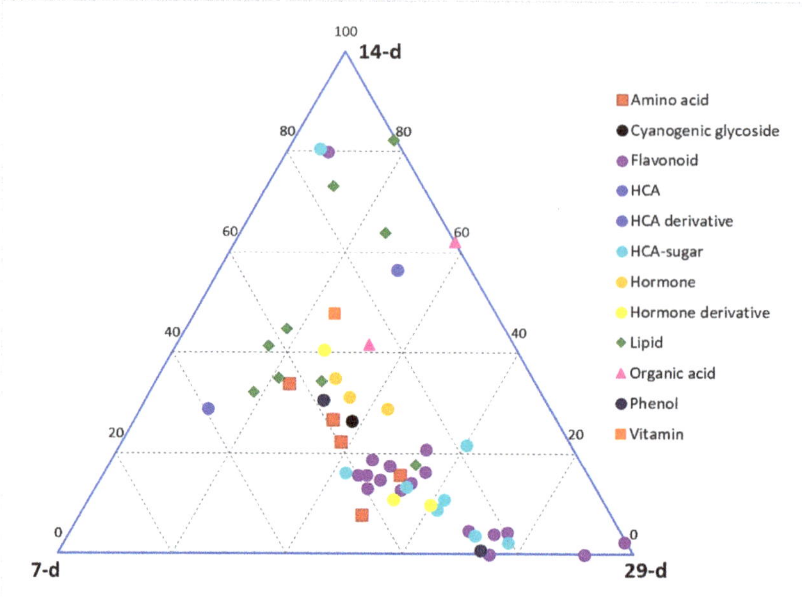

Figure 2. Ternary plot showing all classes of annotated metabolites present in hydromethanolic extracts of developing *Sorghum bicolor* seedlings. Seedlings were harvested at early (7 d), middle (14 d), and late (29 d) growth stages. The plot was constructed based on annotated metabolites across all time points, as listed in Table S1.

3.2. Multivariate Data Analysis and Chemometric Modelling

Chemometric modelling was applied to 46 annotated metabolites that exhibited variation in peak intensities (Table S1) to understand and describe the developmental changes through the lenses of the annotated metabolic profiles. PCA modelling was performed to observe the natural structures in the dataset. PCA is an unsupervised method that depicts similarities and differences (intrinsic interconnectedness) between samples within a dataset by reducing the dimension or complexity of the data, thus allowing for an interpretable visualization and analysis. As such, the PCA methods provide a qualitative representation of the similarities and differences (variation) between and within the samples [28]. The first two principal components (PC1 and PC2) of the generated PCA model explained 47.7% of the total variation and revealed a time-related trend, which reflects the differential metabolome changes across the developmental stages (7, 11, 14, 18, 22, 25, and 29 d) of the seedlings (Figure 3).

While the ternary plot and PCA provided a global view of the metabolome changes in terms of metabolite classes at different developmental stages, orthogonal projection to latent structures discriminant analysis (OPLS-DA) was applied to 43 annotated metabolites for the binary classification of samples to further investigate these changes. OPLS-DA performs sample classification based on linear regression, where differences among groups are modelled; it also identifies discriminant molecules specific to those groups. Supervised methods use predictive models to identify biological responses relating to certain variables, thereby identifying independent and dependent variables in a dataset. The calculated OPLS-DA models were thus computed to separate multivariate relationships into predictive variation and orthogonal variation. The binary classification of the seedlings at the 14 d stage and the 29 d stage, showing the clear separation of the two groups, is represented by the OPLS-DA scores plot in Figure 4A (14 d vs. 29 d). The corresponding figures for 7 d vs. 14 d and 7 d vs. 29 d are presented as Figure S3A,B, respectively.

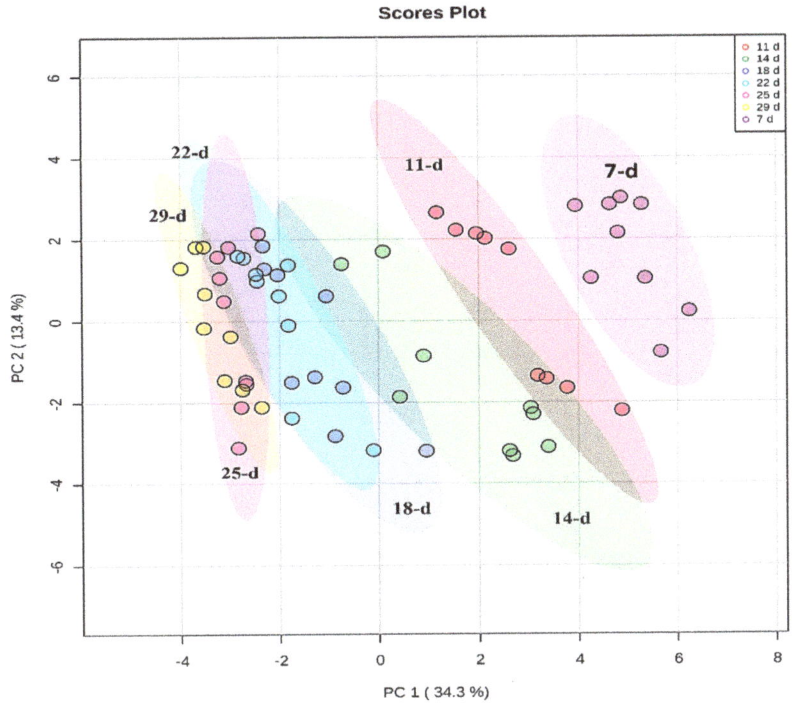

Figure 3. Unsupervised chemometric analysis of annotated metabolites present in extracts from *Sorghum bicolor* seedlings at different developmental stages. Principal component analysis (PCA) plots were constructed using MetaboAnalyst: A scores scatter PCA plot (PC1 vs. PC2) of log-transformed and Pareto-scaled MS data from seedlings on days 11, 14, 18, 22, 25, and 29 post-germination. The clusters are colored based on the different developmental stages (purple = 7 d, red = 11 d, green = 14 d, blue = 18 d, turquoise = 22 d, pink = 25 d, and yellow = 29 d). The PC analysis generated 9 principal components, of which 64.9% of the total component variation was captured.

The OPLS-DA models were validated by diagnostic statistics to ensure the reliability of the models and to prevent the over-fitting of the models to the data. As indicated in the experimental section, the cross-validation (CV) method and permutation tests were used in this research [15]. The permutation test performed was with 100 iterations (n = 100). Permutation tests, as shown in Figure 4B, are randomization-based validation methods that are employed to validate the predictive power of OPLS-DA models, comparing the R^2 and the Q^2 of both the permuted and original models.

The significant metabolites characterizing the early, middle, and late (corresponding to days 7, 14, and 29) development stages were then selected using OPLS-DA S-plots in Figures 4C and S4A,B. These loading S-plots identify the variables that contribute the most to the pattern changes observed on the OPLS-DA scores plots. The variables with the highest absolute values of p[1] and p(corr)[1] are those identified as the discriminant markers and these express/represent the differences between the different groups [29].

The p-value obtained from the OPLS-DA model in Figure 4A was 0.0000, determined by the CV-ANOVA (analysis of variance of cross-validated parameters) test, which indicates that there is a statistically significant difference between the two groups and that the null hypothesis (that there are no differences between the groups) can be rejected. In addition, a VIP plot metric was used for validation of selected metabolites from the S-plots. With VIP, metabolites are scored as a measure of how much they contribute to the model, and the variables that are of significance are those with a VIP score > 1. A higher VIP score is directly related to the significance of a variable. The VIP plot in Figure 4D shows the top 18 most significant metabolites, differentially expressed in the metabolomes of the 14 d and 29 d stages of seedling growth. The metabolites highlighted in red are those correlating to the late (29 d) stage, while the rest are those corresponding to the mid (14 d) stage.

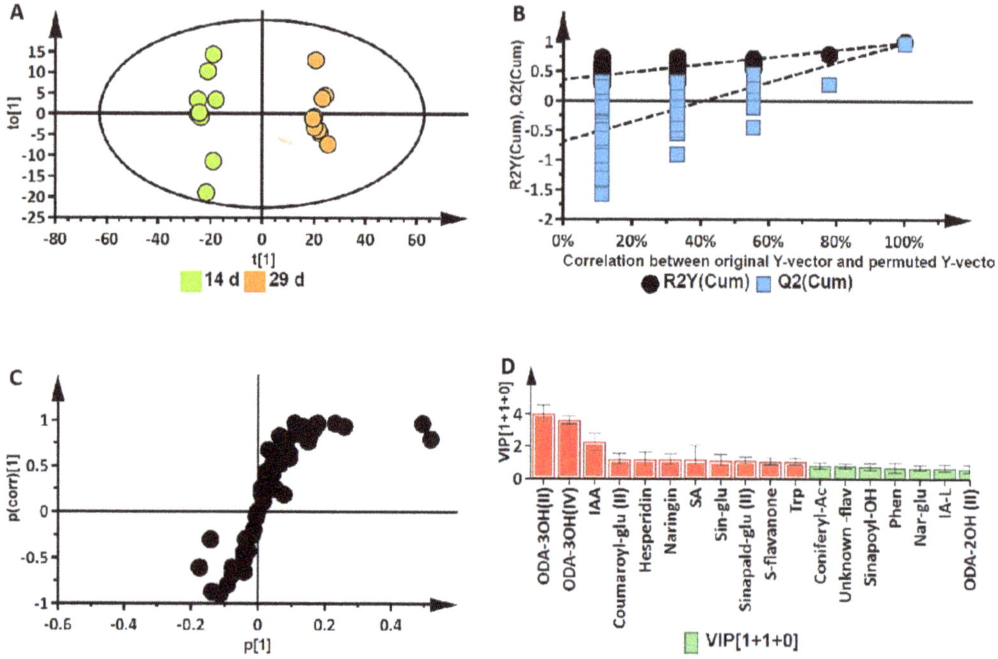

Figure 4. Supervised multivariate data analysis of annotated metabolites in *Sorghum bicolor* seedlings following UHPLC-MS analysis of the 14 d vs. 29 d group samples. (**A**) An OPLS-DA scores plot of the predictive component t[1] and the first orthogonal component t0[1], R^2(cum): 0.989; Q^2(cum): 0.980; R^2 permutation: 0.259; Q^2 permutation: −0.541; *p*-value: 0.0000; and components: (1 + 1 + 0). The ellipse indicates the 95% limit of the Hotelling T^2 distribution for the model. (**B**) Permutation analysis plotting R^2 and Q^2 (black and blue dots, respectively) from n = 100 permutation tests in the OPLS-DA model. The y-axis shows R^2 and Q^2, whereas the x-axis shows the correlation coefficient of permuted and observed data. The cluster of points on the left represents 100 permuted R^2s and Q^2s, and the two points on the right represent the observed R^2(cum) and Q^2(cum). Dashed lines denote corresponding fitted regression lines for observed and permutated R^2 and Q^2. (**C**) OPLS-DA S plot with discriminant biomarkers at each end of the S-plot. The covariance (variable magnitude) and correlation (reliability) of the variables in the model (indicated with black dots) are represented on the axes as p[1] and p(corr)[1], respectively. The features located at the extreme ends of the plot show a positive association (high magnitude and high reliability) to the respective conditions being compared, while those in the middle can be regarded as shared features. (**D**) A VIP plot where the metabolites of a VIP score of >1 is significant to metabolomic differences corresponding to developmental changes. (Metabolite abbreviations: ODA-3OH(II) = trihydroxy-octadecadienoic acid; ODA-3OH(IV) = 9,12,13-trihydroxy-10-octadecenoic acid; IAA = indole-3-acrylic acid; coumaroyl-glu II = coumaroyl glucose; SA = salicylic acid; sin-glu = sinapaldehyde glucoside I; sinapald-glu II = sinapaldehyde glucoside II; S-flavanone = sophoraflavanone; trp = tryptophan; coniferyl-Ac = coniferyl acetate; sinapoyl-OH = sinapoyl-alcohol; phe = phenylalanine; nar-glu = naringenin 7-O-beta-D-glucoside; IA-L = indole-3-acetyl-leucine; ODA-2OH (II) = 9,10-dihydroxy-12-octadecenoic acid).

3.3. Deriving Biochemical Insights from Metabolomics Data

Hierarchical cluster analysis (HiCA) was applied to further explore the time-related trends and groupings revealed by the PC analysis. The HiCA shows the statistical correlation amongst all the samples analyzed and then builds a hierarchy from them. Consequently, similar samples consisting of similar metabolomes form clusters together, generating groups formed on the basis of cluster similarity [30]. Furthermore, the HiCA heatmap also shows the relative abundance of the identified metabolites, thereby displaying the variation of metabolic features present at the different developmental stages of the seedlings.

This HiCA heatmap (Figure 5) shows the average levels of each of the top 25 metabolites at each developmental stage. These patterns suggest that differential reprogramming has occurred over time. This can take the form of high or low accumulation at specified time points, indicating early, late, or oscillatory responses. Where molecules show increasing or decreasing trend patterns, it could be suggestive of increased biosynthesis followed by interconversion, conjugation, degradation, or incorporation into insoluble polymers such as lignin [23,31].

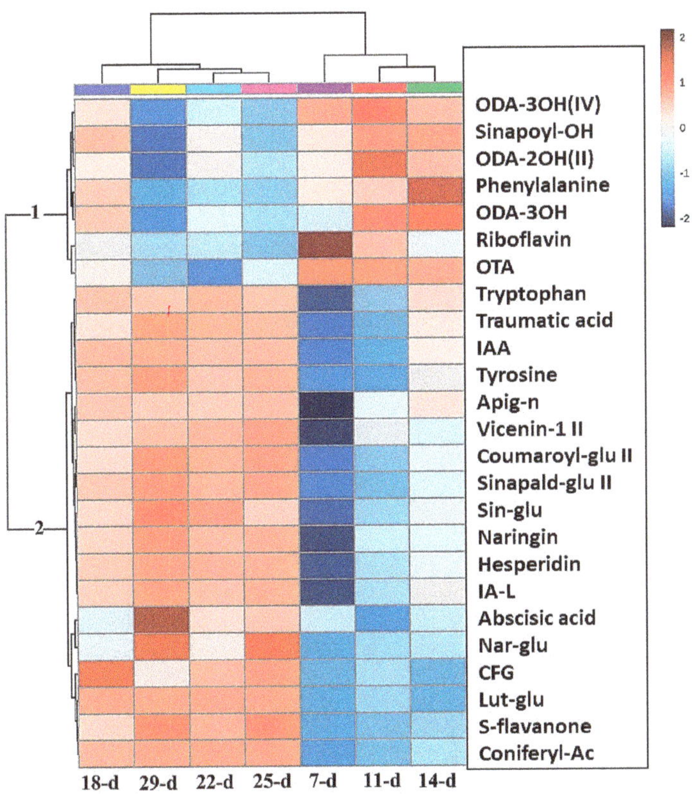

Figure 5. Hierarchical cluster analysis (HiCA) heat map of the top 25 metabolites present in the averaged extracts ($n = 9$) showing the most and least abundant metabolites at each developmental stage (red/blue = high/low concentrations, respectively). The metabolites listed are as follows, respectively: 9,12,13-trihydroxy-10-octadecenoic acid (ODA-3OH(IV)), sinapoyl alcohol (sinapoyl-OH), 9,14-dihydroxy-10,12-octadecadienoic acid (ODA-2OH(II)), phenylalanine, trihydroxy-octadecadienoic acid (ODA-3OH(II)), riboflavin, octadecatetraenoic acid (OTA), tryptophan, traumatic acid, indole-3-acrylic acid, tyrosine, apigenin 7-O-neohesperidoside (rhoifolin) (apig-n), apigenin 6-C-xyloside-8-C-glucoside (vicenin-1), coumaroyl glucose (coumaroyl-glu II), sinapaldehyde glucoside II (sinapald-glu II), sinapaldehyde glucoside I (sin-glu), naringenin 7-O-neohesperidoside (naringin), hesperidin, indole-3

-acetyl-leucine (IA-L), abscisic acid (ABA), naringenin 7-O-beta-D-glucoside (nar-glu), 1,3-O-coumaroyl-feruloylglycerol (CFG), luteolin 7-O-glucoside (Lut-glu), sophoraflavanone G (S-flavanone), coniferyl acetate (coniferyl-Ac).

Two significant groups (1 and 2) are observed in the map that display different patterns of metabolite abundance at each of the time points. The metabolites in group 1 show increased levels in the early stages of development (days 7 to 14) compared to the late stages (days 18 to 29), where the levels of these metabolites decreased. Conversely, the metabolites in group 2 show a linear increase in their levels as the seedlings progress in growth from the early stages to the late stages of development, i.e., from days 7 to 29. The metabolites of group 1 are mostly lipids (ODA-2OH II, ODA-3OH, ODA-3OH IV, and OTA; abbreviations are defined in Figure 5). The other metabolites in this group are phenylalanine, sinapoyl alcohol, and riboflavin, which are from the amino acid, HCA, and vitamin classes, respectively. Significant amounts of lipids are stored in the seeds of plants, that are activated after seed germination. This aids in the growth and development of the seedlings before photosynthesis is initiated. Most of these lipids, such as octadecatetraenoic acid (OTA), act as signalling molecules in plant defense systems [32]; therefore, they are at high levels in the early stages of plant growth, as shown in Figure 5. In contrast, group 2 contains mostly flavonoids. These are not essential for the early survival of seedlings, but more important for the phototrophic response of the seedlings when photosynthesis is initiated; therefore, these metabolites are at low levels in the early growth stages and at high levels in the late stages [33].

The discriminant metabolites identified by the OPLS-DA S-plots that were positively correlated with the 14 d stage were phenylalanine, sinapoyl alcohol, 9,14-dihydroxy-10,12-octadecadienoic acid (ODA-2OH(II)), trihydroxy-octadecadienoic acid (ODA-3OH(II)), and 9,12,13-trihydroxy-10-octadecenoic acid (ODA-3OH(IV)). From the same OPLS-DA S-plot, coniferyl-acetate, 1-O-coumaroyl-beta-D-glucose, sinapaldehyde glucoside, sophoraflavanone G, naringenin 7-O-neohesperidoside, an uncharacterized flavonoid, hesperidin, indole-3-acrylic acid, and tryptophan showed a positive correlation with the 29 d stage. In the growth condition of 7 d vs. 14 d, as shown in Figure S4A, isocitric acid, 15-hydroxylinoleic acid, 9,12,13-trihydroxy-10-octadecenoic acid, and riboflavin showed a positive correlation with the 7 d stage. In the 7 d vs. 29 d growth stage condition, depicted in Figure S4B, the metabolites that were positively correlated to the 7 d stage were trihydroxy-octadecadienoic acid, 15-hydroxylinoleic acid, 9,12,13-trihydroxy-10-octadecenoic acid, and riboflavin, which is similar for the 7 d vs. 14 d condition.

3.4. Pathway Enrichment Analysis Indicates Importance of the Phenylpropanoid Pathway

Analysis of metabolic pathways potentially provides insight into the most significant biochemical and physiological processes occurring in the developing seedlings. These pathways are recognized by identifying groups of metabolites belonging to the same metabolic network.

The KEGG IDs of each compound were used to map these metabolites into principal metabolic pathways. The most significant primary pathways identified by this approach, (listed in Table S2) include phenylalanine, tyrosine, and tryptophan biosynthesis, riboflavin metabolism, and glyoxylate and dicarboxylate metabolism. The most significant pathways identified from the analysis include isoquinoline alkaloid biosynthesis, phenylalanine metabolism, phenylpropanoid biosynthesis, and flavone and flavonoid biosynthesis. Flavone and flavonol biosynthesis had the most hits when arranged by the pathway impact ($\geq$0.1) and when arranged by FDR ($\leq$0.5); phenylpropanoid biosynthesis had the most hits. This was followed by aromatic amino acid biosynthesis (Trp, Tyr, and Phe, with Phe feeding into the phenylpropanoid pathway). Some of the most abundant specialized metabolites in plants originate from these aromatic amino acids.

Biosynthesis of specialized metabolites is based on the formation of precursors of specialized metabolites produced by primary metabolism [34]. These precursors are mostly produced from glycolysis, the tricarboxylic acid (TCA) cycle, etc. Isoquinoline alkaloid biosynthesis produces precursors of alkaloids by decarboxylation of tyrosine. Similarly, they can also be derived from the amino acids Phe, Lys, Trp, and Tyr. The precursor compound, 4-coumaroyl CoA, which supports the production of flavonoids, is itself produced in the phenylpropanoid pathway. Intermediates emerging from the shikimate pathway act as precursors for the phenylpropanoid pathway. Together, most of the sorghum specialized metabolites are derived from these two pathways [35].

Pathway enrichment analysis is based on the quantitative information of the statistically significant compounds and contributes to reducing the complexity of the metabolomes. Its application identified enriched pathways in the developing seedlings and compared pathway functionalities at the different stages when growth is rapid and associated with dynamic changes. Furthermore, pathway topology analysis was applied to identify the connections between the metabolites within the metabolic pathways [19].

The most significant metabolic pathways are represented by the pathway analysis in Figure 6A. Topologically, the phenylpropanoid biosynthesis metabolic pathway is represented in Figure 6B, where four of the metabolites annotated from the OPLS-DA data and matched to the pathway are found. These are phenylalanine, 4-coumarate, sinapoyl alcohol, and sinapaldehyde glucoside.

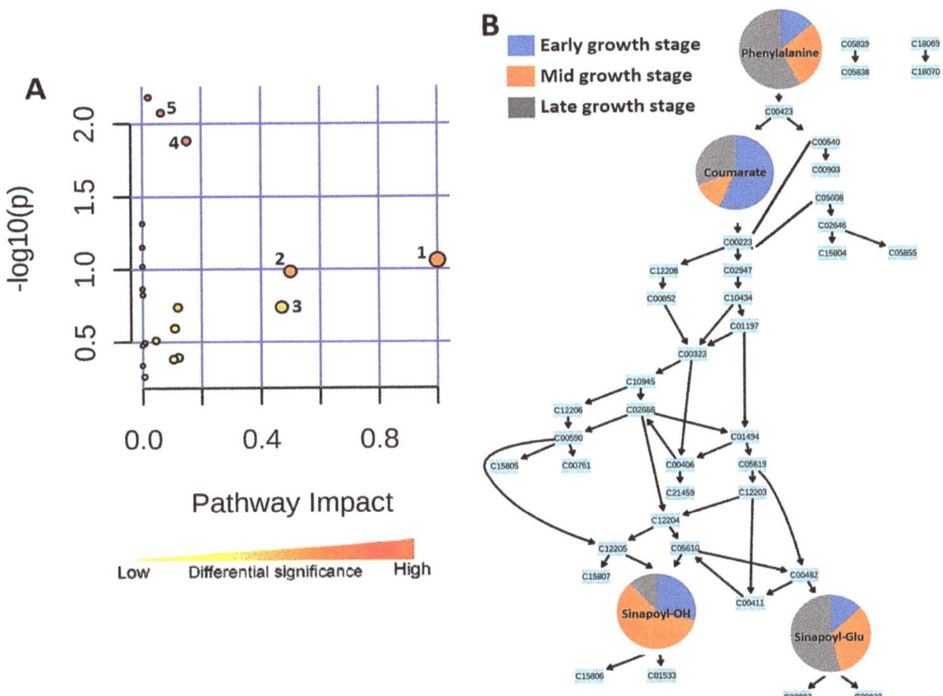

Figure 6. Summarized representation of metabolic pathway analysis. (**A**) Pathways are arranged by increasing pathway impact, which represents differential significance, on the x-axis and by increasing *p*-value on the y-axis. Each pathway is represented by a circle in which size/radius increases with increasing impact, and the color of the circle also becomes more red as the *p*-value increases. 1, Biosynthesis of specialized metabolites. 2, Isoquinoline alkaloid biosynthesis. 3, Phenylalanine metabolism. 4, Flavone and flavonoid biosynthesis. 5, Phenylpropanoid biosynthesis. (**B**) A representation of the topological properties of the phenylpropanoid pathway. The changes in the four matched metabolites in their early, middle, and late seedling growth stages are represented by pie charts, indicating stage-specific quantitative variation.

In the early stage of sorghum seedlings (day 7), phenylalanine is present at low levels. This could be due to the amino acid being converted to 4-coumarate at the start of the pathway. However, phenylalanine accumulates over the period of investigation to day 29, reflecting the increase in flavonoid synthesis and the demand for lignin precursors associated with cell wall synthesis. Coumarate then gives rise to sinapoyl alcohol and sinapaldehyde glucoside further down the metabolic pathway. Sinapoyl alcohol is highly abundant in the middle 14 d stage of seedling growth and is at its lowest level in the late stage. However, sinapylaldehyde glucoside is at its highest level in the late growth stage and is lowest in the early stage, indicative of the glycosylation of the former to the latter.

3.5. *Relative Quantification of Selected Discriminant Metabolites*

In order to gain insights into the dynamics of the changes occurring during the investigated growth period, relative quantification was performed on the selected discriminant metabolites. The bar graphs in Figure 7 show the changes in the levels of the different metabolites in the seedlings, as determined at the 7 d, 14 d, and 29 d stages.

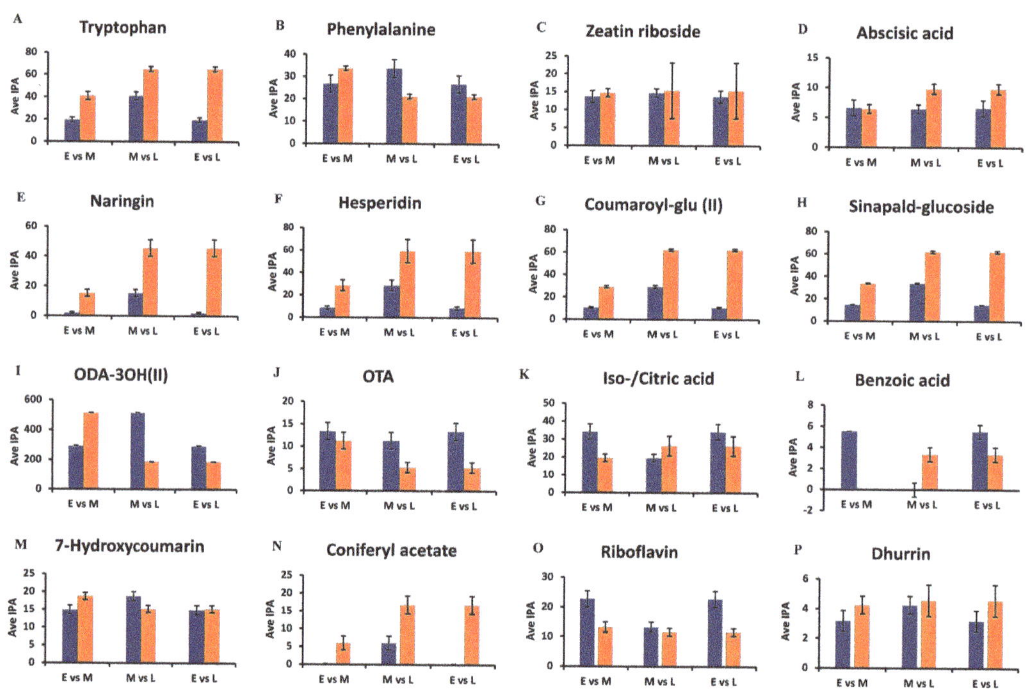

Figure 7. Representative bar graphs showing the semi-quantification of discriminant metabolites retrieved from the OPLS-DA S plots. The average integrated peak area values (y-axis, Ave IPA, n = 9) of metabolites in the early (E = 7 d), middle (M = 14 d), and late (L = 29 d) growth stages are compared to illustrate developing trends within the metabolomes. In the paired comparisons, blue indicates the earlier time point (E or M), while orange indicates the later time point (M or L). (**A**,**B**) Tryptophan and phenylalanine of the amino acid class. (**C**,**D**) Zeatin riboside and abscisic acid of the hormone class. (**E**,**F**) Naringenin 7-O-neohesperidoside (naringin) and hesperidin of the flavonoid class. (**G**,**H**) Coumaroyl glucose and sinapylaldehyde glucoside of the HCA derivatives. (**I**,**J**) Trihydroxy-octadecadienoic acid and octadecatetraenoic acid of the lipid class. (**K**,**L**) Iso-/citric acid and benzoic acid of the organic acid class. (**M**,**N**) 7-Hydroxycoumarin and coniferyl acetate of the phenol class. (**O**) Riboflavin of the vitamin class and (**P**) Dhurrin of the cyanogenic acid class. The error bars indicate the standard deviation.

The levels of tryptophan increase from the early 7 d stage, through to the middle 14 d stage, and finally to the late 29 d growth stage (Figure 7A). Important specialized plant metabolites are derived from tryptophan [36], such as alkaloids and indole-3-acetic acid (IAA), which contribute to plant defense. Therefore, as the seedlings grow, defensive metabolites also increase in concentration [36]. Phenylalanine, on the other hand, increased from the 7 d stage and was at its highest concentration at the 14 d stage, followed by a decrease at the 29 d stage (Figure 7B).

Phenylalanine connects primary and secondary metabolism in plants through the induced action of phenylalanine ammonia lyase (PAL), in response to developmental cues and stress triggers. PAL generates trans-cinnamic acid, which feeds into the early phenylpropanoid pathway, which functions in plant defense and structural support through the synthesis of lignin precursors [37]. In turn, phytohormones, such as abscisic acid and zeatin riboside function in plant growth and plant responses to environmental stress [38]. It was observed that the concentrations of both of these compounds increased from the early 7 d stage to the late 29 d stage of the seedlings (Figure 7C,D).

Plant phenolic compounds (e.g., HCAs, hydroxybenzoic acids, and flavonoids) often exhibit antioxidant properties that allow for adaptation to changing environmental conditions [39]. However, compounds that contain phenolic groups might be active as antimicrobial agents as well, causing a notable correlation between the total phenolic content and antioxidant activity [6,40]. Moreover, some overlap/correlation between antioxidant activity and antimicrobial activity might be found [6,39]. These effects might act synergistically to create an enhanced antimicrobial environment in planta. Unfortunately, little information is available regarding sorghum seedlings [7].

Two important sub-classes of phenolics are the flavonoids and the HCA derivatives. Flavonoids form the largest class of specialized metabolites in most plants and perform several critical functions related to development and environmental adaptation, including plant–microbe interactions [41,42]. Both naringenin 7-O-neohesperidoside (naringin) and hesperidin increased as the seedling progressed in growth from the 7 d stage to the 29 d stage (Figure 7E,F). HCA derivatives and conjugates are compounds derived from core phenylpropanoids that act as antioxidants [43]. In addition, members of this class display antimicrobial activity. In plants, their major role is to provide a measure of defense and thus resistance against pathogens. This property is essential to vulnerable plants to allow seedlings to adapt to the environment in their early stages of growth and development. The HCA derivatives, coumaroyl glucose, and sinapylaldehyde glucoside (Figure 7G,H) similarly display the same pattern of increasing concentrations from the early to the late growth stages. These increases are due to the biosynthesis of HCAs in the early phenylpropanoid pathway, which are precursors for various mono-lignols. The compounds contribute to the lignification of cell walls of newly formed cells, thereby supporting the progressive growth and development of plants and strengthening the cell walls, providing plant defense [44].

Fatty acids are essential members of the lipidome and lipid metabolism and contribute to the synthesis of cell membranes with the associated structural and functional properties. Moreover, they can act as signalling molecules or lipid mediators in plants [32,45]. The oxygenated fatty acids, trihydroxy-octadecadienoic acid and octadecatetraenoic acid in Figure 7I,J, mostly act as signalling molecules [46]. The concentration of trihydroxy-octadecadienoic acid increases from 7 d to 14 d, then decreases at the 29 d growth stage. Octadecatetraenoic acid decreases in its concentration from the 7 d stage to the 29 d stage.

Organic acids contribute to carbon metabolism in plants and take part in the biochemical pathways in plant cells, such as the TCA cycle [47]. Isocitric acid decreases in its levels from the early growth stage to the late growth stage, as depicted by Figure 7K,L, with 14 d having the lowest level of isocitric acid metabolites. Together with the amino acid data, these changes might reflect C and N metabolism and link the related metabolic cycles of anaplerotic reactions to counteract depletion of the TCA cycle by biosynthetic demands. Relatedly, shifts in carboxylic acids levels were reported to be perceived in plants during stress and it was proposed that the tricarboxylates could modulate signal transduction events linked to plant defense [48].

The benzoic acid levels are almost absent at the 14 d stage but are at their highest level at the 7 d stage, decreasing at the 29 d stage. The metabolic changes for the phenolic metabolites 7-hydroxycoumarin and coniferyl acetate are depicted in Figure 7M,N. According to the figures, 7-hydroxycoumarin (umbelliferone) shows minor fluctuations around the middle growth stage. In contrast, coniferyl acetate is at extremely low levels at the 7 d stage but increases at the late 29 d growth stage.

The lowest abundant metabolites are riboflavin and dhurrin, a cyanogenic acid, shown in Figure 7O,P. Riboflavin, a cofactor for diverse metabolic processes and inducer of plant resistance, decreases from the 7 d stage to the 29 d stage. Conversely, durrhin, which is a wound metabolite and source of resistance of sorghum seedlings to fungal infection [49], is already present at the 7 d stage and increased in concentration at the 29 d stage.

3.6. *Quantitative Determination of Flavones*

As a phytochemical class, flavonoids are composed of flavonols, flavanones, flavanols, flavones, isoflavones, and their respective derivatives [42]. Flavone O-conjugates and their aglycones, (e.g., apigenin and luteolin) are constitutive metabolites in stem and leaf tissues of grasses. Apigenin and luteolin are generally O-glycosylated; however, flavone C-glycosides, which contain sugar residues covalently linked to C-6 and/or C-8 in the flavonoid A-ring, co-exist with flavone O-conjugates as predominant flavonoids [33,50].

In particular, flavonoids are considered as plant UV protectants due to their epidermal accumulation and radical scavenging properties [35,42]. In grasses, accumulation of flavone aglycones and their O-conjugates in grasses can be triggered by biotic stresses. For example, apigenin and luteolin are defense-related metabolites in sorghum infected by the anthracnose fungus *Colletotrichum sublineola* [10,51]. These pathogen-inducible flavones inhibit the in vitro spore germination of *C. sublineola*, implicating a functional role in chemical defense [51]. In addition, apigenin and luteolin and their O- and C-glycosylated derivatives have been reported to possess direct antimicrobial activity [8,52] and were also found as discriminant metabolites in sorghum seedlings responding to *Burkholderia andropogonis* [9] and *C. sublineola* [10].

In addition to other flavonoids, Table S1 lists the following derivatives of two flavones, apigenin and luteolin (3′-hydroxyapigenin), as discriminant metabolites: apigenin 8-C-glucoside (vitexin), apigenin 7-O-glucoside (apigetrin), apigenin 6-C-glucosyl-8-xyloside (vicenin-3), apigenin 6-C-xylosyl-8-C-glucoside (vicenin-1), apigenin 7-O-neohesperidoside (rhoifolin), luteolin 7-O-glucoside (luteoloside), luteolin 7-O-neohesperidoside (lonicerin), with the flavanone naringenin as precursor (naringenin 7-O-beta-D-glucoside and naringenin 7-O-neohesperidoside or naringin) (Figure S5). Quantitative studies of selected flavonoids can play a significant role in deciphering data acquired from untargeted and targeted metabolomic studies. Their concentrations were thus further investigated using quantitative MRM analysis during the same time periods as used for the metabolomics analysis [9]. The mean peak areas from the UHPLC-MS/MS data along with the respective standard curve equations were used to determine the concentration values in ng/g (Figure 8).

The flavones luteolin, luteoloside, vicenin-2, vicenin-3, and vitexin were detected at varying concentrations at different d.p.g., but apigetrin and isovitexin were below the LOD. From day 11, the dominant flavones were vicenin-2, vicenin-3 (with apigenin as aglycone), and to a lesser extent, luteoside (with luteolin as aglycone). Vicenin-2 and vicenin-3 were the most abundant at all stages of development, which may suggest that they play a significant role in the preparation of the plants' defense against stresses during early development. Of interest is the profile of vitexin, which is initially relatively high (day 7, decreasing to day 11), followed by its apparent disappearance from the samples (day 14 onwards). Vitexin is known to occur in sorghum seeds [53] and it is possible that some of the metabolite could have originated from the germinating seed. Vitexin is an 8-C-mono-glycosylated form of apigenin that is metabolized to the diglycosylated vicenins (Figure S5). The consistent low level of apigenin can be explained in view of its role as precursor compound (aglycone) of the vicenins, thus expanding the multi-metabolite apigenin-based chemical space in the seedlings. A similar situation my apply in the case of luteolin and luteoloside.

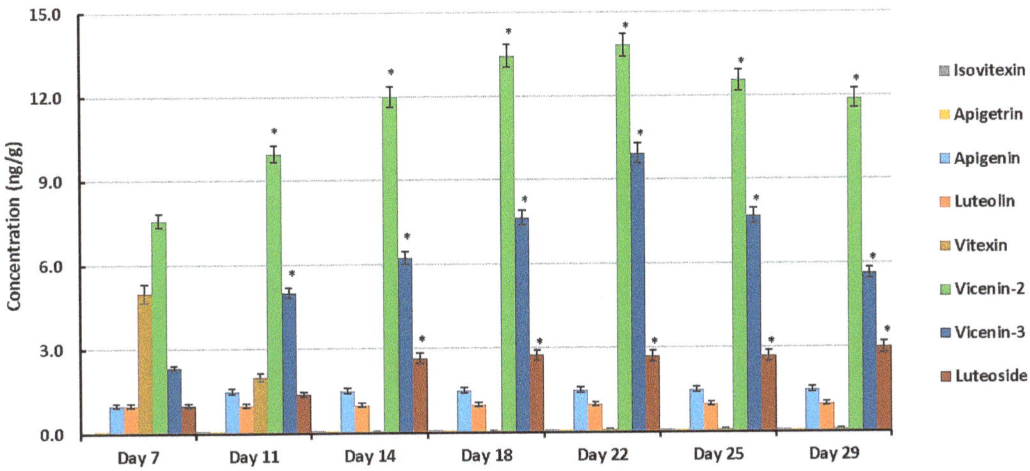

Figure 8. The varying concentrations of flavones apigenin and luteolin and the glycosylated derivatives vitexin, vicenin-2, and vicenin-3 and luteoside in developing seedlings of the sorghum cultivar NS5511 at 7, 11, 14, 18, 22, 25, and 29 days post-germination, as determined by UHPLC-3Q-MS/MS using optimized MRM conditions. All concentrations (y-axis, ng/g fresh weight) are mean concentration values with $n = 9$, and error bars indicate the standard error. Isovitexin and apigetrin were below the limit of detection. An asterisk (*) indicates the statistical significance (ANOVA followed by Tukey post hoc test at a p-value < 0.05 when comparing mean values to that of day 7).

Previous investigations on metabolomic profiling of the response of sorghum seedlings to pathogen attack have revealed the reprogramming of pathways that synthesize flavonoids (especially flavones) in the sorghum metabolome post-infection with *B. andropogonis* and *C. sublineola*, respectively [9,10]. The flavones upregulated in both pathogen responses were identified as apigetrin, apigenin, vitexin, isovitexin, luteolin, luteoloside, vicenin-2, and vicenin-3, indicating a positive correlation of the glycosylated derivatives of apigenin and luteolin with the defense response of sorghum seedlings [54]. Although glycosylated flavonoids appear to have a reduced antimicrobial effect compared to the respective aglycones [55], they can serve as storage forms of constitutive anti-microbial metabolites for the rapid release of the aglycones upon pathogen attack.

4. Conclusions

The seedling stage of plant development is the most crucial and vulnerable to varying environmental conditions. Various metabolic pathways occur within sorghum seedlings to support biological functions essential to the growth and development of the plants as well as to support innate immune responses in defense against attempted pathogen attacks. The study of the development-related metabolomes provided superior insights into these processes as the seedlings progressed in their initial post-germination growth. Changes in the metabolomes, as evident from PCA plots, showed distinct clustering. Significant changes in the metabolic patterns of these metabolomes were focused in the early, middle, and late growth stages of the seedlings. In turn, OPLS-DA was utilized to identify discriminatory markers associated with specific developmental stages. Nine classes of specialized metabolites associated with the changing metabolomes were identified from the annotated UHPLC-MS datasets.

Pathway analysis revealed five metabolic pathways, with the phenylpropanoid pathway being the most pronounced and flavonoids the most abundant class of metabolites present in the samples. Flavonoids and HCA derivatives allow the seedlings to survive post-germination, enabling them to be established in the environment and overcome future obstacles during the growth and development of the plant. This offers an explanation as to why these phenylpropanoids are important/at high levels during the early stages of seedling growth, i.e., to act as antimicrobial phytoanticipins. Site-specific hydroxylation, methylation, and glycosylation reactions not only lead to structural diversification within a class of specialized metabolites (and thus increase the 'chemical space' thereof) but may also broaden the functional diversification of the molecules. In the later stages, when the plants are more developed and less vulnerable to detrimental environmental conditions, the same metabolic pathways can be (re)activated as part of inducible immunity to provide plant defense responses when needed. Profiling of flavone-based phytoanticipins in juvenile plants may thus be developed as a potential tool for the identification of host resistance.

Overall, the metabolomics investigation substantially extends the knowledge of the metabolite dynamics of early seedling development in sorghum. This affords new insights into the involved metabolic pathways and their regulation through mechanisms not yet documented in the literature that require further investigation.

Supplementary Materials: The following supporting information can be downloaded at: https://www.mdpi.com/article/10.3390/metabo14020112/s1, Figure S1. Growth of *Sorghum bicolor* seedlings; Figure S2. UHPLC-MS BPI chromatograms of methanolic extracts derived from sorghum in ESI negative (A) and ESI positive ionization modes (B); Figure S3. OPLS-DA scores plots of differentially occurring metabolites in extracts of *Sorghum bicolor* seedlings following UHPLC-MS analysis; Figure S4. OPLS-DA S plots of discriminant biomarkers occurring in extracts from *Sorghum bicolor* seedlings following UHPLC-MS analysis; Figure S5. Structures of the flavone, apigenin, and mono- and diglycosylated derivatives thereof. Table S1. Classification and annotation of metabolites from extracts of *Sorghum bicolor* seedlings from different development stages. Table S2. Pathway enrichment analysis based on the presence of metabolites present in hydromethanolic extracts from developing *Sorghum bicolor* seedlings; Table S3. Retention times and MRM/MS data of the precursor ions, product ions, dwell times and collision energies of the standard compounds; Table S4. Standard curve equations and R^2 values of the standard compounds.

Author Contributions: Conceptualization, I.A.D.; methodology, P.A.S., F.T. and I.A.D.; validation, L.P.N.; investigation, L.P.N. and P.A.S.; resources, P.A.S. and I.A.D.; data curation, L.P.N. and F.T.; writing—original draft preparation, L.P.N.; writing—review and editing, F.T. and I.A.D.; visualization, I.A.D.; supervision, I.A.D.; project administration, I.A.D.; All authors have read and agreed to the published version of the manuscript.

Funding: This research received no external funding.

Institutional Review Board Statement: Not applicable.

Informed Consent Statement: Not applicable.

Data Availability Statement: Data are contained within the article or supplementary material.

Acknowledgments: C. Mareya, N. Mokoena, and R. Motsatsi assisted with sample preparation, data processing, and quantitative analysis, respectively, and are acknowledged for technical support. N. Buthelezi is acknowledged for analytical method development.

Conflicts of Interest: The authors declare no conflicts of interest.

References

1. Ramatoulaye, F.; Mady, C.; Fallou, S.; Amadou, K.; Cyril, D.; Massamba, D. Production and use sorghum: A literature review. *J. Nutr. Health Food Sci.* **2016**, *4*, 1–4. [CrossRef]
2. McLaren, N.; Rothmann, L. Management of Sorghum Diseases. *Grain SA* 2019, February 2019. Available online: https://www.grainsa.co.za/management-of-sorghum-diseases (accessed on 13 November 2023).

3. Isah, T. Stress and defense responses in plant secondary metabolites production. *Biol. Res.* **2019**, *52*, 39. [CrossRef]
4. Li, Q.; Duncan, S.; Li, Y.; Huang, S.; Luo, M. Decoding plant specialized metabolism: New mechanistic insights. *Trends Plant Sci.* **2023**, *9*. [CrossRef]
5. Awika, J.; Rooney, L. Sorghum phytochemicals and their potential impact on human health. *Phytochemistry* **2004**, *65*, 1199–1221. [CrossRef] [PubMed]
6. Kozłowska, M.; Scibisz, I.; Przybył, J.L.; Laudy, A.E.; Majewska, E.; Tarnowska, K.; Małajowicz, J.; Ziarno, M. Antioxidant and antibacterial activity of extracts from selected plant material. *Appl. Sci.* **2022**, *12*, 9871. [CrossRef]
7. Kil, H.Y.; Seong, E.S.; Ghimire, B.K.; Chung, I.-M.; Kwon, S.S.; Goh, E.J.; Heo, K.; Kim, M.J.; Lim, D.J.; Lee, D.; et al. Antimicrobial activities of crude sorghum extract. *Food Chem.* **2009**, *115*, 1234–1239. [CrossRef]
8. Karpiński, T.M.; Adamczak, A.; Ożarowski, M. Antibacterial activity of apigenin, luteolin, and their C-glucosides. In Proceedings of the 5th International Electronic Conference on Medicinal Chemistry, Basel, Switzerland, 1–30 November 2019. [CrossRef]
9. Mareya, C.; Tugizimana, F.; Piater, L.; Madala, N.; Steenkamp, P.; Dubery, I. Untargeted metabolomics reveal defensome-related metabolic reprogramming in *Sorghum bicolor* against infection by *Burkholderia andropogonis*. *Metabolites* **2019**, *9*, 8. [CrossRef] [PubMed]
10. Tugizimana, F.; Djami-Tchatchou, A.; Steenkamp, P.; Piater, L.; Dubery, I. Metabolomic analysis of defense-related reprogramming in *Sorghum bicolor* in response to *Colletotrichum sublineolum* infection reveals a functional metabolic web of phenylpropanoid and flavonoid pathways. *Front. Plant Sci.* **2019**, *9*, 1840. [CrossRef] [PubMed]
11. Hamany Djande, C.Y.; Pretorius, C.; Tugizimana, F.; Piater, L.A.; Dubery, I.A. Metabolomics: A tool for cultivar phenotyping and investigation of grain crops. *Agronomy* **2020**, *10*, 831. [CrossRef]
12. Sumner, L.W.; Amberg, A.; Barrett, D.; Beale, H.M.; Beger, R.; Daykin, C.A.; Fan, T.W.-M.; Fiehn, O.; Goodacre, R.; Griffin, J.R.; et al. Proposed minimum reporting standards for chemical analysis: Chemical Analysis Working Group (CAWG) Metabolomics Standards Initiative (MSI). *Metabolomics* **2007**, *3*, 211–221. [CrossRef] [PubMed]
13. Brown, M.; Wedge, D.C.; Goodacre, R.; Kell, D.B.; Baker, P.N.; Kenny, L.C.; Mamas, M.A.; Neyses, L.; Dunn, W.B. Automated workflows for accurate mass-based putative metabolite identification in LC/MS-derived metabolomic datasets. *Bioinformatics* **2011**, *27*, 1108–1112. [CrossRef]
14. Carlson, R.; Tugizimana, F.; Steenkamp, P.A.; Dubery, I.A.; Labuschagne, N. Differential metabolic reprogramming in *Paenibacillus alvei*-primed *Sorghum bicolor* seedlings in response to *Fusarium pseudograminearum* infection. *Metabolites* **2019**, *9*, 150. [CrossRef] [PubMed]
15. Fallah, N.; Pang, Z.; Dong, F.; Zhou, Y.; Lin, W.; Fabrice, K.M.A.; Hu, C.; Yuan, Z. Niche differentiation modulates metabolites abundance and composition in silicon fertilizer amended soil during sugarcane growth. *BMC Plant Biol.* **2022**, *22*, 497. [CrossRef]
16. Benton, P.H.; Ivanisevic, J.; Rinehart, D.; Epstein, A.; Kurczy, M.E.; Boska, M.D.; Gendelman, H.E.; Siuzdak, G. An Interactive Cluster Heat Map to Visualize and Explore Multidimensional Metabolomic Data. *Metabolomics* **2015**, *11*, 1029–1034. [CrossRef]
17. Xia, J.; Wishart, D.S. Metabolomic data processing, analysis, and interpretation using MetaboAnalyst. *Curr. Protoc. Bioinform.* **2011**, *14*, 14.10.1–14.10.48. [CrossRef]
18. Trivedi, D.; Iles, R. The application of SIMCA P+ in shotgun metabolomics analysis of ZIC®HILIC-MS spectra of human urine-experience with the Shimadzu IT-ToF and profiling solutions data extraction software. *J. Chromatogr. Sep. Tech.* **2012**, *3*, 145. [CrossRef]
19. Liu, G.; Lee, D.P.; Schmidt, E.; Prasad, G.L. Pathway analysis of global metabolomic profiles identified enrichment of caffeine, energy, and arginine metabolism in smokers but not moist snuff consumers. *Bioinform. Biol. Insights* **2019**, *13*, 1177932219882961. [CrossRef] [PubMed]
20. Fukushima, A.; Kusano, M. Recent progress in the development of metabolome databases for plant systems biology. *Front. Plant Sci.* **2013**, *4*, 73. [CrossRef] [PubMed]
21. Rodrigues, A.M.; António, C. Standard key steps in mass spectrometry-based plant metabolomics experiments: Instrument performance and analytical method validation. *Methods Mol. Biol.* **2018**, *1778*, 19–31. [CrossRef]
22. De-la-Cruz Chacón, I.; Riley-Saldaña, C.; González-Esquinca, A. Secondary metabolites during early development in plants. *Phytochem. Rev.* **2012**, *12*, 47–64. [CrossRef]
23. Mhlongo, M.I.; Piater, L.A.; Steenkamp, P.A.; Labuschagne, N.; Dubery, I.A. Metabolomic evaluation of tissue-specific defense responses in tomato plants modulated by PGPR-priming against *Phytophthora capsici* infection. *Plants* **2021**, *8*, 1530. [CrossRef] [PubMed]
24. Moënne-Loccoz, Y.; Mavingui, P.; Combes, C.; Normand, P.; Steinberg, C. Microorganisms and Biotic Interactions, Chapter 11. In *Environmental Microbiology: Fundamentals and Applications: Microbial Ecology;* Bertrand, J.C., Caumette, P., Lebaron, P., Matheron, R., Normand, P., Sime-Ngando, T., Eds.; Springer Science: Dordrecht, The Netherlands, 2015. [CrossRef]
25. Aksenov, A.A.; da Silva, R.; Knight, R.; Lopes, N.; Dorrestein, P.C. Global chemical analysis of biology by mass spectrometry. *Nat. Rev. Chem.* **2017**, *1*, 0054. [CrossRef]
26. Saftić, L.; Peršurić, Ž.; Fornal, E.; Pavlešić, T.; Kraljević Pavelić, S. Targeted and untargeted LC-MS polyphenolic profiling and chemometric analysis of propolis from different regions of Croatia. *J. Pharm. Biomed. Anal.* **2019**, *165*, 162–172. [CrossRef] [PubMed]

27. De Vos, R.; Moco, S.; Lommen, A.; Keurentjes, J.; Bino, R.; Hall, R. Untargeted large-scale plant metabolomics using liquid chromatography coupled to mass spectrometry. *Nat. Protoc.* **2007**, *2*, 778–791. [CrossRef] [PubMed]
28. Manickam, S.; Rajagopalan, V.R.; Kambale, R.; Rajasekaran, R.; Kanagarajan, S.; Muthurajan, R. Plant Metabolomics: Current Initiatives and Future Prospects. *Curr. Issues Mol. Biol.* **2023**, *45*, 8894–8906. [CrossRef] [PubMed]
29. Sanz-Cortés, M.; Carbajo, R.; Crispi, F.; Figueras, F.; Pineda-Lucena, A.; Eduard, G. Metabolomic profile of umbilical cord blood plasma from early and late intrauterine growth restricted (IUGR) neonates with and without signs of brain vasodilation. *PLoS ONE* **2013**, *8*, e80121. [CrossRef] [PubMed]
30. Granato, D.; Santos, J.; Escher, G.; Ferreira, B.; Maggio, R. Use of principal component analysis (PCA) and hierarchical cluster analysis (HCA) for multivariate association between bioactive compounds and functional properties in foods: A critical perspective. *Trends Food Sci. Technol.* **2018**, *72*, 83–90. [CrossRef]
31. Pretorius, C.J.; Steenkamp, P.A.; Tugizimana, F.; Piater, L.A.; Dubery, I.A. Metabolomic characterisation of discriminatory metabolites involved in halo blight disease in oat cultivars caused by *Pseudomonas syringae* pv. *coronafaciens*. *Metabolites* **2022**, *12*, 248. [CrossRef]
32. Seth, T.; Asija, S.; Umar, S.; Gupta, R. The intricate role of lipids in orchestrating plant defense responses. *Plant Sci.* **2024**, *338*, 111904. [CrossRef]
33. Tohge, T.; de Souza, L.; Fernie, A. Current understanding of the pathways of flavonoid biosynthesis in model and crop plants. *J. Exp. Bot.* **2017**, *68*, 4013–4028. [CrossRef]
34. Pott, D.; Osorio, S.; Vallarino, J. From central to specialized metabolism: An overview of some secondary compounds derived from the primary metabolism for their role in conferring nutritional and organoleptic characteristics to fruit. *Front. Plant Sci.* **2019**, *10*, 835. [CrossRef]
35. Falcone Ferreyra, M.; Rius, S.; Casati, P. Flavonoids: Biosynthesis, biological functions, and biotechnological applications. *Front. Plant Sci.* **2012**, *3*, 222. [CrossRef]
36. Ishihara, A.; Matsuda, F.; Miyagawa, H.; Wakasa, K. Metabolomics for metabolically manipulated plants: Effects of tryptophan overproduction. *Metabolomics* **2007**, *3*, 319–334. [CrossRef]
37. Pascual, M.; El-Azaz, J.; de la Torre, F.; Cañas, R.; Avila, C.; Cánovas, F. Biosynthesis and metabolic fate of phenylalanine in conifers. *Front. Plant Sci.* **2016**, *7*, 1030. [CrossRef] [PubMed]
38. Adolfsson, L.; Nziengui, H.; Abreu, I.; Šimura, J.; Beebo, A.; Herdean, A.; Aboalizadeh, J.; Široká, J.; Moritz, T.; Novák, O.; et al. Enhanced secondary- and hormone metabolism in leaves of arbuscular mycorrhizal *Medicago truncatula*. *Plant Physiol.* **2017**, *175*, 392–411. [CrossRef] [PubMed]
39. Sartini, S.; Djide, M.N.; Nainu, F.L. Correlation Phenolic Concentration to Antioxidant and Antibacterial Activities of Several Ethanolic extracts from Indonesia. *J. Phys. Conf. Ser.* **2019**, *1341*, 072009. [CrossRef]
40. Daglia, M. Polyphenols as antimicrobial agents. *Curr. Opin. Biotechnol.* **2012**, *23*, 174–181. [CrossRef]
41. Mathesius, U. Flavonoid functions in plants and their interactions with other organisms. *Plants* **2018**, *7*, 30. [CrossRef]
42. Wu, J.; Lv, S.; Zhao, L.; Gao, T.; Yu, C.; Hu, J.; Ma, F. Advances in the study of the function and mechanism of the action of flavonoids in plants under environmental stresses. *Planta* **2023**, *257*, 108. [CrossRef]
43. Khawula, S.; Gokul, A.; Niekerk, L.-A.; Basson, G.; Keyster, M.; Badiwe, M.; Klein, A.; Nkomo, M. Insights into the Effects of Hydroxycinnamic Acid and Its Secondary Metabolites as Antioxidants for Oxidative Stress and Plant Growth under Environmental Stresses. *Curr. Issues Mol. Biol.* **2024**, *46*, 81–95. [CrossRef]
44. Behr, M.; Sergeant, K.; Leclercq, C.; Planchon, S.; Guignard, C.; Lenouvel, A.; Renaut, J.; Hausman, J.; Lutts, S.; Guerriero, G. Insights into the molecular regulation of monolignol-derived product biosynthesis in the growing hemp hypocotyl. *BMC Plant Biol.* **2018**, *18*, 1. [CrossRef]
45. Kim, H. Lipid metabolism in plants. *Plants* **2020**, *9*, 871. [CrossRef]
46. Pretorius, C.J.; Zeiss, D.R.; Dubery, I.A. The presence of oxygenated lipids in plant defence in response to biotic stress: A metabolomics appraisal. *Plant Signal. Behav.* **2021**, *16*, 1989215. [CrossRef]
47. Drincovich, M.; Voll, L.; Maurino, V. Editorial: On the diversity of roles of organic acids. *Front. Plant Sci.* **2016**, *7*, 1592. [CrossRef]
48. Balmer, A.; Pastor, V.; Glauser, G.; Mauch-Mani, B. Tricarboxylates induce defense priming against bacteria in *Arabidopsis thaliana*. *Front. Plant Sci.* **2018**, *9*, 1221. [CrossRef] [PubMed]
49. Nicholson, R.; Jamil, F.; Snyder, B.; Lue, W.; Hipskind, J. Phytoalexin synthesis in the juvenile sorghum leaf. *Physiol. Mol. Plant Pathol.* **1988**, *33*, 271–278. [CrossRef]
50. Lam, L.P.Y.; Wang, L.; Lui, A.C.W.; Liu, H.; Umezawa, T.; Tobimatsu, Y.; Lo, C. Flavonoids in major cereal grasses: Distribution, functions, biosynthesis, and applications. *Phytochem. Rev.* **2023**, *22*, 1399–1438. [CrossRef]
51. Du, Y.; Chu, H.; Wang, M.; Chu, I.K.; Lo, C. Identification of flavone phytoalexins and a pathogen-inducible flavone synthase II gene (SbFNSII) in sorghum. *J. Exp. Bot.* **2010**, *61*, 983–994. [CrossRef] [PubMed]
52. Adamczak, A.; Ożarowski, M.; Karpiński, T.M. Antibacterial activity of some flavonoids and organic acids widely distributed in plants. *J. Clin. Med.* **2020**, *9*, 109. [CrossRef] [PubMed]
53. Przybylska-Balcerek, A.; Frankowski, J.; Stuper-Szablewska, K. Bioactive compounds in sorghum. *Eur. Food Res. Technol.* **2018**, *245*, 1075–1080. [CrossRef]

54. Poloni, A.; Jan Schirawski, J. Red Card for Pathogens: Phytoalexins in Sorghum and Maize. *Molecules* **2014**, *19*, 9114–9133. [CrossRef] [PubMed]
55. Xie, Y.; Chen, J.; Xiao, A.; Liu, L. Antibacterial activity of polyphenols: Structure-activity relationship and influence of hyperglycemic condition. *Molecules* **2017**, *22*, 1913. [CrossRef] [PubMed]

metabolites

MDPI

Review

Deleterious Effects of Heat Stress on the Tomato, Its Innate Responses, and Potential Preventive Strategies in the Realm of Emerging Technologies

Qaisar Khan, Yixi Wang, Gengshou Xia, Hui Yang, Zhengrong Luo and Yan Zhang *

Department of Landscape and Horticulture, Ecology College, Lishui University, Lishui 323000, China; qaisar.khan@yahoo.com (Q.K.); yxwangls@163.com (Y.W.); lsxyxgs@163.com (G.X.); lsxyyh@126.com (H.Y.); zrluo@126.com (Z.L.)
* Correspondence: yzhang@lsu.edu.cn; Tel.: +86-18867805285

Abstract: The tomato is a fruit vegetable rich in nutritional and medicinal value grown in greenhouses and fields worldwide. It is severely sensitive to heat stress, which frequently occurs with rising global warming. Predictions indicate a 0.2 °C increase in average surface temperatures per decade for the next three decades, which underlines the threat of austere heat stress in the future. Previous studies have reported that heat stress adversely affects tomato growth, limits nutrient availability, hammers photosynthesis, disrupts reproduction, denatures proteins, upsets signaling pathways, and damages cell membranes. The overproduction of reactive oxygen species in response to heat stress is toxic to tomato plants. The negative consequences of heat stress on the tomato have been the focus of much investigation, resulting in the emergence of several therapeutic interventions. However, a considerable distance remains to be covered to develop tomato varieties that are tolerant to current heat stress and durable in the perspective of increasing global warming. This current review provides a critical analysis of the heat stress consequences on the tomato in the context of global warming, its innate response to heat stress, and the elucidation of domains characterized by a scarcity of knowledge, along with potential avenues for enhancing sustainable tolerance against heat stress through the involvement of diverse advanced technologies. The particular mechanism underlying thermotolerance remains indeterminate and requires further elucidatory investigation. The precise roles and interplay of signaling pathways in response to heat stress remain unresolved. The etiology of tomato plants' physiological and molecular responses against heat stress remains unexplained. Utilizing modern functional genomics techniques, including transcriptomics, proteomics, and metabolomics, can assist in identifying potential candidate proteins, metabolites, genes, gene networks, and signaling pathways contributing to tomato stress tolerance. Improving tomato tolerance against heat stress urges a comprehensive and combined strategy including modern techniques, the latest apparatuses, speedy breeding, physiology, and molecular markers to regulate their physiological, molecular, and biochemical reactions.

Keywords: heat stress; reactive oxygen species; heat shock proteins; stress signaling; genome editing; omics; heat tolerance pyramiding; genetic resources

1. Introduction

The tomato, scientifically known as *Solanum lycopersicum* in the Solanaceae family, is cultivated in diverse environmental circumstances and geographical regions ranging from tropical to temperate environments. The tomato arrived in Europe during the Renaissance and was scattered to the Mediterranean region [1]. The tomato is a fruit vegetable among the most cultivated crop plants on the earth and is grown in greenhouses and fields worldwide [2]. It is rich in medicinal and nutritional contents, including lycopene, the valuable compound having anti-oxidative and anti-cancer properties, vitamins A and C, β-carotene, iron, phosphorus, flavonoids, ferulic acid, hydroxycinnamic acid, chlorogenic acid, homovanillic acid, folate, and low calories [3–6]. Around 80% of tomatoes are used as processed food like ketchup, soup, paste, sauces, and juices [7,8]. Globally, China is the biggest producer of tomatoes, followed by India and Turkey (FAO-2021) [9].

Earlier researchers have studied and discussed numerous facets of heat stress on tomatoes, which include plant growth, leaf morphology, photosynthesis, and reproductive performance, including fruit sets, root growth, ROC species, pollen viability, pollen numbers, and inflorescence numbers, focusing on individual aspects. In the context of global warming, this current review provides a thorough critical analysis of heat stress on tomatoes, covering all major aspects, including seed germination, growth, and development and physiological, biochemical, genetic, and molecular reactions. Furthermore, it offers comprehensive information about the available technologies and potential approaches for creating imminent heat-tolerant cultivars. This present review provides complete insight into all significant negative aspects of heat stress on tomatoes, their morphological, physiological, biochemical, and molecular responses, analytical methodologies, and strategies for developing heat-tolerant tomato cultivars.

2. Heat Stress

The undesirable influence of non-living dynamics and factors on living organisms in a specified environment is termed abiotic stress [10]. Several abiotic stresses, such as heat, flood, drought, and salt, reduce the production and yield of tomato crops by up to 75%; particularity is subjected to the severity of stresses [11]. Generally, heat stress is defined as an increase in temperature beyond tolerance for an unknown duration, adequate to trigger irretrievable impairment in plant growth and development. In contrast, heat tolerance is defined as a plant's capability of growth and production to an economic yield level under high temperatures [12,13]. In the context of tomato cultivation, heat stress is commonly classified as moderate heat stress, ranging from 32 °C to 37 °C, and severe heat stress, ranging from 38 °C to 45 °C [14]. Climate changes drastically affect tomato crop production and yield, particularly in Asian countries [15]. It is a common opinion that soaring temperatures will enhance the average temperature of the earth's surface by 0.2 °C every ten years in the coming thirty years, increasing extreme weather and, consequently, negatively affecting tomato plant growth and development and severely reducing its production and yield [16,17].

3. Negative Effects of Heat Stress on Growth and Development

Tomato plants can typically grow and develop reproductive organs, pollen grains, and fruit sets at an optimum temperature between 15 °C and 32 °C; however, temperatures beyond 35 °C badly stress sexual and asexual development [18,19]. Being sessile, tomato plants often face erratic high-temperature conditions, which adversely influence them as temperatures go beyond the optimal ranges. Studies have revealed that high temperatures increase the frequencies of hot and dry days, affecting tomato plant growth, biomass, phenology, agronomic traits, production, and yield [20,21]. High temperature significantly disrupted physiological characteristics such as leaf water content, membrane stability, canopy temperature drop, photosynthesis, stomatal conductance, chlorophyll content, and fluorescence [22,23]. High heat stress negatively influences the metabolic processes involved in growth and development [24]. It produces reactive oxygen species, like hydrogen peroxide (H_2O_2), superoxide, hydroxyl radical (OH), and singlet oxygen 1[O2], which adversely disturb cellular homeostasis [25–27].

In numerous crops, particularly tomatoes, reproductive growth is highly prone to heat stress compared with vegetative growth [28]. Seeds are a significant part of plants, which carry genetic information to descending generations [29]. However, higher temperatures seriously threaten seed germination, seedling physiology, and phenotypic expression [30,31]. Seed germination tested at a constant range of temperatures from 24 °C to 37 °C for 8 days showed that the rate of seed germination started reducing after 28 °C and entirely ceased at 36 °C. Cotyledon size reduced at a temperature higher than 24 °C but the seedling's hypocotyl length increased by 1.9 cm (24 °C), 4.1 cm (28.5 °C), and 2.6 cm (31.5 °C), which shows that temperatures higher than 28.5 °C also affect hypocotyl length negatively. In the same study, tomato seedlings aged 12 days (germination: 24 °C) were exposed to 37 °C for 24 h, and a 1 h heat wave (45 °C) damaged the seedling's recovery ability. Exposure to a 45 °C heat wave for 1 h, 3 h, 6 h, and 12 h showed that the seedlings started drying at 6 h and lost recovery capability at 12 h. The number of lateral roots was reduced, but the growth of the primary root was stopped at 37 °C. Although 45 °C did not affect lateral roots significantly, it halted the growth of the primary root [32,33]. High temperature reduces tomato root growth and nutrient uptake, affecting root–shoot source–sink relationships that affect fruit yield and quality [28,34]. The 30-day-old seedlings of two tomato cultivars (Dafnis and Minichal) were subjected to heat stress of 40 °C for 7 days in a growth chamber, and the results indicated that the effects of high temperature on tomato leaves started to appear on the second day. However, a big difference was noticed on the seventh day. The damage to the leaves of the Dafnis cultivar was over 60%, but Minichal showed resistance [35], which suggests that heat is a serious problem for tomato plants, and the creation of heat-resistant varieties is very important to avoid economic losses.

4. Adverse Impacts of Heat Stress on Photosynthetic Parameters

Exposure of tomato plants to higher temperatures leads to significant disruption of the chloroplast, which produces adenosine triphosphate (ATP) and phytochemicals. A good performance of the photosynthetic apparatus under high-temperature stress shows the ability of a plant to tolerate and adapt to stressful conditions [36,37]. However, HS negatively affects several important components of photosynthesis (Figure 1). Heat stress inhibits chlorophyll formation; hence, measuring chlorophyll (a, b) concentrations can be a parametric indication for identifying heat-resistant plants. Under directly applied high-temperature stress of 45 °C (severe stress) for 2 h, a heat-resistant tomato cultivar showed a decline in the ratio of chlorophyll (a:b) and an increase in the ratio of chlorophyll to carotenoid in contrast to the control condition of 25/20 °C (day/night). Heat-sensitive cultivars, on the other hand, showed a decrease in the CO_2 assimilation rate (A), the net photosynthetic rate (Pn), and photosystem II efficiency (Fv/Fm), which represents the highest quantum efficacy of photosystem II (PSII) and is used to assess chloroplasts' normal or superior functioning under heat stress conditions [38–42].

Photosynthesis in plants is a heat-sensitive physiological process that influences chlorophyll content, CO_2 integration, D1 and D2 protein turnover, chloroplast components, and heat-responsive protein deactivation [43]. Plant growth, development, production, yield, and future food security are deeply connected with photosynthesis [44,45]. Persistent higher heat stress inhibits photosynthetic activities, which affect the growth and production of plants. Photosystems I and II (PSI, II), chlorophyll, the electron transport chain, and CO_2 assimilation are among the significant photosynthesis process components, so damage to any of them retard the photosynthetic mechanism [46]. A study by reference [47] revealed that the production of protochlorophyllide (Pchlide), an intermediate in the biosynthetic pathway of chlorophyll, was repressed by 70% at high temperature (42 °C) compared with a control (25 °C), which reduced chlorophyll manufacture to 60% in cucumber seedlings. Similarly, in the same study, the activities of the 5-aminolevulinic acid dehydratase (ALAD) enzyme, which is responsible for converting 5-aminolevulinic acid (ALA) into porphobilinogen (PBG), an intermediate in chlorophyll biosynthesis, and porphobilinogen deaminase (PBGD), which is necessary for converting PBG into urogen, were reduced by 45% and 28% at a higher temperature compared with a control. The PSII electron transport system is highly vulnerable to high temperature because it increases thylakoid membrane fluidity, which knockouts the PSII light-garnering system from the thylakoid membrane and, consequently, destroys PSII integrity [48,49]. High temperature severely disturbs tomato plants' photosynthetic activities, specifically in susceptible tomato varieties [50].

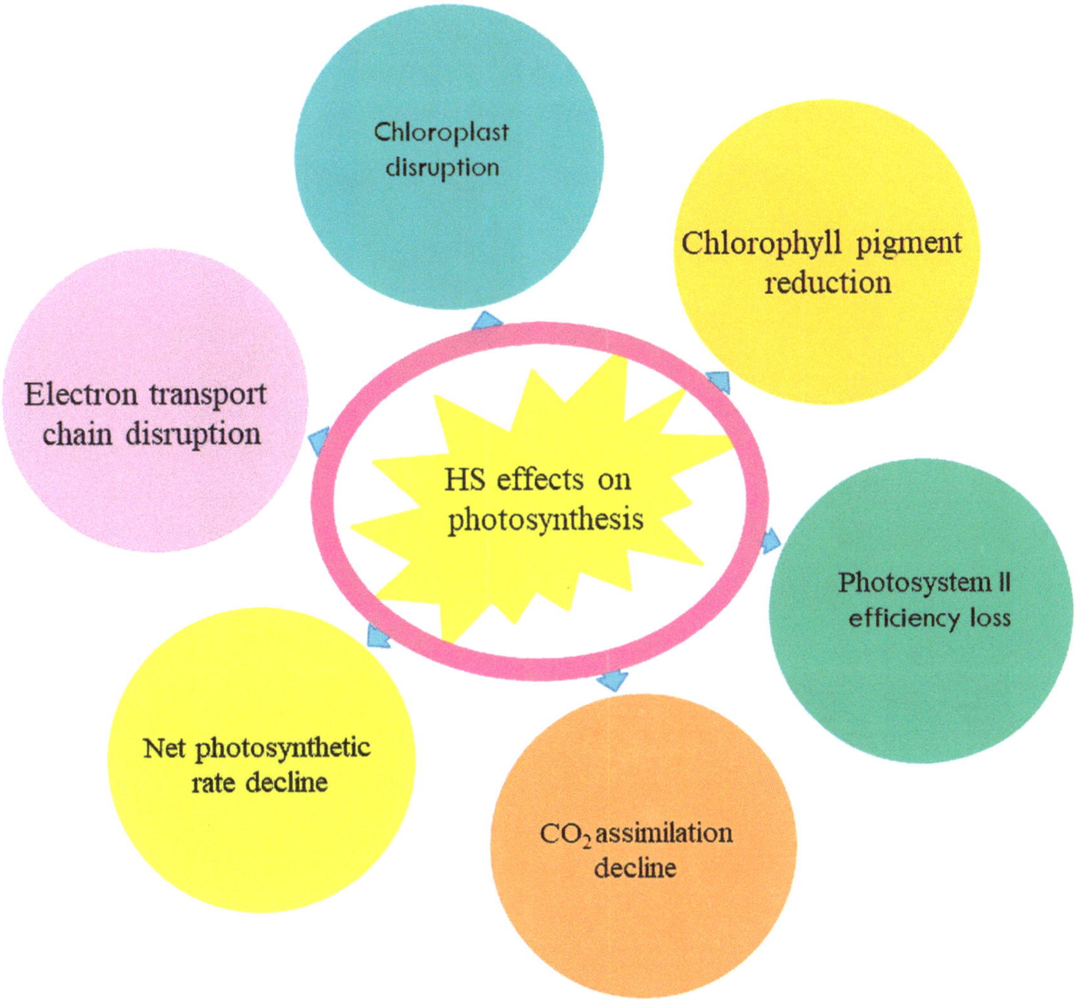

Figure 1. The negative impacts of heat stress on photosynthetic parameters.

5. Heat Stress Represses Reproductive Performance

Cultivated tomatoes are autogamous plants, and high temperatures negatively impact their pollination [51]. Under high heat stress, the tomato style, which is the female reproductive part of the flower gynoecium holding the stigma, extends abnormally and goes out the antheridial cones, minimizing the chances of pollination and, consequently, reducing fruit sets (Figure 2) [52–55]. Heat stress distorts pollen grain development by reducing the amount of carbohydrates at the early stages of development, reducing the sugar concentration in mature pollens, and resulting in slashed pollen viability [56,57]. The responses of heat-resistant and susceptible genotypes of *Lycopersicon esculentum* Mill. and *L. pimpinellifolium* Mill. to heat stress was evaluated by subjecting plants to optimal (27/23 °C, day/night) and high-temperature (35/23 °C) regimes in a greenhouse. The heat tolerance ratings of the genotypes were determined by calculating the percentage of fruit that successfully developed under high and optimal temperatures. The fruit sets varied from 41% to 84% in the temperature-sensitive genotypes and 45% to 91% in the heat-tolerant genotypes at optimal temperatures. The genotypes with great heat sensitivity did not yield any fruit. In contrast, the genotypes that could withstand high temperatures produced fruit set rates ranging from 45% to 65% [58]. Stigma and stylar exsertion negatively affect fruit set forming capabilities because of elevated temperatures [59]. It is essential to create heat-resistant tomato varieties with higher fruit sets as these varieties will benefit tomato crop yield in areas where the growing season's average temperature is 35 °C or higher.

Figure 2. Phenotypic changes in the tomato (cv. Saladette) flowers subjected to heat stress. (**a**,**b**) are young flower buds and flowers at the blooming stage under normal temperatures (26/19 °C). (**c**,**d**) are flower buds and opened flowers under heat stress (36/26 °C).

6. Negative Impacts of Heat Stress on Agronomic Traits

Agronomic traits refer to the characteristics of plants that exert influence on their productivity, quality, and ability to cope with biotic and abiotic stressors. The adverse effects of heat stress hinder the overall capacity of tomatoes to reach the desired agronomic performance. Several investigations have been carried out to assess the negative impacts of heat stress on different aspects of tomato leaves, such as fresh mass, the leaf area, the leaf area ratio, the specific leaf area, and plant height and stem diameter under multiple heat stress levels [60,61]. The total area of all leaves on a single plant is referred to as the leaf area (LA) [62]. The specific leaf area (SLA) is a crucial statistic for plant growth modelers as it specifies the amount of fresh leaf area to allocate for each unit of biomass produced; it is calculated by dividing the leaf area by the leaf mass (LA/LM) [63]. Heat stress negatively affects plant leaves in several other ways, including reducing their capacity to retain water and early leaf mortality [64,65]. Heat stress causes glucose reserve shortages because it impedes starch accumulation, which results in a decrease in soluble sugar concentration obtained from the decomposition of starch in fully developed pollen grains [66]. These incidents can potentially decrease tomato pollen fertilization capacity [67]. An increase in diurnal temperature over 25 °C adversely impacted fruit quantity, weight, and seed count per fruit markedly [68].

Heat Stress and Heat Shock Combined Effects

The tomato cultivars Kervic F1 (heat-resistant) and UC 82-B (heat-susceptible) at the age of 35 days after heat shock at 50 °C for 30 s were subjected to a heat stress of 35/27 °C (day/night) compared to control 26/20 °C (day/night) conditions to study agronomic traits including the leaf area (LA), leaf area ratio (LAR), specific leaf area (SLA), number of pollen grains per flower (NPGF), number of fruits per plant (NFP), and fruit fresh mass per plant (FFMP) [69]. Heat stress and heat shock negatively influenced the agronomic traits of tomatoes, particularly the leaf area, pollen grains, fruit sets, and fruit weight (Figure 3). The heat stress repercussions mentioned herein hinder the overall capacity of tomatoes to perform better agronomically. Therefore, it is crucial to extensively examine all physiological and agronomic characteristics to address heat stress issues effectively.

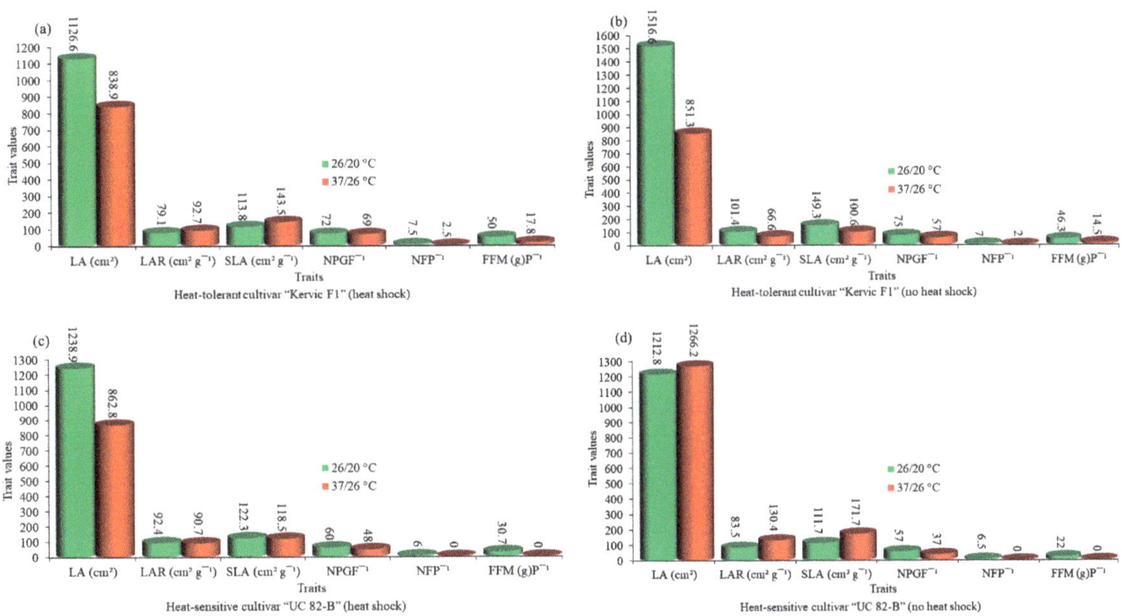

Figure 3. Effects of heat stress and heat shock on the agronomic parameters of resistant (Kervic F1) and sensitive (UC 82-B) tomato cultivars. (**a**) Heat-resistant cultivar (Kervic F1) under heat stress after heat shock stress. (**b**) Heat-resistant cultivar (Kervic F1) under heat stress without heat shock stress. (**c**) Heat-sensitive cultivar (UC 82-B) under heat stress after heat shock stress. (**d**) Heat-sensitive cultivar (UC 82-B) under heat stress without heat shock stress. The X-axis indicates the types of agronomic parameters investigated, and the Y-axis shows values of changes in agronomic parameters under heat stress.

7. Over Production of Reactive Oxygen Species (ROS)

An equilibrium among numerous pathways in diverse cell compartments maintains cellular homeostasis under an optimal temperature. The sustainability of homeostasis cannot be guaranteed when temperatures go beyond the optimal level because various pathways have diverse optimum temperatures within the cell, and heat stress upsets this functional balance between different pathways [70]. ROS are over-produced in response to high-temperature stress and other harmful factors that affect several intracellular pathways [71,72]. ROS include free and non-free radicals that contain oxygen and are capable of self-regulating survival with one or more unpaired electrons (Figure 4). Free-radical ROS, like hydroxyl ion radical (OH•), superoxide anion radical ($O_2^{\bullet -}$), and alkoxyl ($RO^{\bullet}$), carbonate ($CO_3^{\bullet -}$), peroxyl ($RO_2^{\bullet}$), hydroperoxyl ($HO_2^{\bullet}$) molecular oxygen (O_2), and non-free radical species such as ozone (O_3), hydrogen peroxide (H_2O_2), singlet oxygen ($^1O_2$), hypobromous acid (HOBr), hydroperoxy (ROOH), hypoiodous acid (HOI), hypochlorous acid (HOCl), are severely toxic to plant growth and development [73–76]. In response to heat stress, ROS are generated in different cellular parts, like the plasma membrane, mitochondria, cell wall, chloroplast, peroxisome, endoplasmic reticulum, and apoplast [77,78]. Excessive production of ROS damages molecules and compounds in plant cells like lipids, deoxyribonucleic acid (DNA), ribonucleic acid (RNA), proteins, and carbohydrates [79–81].

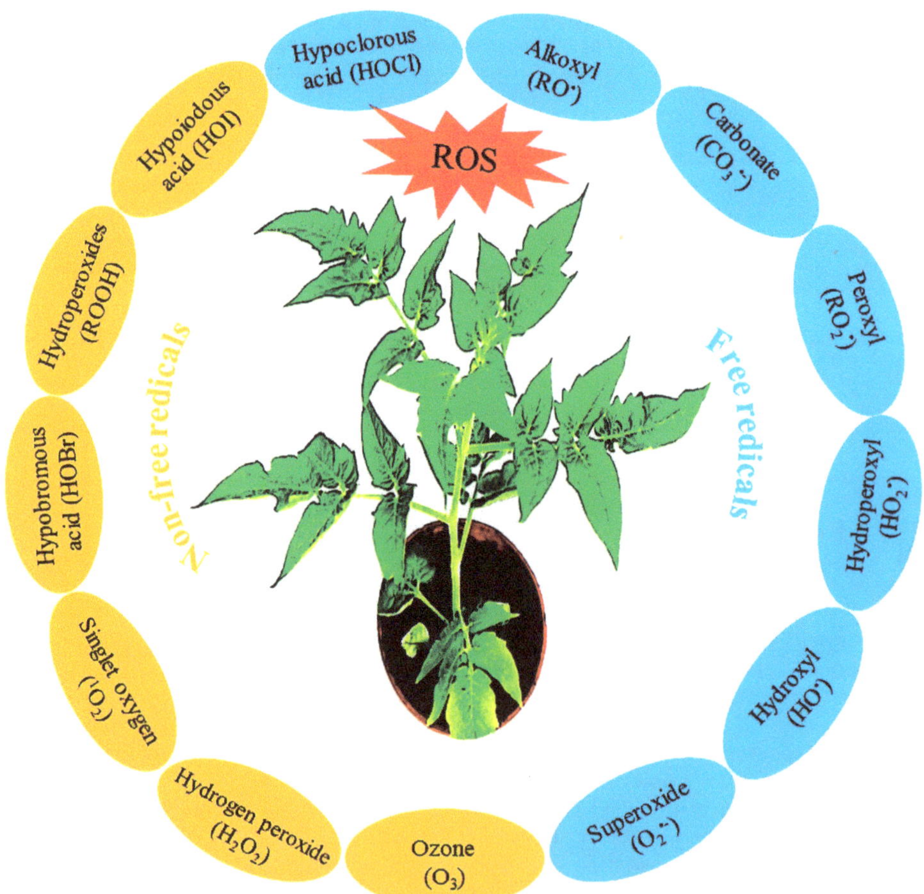

Figure 4. A display showing various types of reactive oxygen species functioning in tomato plants.

8. Heat Stress Causes Oxidative Stress

Tomato plants are sensitive to high temperatures, even a little beyond optimal, which causes the overproduction of reactive oxygen species (ROS) [82]. An equivalence between ROS production and antioxidants is essential for the proper growth and development of plants [83], but high temperatures are known to disrupt this equivalence in tomato plants. The tomato variety "Tmknvf$_2$" was subjected to oxidative metabolism analysis at an optimal temperature of 25 °C and a high temperature of 35 °C by reference [84], focusing on superoxide dismutase (SOD), ascorbate peroxidase (APX), dehydroascorbate peroxidase (DHAR), guaiacol peroxidase (GPX), catalase (CAT), ascorbate (AsA), glutathione reductase (GR), hydrogen peroxide (H_2O_2), dehydroascorbate (DHA), glutathione (GSH), oxidized glutathione (GSSG), total ascorbate, total glutathione, and dry weight (DW). Generally, the activities of the CAT, APX, DHAR, GR, and GPX enzymes are enhanced in response to higher temperatures [85,86]; however, in the case of reference [84], their activities were reduced because high temperatures denatured these proteins. The concentrations of GSH, GSSG, DHA, AsA, total ascorbate, and glutathione antioxidant compounds were higher at 35 °C than at 25 °C. These substrates are utilized by CAT, APX, DHAR, GR, and GPX in the ascorbate–glutathione cycle, but their activities were reduced by higher temperatures (35 °C), resulting in a higher accumulation of these substrates (Figure 5) and increased hydrogen peroxide (H_2O_2) accumulation in tomato leaves. Overproduction of ROS seriously impairs plant growth, development, and yield [87]. Therefore, it is imperative to investigate oxidative metabolism in tomato plants thoroughly and develop heat-tolerant varieties.

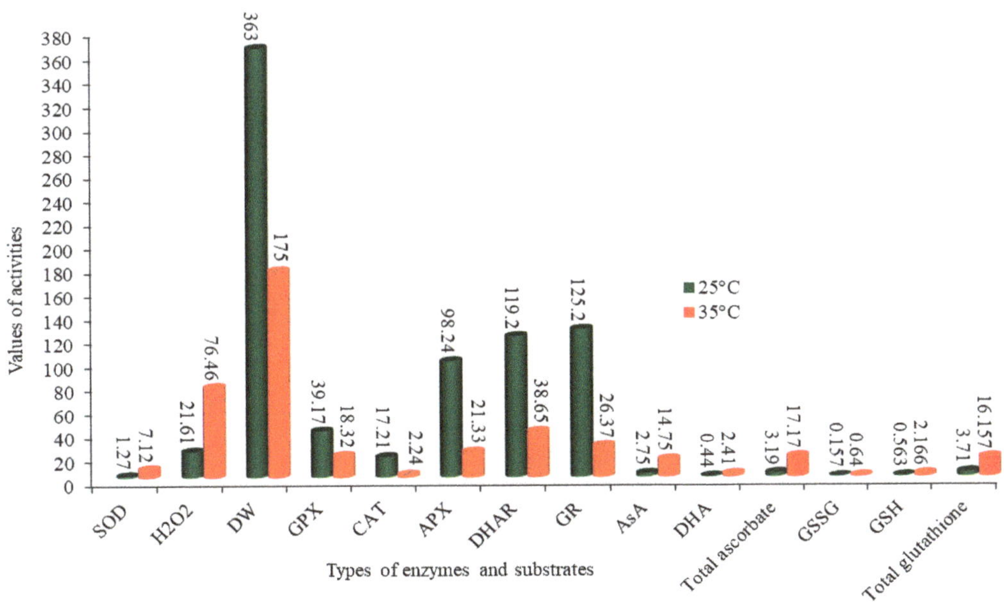

Figure 5. The activities of enzymes and substrates in tomato plants under heat stress. Measurement units, SOD: unit mg protein^{-1} min^{-1}, H_2O_2: mmol g^{-1} (FW), W.D: g plant^{-1}, GPX, CAT, APX, AsA, DHAR, DHA, GR, GSH, total ascorbate, and total glutathione: µmol mg-prot^{-1} min^{-1}.

9. Phenological Modifications in Response to Heat Stress

Plant heat resistance refers to the ability of plants to thrive and produce the required yield under high temperatures, which is specifically linked to the plant species or potentially to the distinct organs and tissues within the same plant. Plant reactions to heat stress depend on the threshold degree, exposure period, and plant nature. The effects of heat stress on a plant's many functioning processes, such as seed germination, development, growth, procreation, and yield, are toxic [88,89]. Under conditions of severely high temperature, serious damage to cells, even complete breakdown of cellular structures, and cell demise might occur rapidly [90]. In response to high temperatures, plants implement several short-term acclimation mechanisms and long-term evolutionary strategies for persistence (Figure 6) [91]. Among these stratagems are stomatal closure, leaf position changes, variations in the lipid configuration of the membrane, larger xylem, reduced water loss, fast maturation, increased transpiration, decreased absorption of radiation, an increase in the number of hairs on the surface, cuticle layer thickening, adoption of paraheliotropism, an increase in wax, late embryogenesis abundant proteins, transcriptional regulation, more vigorous antioxidant defense, signaling cascades stimulation, osmoprotectant, and phenological, morphological, biochemical, anatomical, molecular, and genetic adaptations [92,93]. Numerous heat-inducible genes, often referred to as heat shock genes (HSGs), exhibit upregulation in response to thermal stress. These genes encode HSPs, which are essential for plants to survive in life-threatening heat stress [94,95]. Heat shock proteins (HSPs) are biologically active only during certain plant development and growth stages, including seed germination, embryo microsporogenesis, and fruit ripening [96,97].

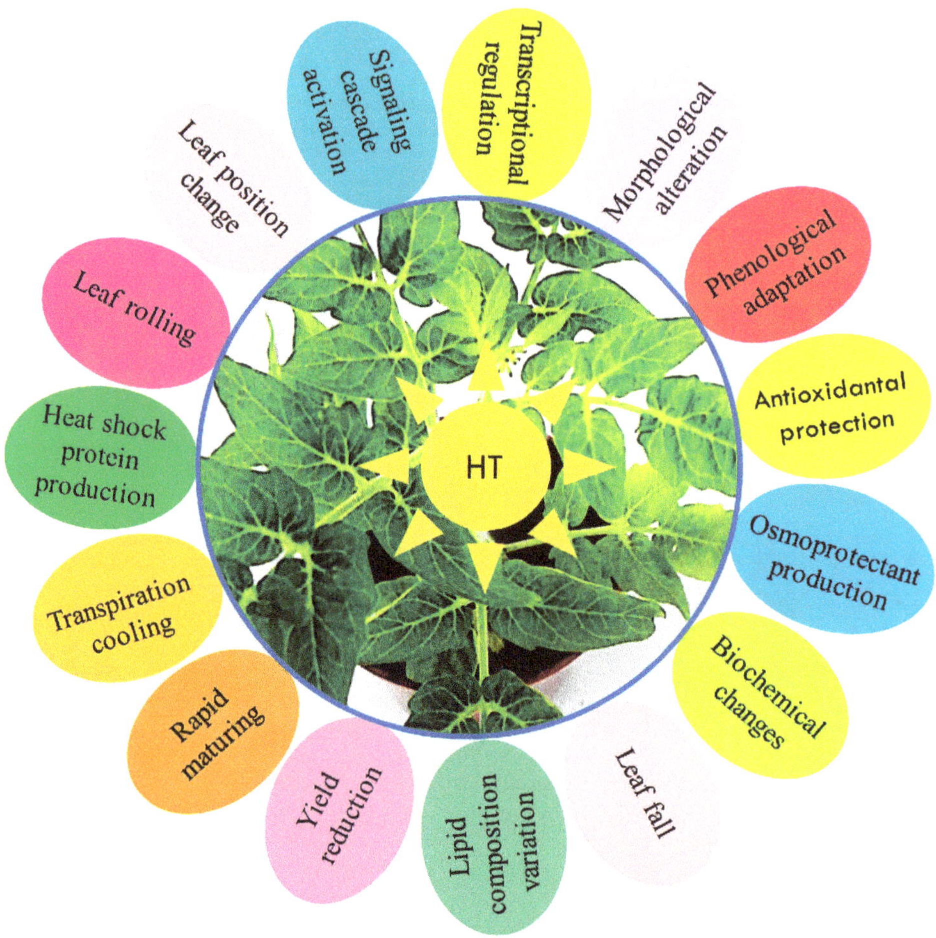

Figure 6. Various long- and short-term phenological changes adopted by tomato plants in response to heat stress.

Under elevated temperatures, tomato plants manifest symptoms including stunted growth, aberrant development, poor photosynthesis, reduced crop output, and even plant mortality [98]. However, it is essential to note that not all genotypes of tomatoes are susceptible to high temperatures [99]. Some studies found that growing tomatoes between 21 °C and 26 °C decreased the entire carotene concentration but did not affect lycopene quantity. In contrast, cultivating tomatoes within the temperature range of 27 to 32 °C reduced ascorbate and lycopene levels while concurrently enhancing the levels of routine caffeic acid derivatives and glucosides [100]. Moreover, tomato fruit firmness and better shelf life were found in F1 hybrids, having mutant genes such as alcobaca (alc), ripening inhibitor (rin), and non-ripening (nor). These hybrids can maintain, to a greater extent, tomato hue, feel, taste, and nutritional value even when exposed to high-heat-stress conditions [101]. High temperatures during fruit development negatively affect assimilation, distribution, and shelf storage. The fruit produces several structural and functional elements throughout the ripening process, including starch and secondary metabolites that affect the interior quality of fruits [102]. The sucrose that fruit receives from the leaves as photosynthesized sugars adds to the fruit's dry matter. A tomato's flavor results from transforming carbs like sucrose into organic acids and aromatic compounds [103]. Environmental parameters, such as temperature, water irradiation, and photosynthesis, affect fruit quality [104]. These problems, because of heat stress, draw our attention to exploring, selecting, and utilizing cultivars of tomatoes capable of enduring high temperatures during the cultivation period. Therefore, it is critical to understand the molecular and genetic mechanisms regulating plants' short- and long-term natural defense strategies in response to heat stress, which could be applied to regulate the heat stress problem in crops, particularly in heat-sensitive plants like tomatoes.

10. Heat Shock Signaling Pathway Modulation

In the face of heat stress, plants have several free and dependent pathways to perceive external and internal signals, which significantly regulate the development of responses to create resistance to cope with the situation [105]. These responses entail the overexpression of several genes and the activation of complex integrated circuits involving various pathways. Cofactors and signaling molecules such as mitogen-activated protein kinase (MAPK/MPKs), sugar compounds, and Ca-dependent protein kinases (CDPKs) play a fundamental role in activating stress-responsive genes [106,107]. However, an intrinsic study must fully elucidate and understand the signaling molecules and pathways involved in developing heat tolerance.

11. Heat Shock Protein (HSP) Production

Heat stress often triggers the activation of heat-inducible genes known as heat shock genes (HSGs), which produce heat shock proteins (HSPs) that are essential for a plant's existence under very high temperatures [108]. HSPs act as chaperones to safeguard intracellular proteins from decomposition and maintain their integrity and functionality by facilitating protein folding [109]. Previously, scientists thought heat stress was the main trigger for HSP formation. However, they have since learned that many biotic and abiotic stimuli could cause HSP formation. They show up- or downregulation responses to biotic and abiotic stress situations, but further research is needed to understand signal recognition and transmission processes fully [6]. Additionally, a plant can overcome these obstacles with the help of post-transcriptional modifications, including alternative splicing and micro RNA (miRNA). Alternative splicing creates many transcripts from a single gene, while miRNA binds to mRNA to inhibit translation or induce mRNA cleavage at any location [110,111]. In plants, heat shock proteins can be divided into five categories, including small HSP20 (sHSP20), HSP60 (GroE), HSP70 (DnaK), HSP90, and (HSP100). Among these HSPs, HSP60 and HSP70 are incredibly conserved, suggesting their crucial function in the heat stress response [112]. HSP20s is a low-molecular-mass (15 to 42 kDa) family with a 90-amino acid alfa-crystallin domain (ACD) that forms a seven-stranded β-sandwich flanked by a variable N-terminal domain (NTD) with fewer to 85 amino acids and a short C-terminal extension (CTE) and is predominantly induced by heat stress in several higher plants [113].

Plants sense heat stress principally at the plasma membrane, leading to the opening of particular calcium channels, permitting calcium ions to enter the cell, and triggering the activation of mitogen-activated and calcium-dependent protein kinases, which in turn activate the heat stress response (HSR) [114–116]. At the time of the HSR, numerous specific genes are upregulated essentially, leading to the accumulation of a significant amount of HSPs in different cellular compartments, which play a crucial role in signaling and heat resistance mechanisms during the HSR; HSPs are commonly regulated by heat shock factors (HSFs) [117,118]. Various pathways transmit heat signals to HSFs, activating HSPs and heat-responsive genes (HRGs) and playing a significant role in plant heat adaption mechanisms, which suggests that the HSF-HSP pathway is critical in governing plant responses to heat stress [119].

12. Heat Shock Factor (HSF) Activation

Heat shock factors (HSFs) are activators that trigger the transcription of heat shock genes and bind to heat shock sequence elements (HSEs) found throughout the genome, which consist of a tandem array of three oppositely orientated "AGAAN" motifs or a variant of them that is less similar [120,121]. The structure of plant HSFs is very conserved and consists of several vital parts such as the oligomerization domain (OD), DNA binding domain (DBD), transcriptional activation motif (AHA), nuclear export signal (NES), and the nuclear localization signal (NLS) [122]. The oligomerization domain (OD) consists of a bipartite heptad pattern of hydrophobic amino acid residues in the HR-A and HR-B regions, and a flexible linker links it to the DNA-binding domain (DBD) [123]. The N-terminal DNA binding domain (DBD) is distinguished by a core helix–turn–helix motif that particularly attaches to the target promoter's heat stress elements (HSEs), activating stress-inducible gene transcription [124]. The plant HSF C-terminal stimulation domain is described by short peptide motifs (AHA) that consist of giant hydrophobic and acidic amino acid residues. These residues are unique to HSFA and are absent in the HSFB and C classes [125]. The nuclear localization signal (NLS) and nuclear export signal (NES) of HSFs play a significant role in forming a nuclear entrance complex consisting of the target protein and the receptor-mediated export complex, including the NES receptor exportin-α [126]. The classification of plant HSFs into HSFA, HSFB, and HSFC is based on the number of amino acid residues inserted into the HR-A and HR-B regions and the linker length area between the DBD and HR-A and HR-B regions [127]. Class A HSFs have the transcriptional activation domain, while classes B and C HSFs lack this specific amino acid motif and cannot promote transcriptional activation alone [128,129]. It is now well-established that several HSF classes modulate HSP expression in tomato plants and have a positive regulatory role in osmotic, oxidative, thermal, anoxia, and stress tolerance, particularly by HSFA [130].

The Heat Shock Factor A1 Class (HSFA1)

Investigations on model crop plants, including tomato [131], *A. thaliana* [132], and soybean [133], have shown that HSFA1-related genes are expressed constitutively under normal circumstances; however, their expression increases rapidly under heat stress, which designates them as significant master regulators of the heat stress response. In tomatoes, there are four members of the class HSFA1, namely, HsfA1a (Solyc08g005170), HsfA1b (Solyc03g097120), HsfA1c (Solyc08g076590), and HsfA1e (Solyc06g072750). Among these members, HSFA1a is the master regulator because of its consistent expression in control and heat stress (HS) conditions across all tissues. On the other hand, HSFA1c and HSFA1e are typically significantly expressed in red ripe fruits, while HSFA1b is strongly stimulated in all fruit stages [134]. HSFA2 plays a crucial role in the priming mechanism of tomato plants, which is responsible for maintaining pollen thermotolerance throughout the process of microsporogenesis [135]. A previous study provided evidence that the expression reduction in HSFA2 resulted in a decrease in the viability and germination rate of pollen exposed to HS during the meiosis and microsporogenesis phases, which supports the notion that it plays a crucial role in maintaining thermotolerance [136]. The expression levels of tomato HSF genes, namely, SlyHSF01, SlyHSF8, SlyHSF9, SlyHSF10, and SlyHSF11, have been observed to be significantly higher in leaf tissues under a high temperature (45 °C) compared with a control (30 °C) situation [137]. The cytoplasm is the site for tomato HSFA3 (Solyc09g009100) expression under a controlled environment, while the nucleus is the site of its expression under HS circumstances [138]. According to reports, tomato HsfA4s (Solyc07g055710, Solyc03g006000, and Solyc02g072000) significantly boost the expression of HS genes, while HSFA5 (Solyc12g098520) is a particular inhibitor of HSFA4 action [139,140]. A reduction in HSFB4a ((Solyc04g078770) expression and a boost in HSFA7 levels regulate thermo-tolerance in tolerant tomato cultivars [141]. The overexpression of SUMO E3 ligase (SlSIZ1) in tomatoes led to an enhanced heat tolerance by regulating the activities of HSFA1 and promoting the accumulation of HSP70 [142]. In a heat-resistant tomato cultivar (CLN1621L), the gene notabilis (Solyc07g056570) and acyl-sugar acyltransferase (Solyc09g014280) exhibit upregulation as positive regulators of HS tolerance, while the gene Pin-II proteinase inhibitor (Solyc03g020030) shows downregulation as a negative regulator of thermotolerance, indicating that the inverse expression of these genes encodes enzymes and proteins that play significant roles in mitigating heat stress [143].

13. Involvement of Omics Approaches

Omics technologies are distinguished by their systematic investigation and analysis of extensive datasets that capture the entirety of a biological system's structure and function at a specific level, which has significantly transformed the approaches used to study biological systems [144]. Multi-omics strategies involve techniques such as transcriptomics, genomics, metabolomics, proteomics, epigenomics, proteogenomics, lipidomics, interactomics, ionomics, phenomics, and bioinformatics, which produce a significant amount of data that can be utilized to understand the physiological and molecular mechanisms functioning in plants under stresses and devise effective strategies for mitigating the adverse impacts of such stresses [145,146]. However, relying exclusively on a single omics approach is inadequate to fully elucidate the complexities of plant responses to abiotic stresses, particularly HS. The utilization and incorporation of multi-omics methodologies are necessary to achieve promising results. Hence, integrating multi-omics methods is essential for satisfactory inferences [147,148].

13.1. Genomics

Genomics research explores a genome's structure, function, evolution, mapping, and changes. At the same time, the latest advances in molecular biology have quickened the rate of high-throughput genome sequencing, genomic characterization, and gene expression analysis [149]. Functional genomics involves the analysis of partial or unbiased genome sequencing data to elucidate gene functions and interactions, which is achieved through a forward approach consisting of investigating randomly obtained mutants of a particular phenotype and identifying the responsible gene or a reverse approach by disrupting a known gene to examine the organism's phenotype [150,151]. Genome-wide association studies (GWASs) involve the comprehensive analysis of a complete genome to uncover DNA changes associated with a particular trait [38]. The main objective of GWASs is to determine genomic regions related to agronomic or morphological characteristics and any phenotypes that may serve as markers, genes, or QTLs for gene identification, introgressive hybridization, and marker-assisted breeding (MAB) [152,153]. GWASs revealed the upregulation of SlTFT6, a gene belonging to the Sl14-3-3 family, which improved thermotolerance in tomato plants [154]. Structural genomics focuses on elucidating the three-dimensional configuration of genes to ascertain their identity, position, and arrangement along the chromosome [155]. Genomic selection represents an innovative approach to enhancing quantitative traits by leveraging marker and phenotypic data obtained from observed populations, thereby evaluating the influence of all genetic loci [156]. Genome sequencing and mapping comprise several systems, such as the Roche 454GS FLX Titanium or Illumina Solexa Genome Analyzer, which are considered next-generation sequencing (NGS) platforms and have significantly reduced the cost and time required for sequencing compared with traditional methods like the Sanger method [157]. These platforms have provided comprehensive information regarding the characteristics of genomes, including coding and non-coding genes, GC contents, repetitive elements, and regulatory sequences, which have facilitated the development of improved crop varieties such as tomato, rice, wheat, maize, sorghum, and soybean [158,159]. Molecular markers, also known as genetic markers, are segments of DNA that may detect changes in a population's DNA or polymorphisms, including deletions, insertions, and substitutions of bases [160]. Various molecular markers, such as random amplified polymorphic DNA (RAPD), simple sequence repeats (SSRs), sequence-tagged sites (STSs), restriction fragment length polymorphism (RFLP), single-nucleotide polymorphism (SNP), and amplified fragment length polymorphism (AFLP), have been recently identified as valuable tools for identifying polymorphisms in plants [161]. The investigation of comparative genomics involves the alignment of biological sequences and the identification of conserved sequences, which reveals significant synteny among related species [162] and enables the detection of small-scale changes within different genomes, including protein-coding regions and their impact on protein structure and function [163].

13.2. Transcriptomics

The term "transcriptome" covers the complete collection of ribonucleic acid (RNA) molecules within an organism or a particular cell type, which mainly ranges from protein-coding messenger RNA (mRNA) to various non-coding RNAs such as transfer RNA (tRNA), long non-coding RNA (lncRNA), ribosomal RNA (rRNA), primary microRNA (pri-miRNA), and small nuclear RNA (snRNA) [164–166]. The transcriptomic approach covers multiple facets of RNA-seq evaluation, especially experimental design, quality control, read alignment, quantification of gene and transcript levels, visualization, differential gene expression, alternative splicing, functional analysis, gene fusion detection, and expression quantitative trait loci (eQTL) mapping [167,168]. The primary focus of transcriptomic research is to examine gene transcripts or RNA linked to a plant's phenotypic expression under various stress conditions [169] by employing a range of techniques, including serial analysis of gene expression (SAGE), DNA microarrays, and high-throughput technologies based on next-generation sequencing (NGS) for conducting digital gene expression (DGE) and RNA sequencing (RNAseq) [170,171]. The transcriptomic analysis of microspores from a heat-tolerant tomato cultivar (cv. Hazera 3042) revealed elevated levels of heat-responsive gene expression, specifically LeHSFA2, LeHSP17.4-CII, homologs of LeHSP90 (*Laternula elliptica*), and AtVAMP725 (*A. thaliana*), compared with a control [172]. The transcriptomic study findings indicated a notable increase in the expression of SAUR (small auxin upregulated RNA) family proteins, MYB (myeloblastosis viral oncogene homolog) transcription factors, and NAC (no apical meristem) domain proteins in response to arid environmental conditions. Furthermore, it was observed that the heat-tolerant line exhibited a significant inclusion of heat shock proteins and proteinase inhibitors [173]. The transcriptomic analysis of tomato plants subjected to heat stress at temperatures of 35/25 °C, in conjunction with specific nitrogen fertilizer levels, showed a significant upregulation of genes, including cell wall invertase (CWINV2; Solyc10g085650.2, Solyc10g085640.1) and sucrose transporter (SUT1; Solyc11g017010.2), while hexokinase 2 (HK2) (Solyc06g066440.3), SWEET2 (Solyc07g062120.4), and SWEET1 (Solyc04g064610.3) exhibited downregulation [174].

13.3. Metabolomics

Metabolomics is the scientific investigation of naturally occurring tiny, low-molecular-weight metabolites, including carbohydrates, fatty acids, amino acids, steroids, and lipids, which play distinctive roles in interpreting cellular biochemistry [175,176]. The function of a metabolite can be significantly altered by minor alterations in its chemical structure and the presence of external abiotic or biotic stimuli [177]. Metabolomics inquiry offers distinct advantages over other omics because metabolites are the downstream products of gene and protein activities, which determine the impact on biological phenotype and other physiologic processes [178]. Plant metabolites can be primary metabolites, which are crucial for growth and significantly impact physiological processes, and secondary metabolites, which are vital for defense mechanisms in response to various stressors [179,180]. A variety of advanced techniques exist for the analysis of plant metabolites, including gas chromatography (GC), high-performance liquid chromatography (HPLC), thin-layer chromatography (TLC), paper chromatography (PC), nuclear magnetic resonance (NMR), metabolic flux analysis (MFA), extracellular flux analysis (EFA), direct-inject mass spectrophotometry (DIMS), Fourier transform infrared spectroscopy (FTIR), capillary electrophoresis (CE), and mass spectrometry (MS), which have proven to be valuable tools for researchers [181,182]. A metabolic investigation of tomatoes under elevated temperatures and relative air humidity revealed the disruption of enzymes involved in sucrose metabolism, resulting in a decrease in the fruit-soluble sugar content. Conversely, an increase in the activities of enzymes associated with phosphopyruvate carboxylase (PEPC), mitochondria aconitase (MDH), and citrate synthetase (CS) led to an elevated content of malic acid [183]. Metatomic analysis has shown a significant association among sucrose, glucose, fructose, the TCA cycle, starch production, and HS tolerance [184]. Liquid chromatography–mass spectrometry (LC-MS) identified an increased accumulation of secondary metabolites, specifically flavonoids, within the pollen microspore of tomatoes under heat stress [185]. A metabolic analysis of tomatoes using gas chromatography–mass spectrometry (GC-MS) revealed that heat treatment mitigated the effects of chilling on fruits by modifying the concentrations of several fruit metabolites, including arabinose, fructose-6-phosphate, valine, and shikimic acid, in the chilled samples as compared with a control [186].

13.4. Proteomics

Proteomics comprehensively explores protein composition, structure, expression, modification status, connections, and interactions among proteins [187]. Basic proteomics techniques include one-dimensional (1D) and two-dimensional (2D) gel electrophoresis (2-DE) methodologies [188]. Several other high-throughput screening technologies such as shotgun proteomics (SP), nanoflow liquid chromatography coupled to tandem mass spectrometry (nLC-MS/MS) [189], stable isotope labeling by amino acids in cell culture (SILAC) [190,191], multidimensional protein identification technology (MudPIT) [192], isobaric tags utilized in relative and absolute quantitation (ITRAQ) [193,194], the Western blot (WB) technique [195], multiple reactions monitoring mass spectrometry (MRM-MS) [196], and tandem mass tags (TMTs) [197,198] are available for utilization according to research objectives. Proteomic analysis of the tomato revealed better pollen tolerance to heat stress following ethephon pre-treatment by increasing protein abundance in processes of protein synthesis, degradation, the tricarboxylic acid cycle, and RNA regulation [199]. Another proteome analysis of tomatoes subjected to high-light-induced stress revealed a notable presence of oxygen-evolving complex and PSII complex proteins, including PsbH, PsbS, PsbR, and Psb28, within the leaf zone that exhibited the maximum damage [200]. Tandem mass tag (TMT)-based analysis of pollen mother cells at the initial anther developmental stage in the Maxifort tomato variety revealed the upregulation of 96 proteins including heat shock proteins, calreticulin, and exocytosis associated with protein folding/refolding/targeting/removal along with the secretion of aggregated and damaged proteins/peptides and the downregulation of 158 proteins active in ubiquitin-mediated protein breakdown, antioxidant mechanisms, and the metabolism of lipids and carbohydrates [201].

14. Genome Editing Strategy Application

Genome editing has emerged as a promising tool in tomato breeding, offering the potential for immense success and fully utilizing genome information and phenotyping technologies. It is divided into two major approaches, first, as site-directed nuclease (SDNs) and, second, as oligonucleotide-directed mutagenesis (ODMs) involved in creating mutations in the genome [202]. Its application could enhance HS resistance by introducing mutations into negative regulatory genes, which have a pivotal role in tomato HS tolerance [203]. It entails the utilization of several DNA-cleaving enzymes, known as nucleases, specifically designed to cleave the DNA at a pre-established site through diverse DNA binding systems. Several techniques can be applied to carry out specific DNA cleavages, such as zinc finger nucleases (ZFNs), mega-nucleases (MNs), clustered regularly interspaced short palindromic repeat (CRISPR)-associated proteins (CRISPR/Cas), and transcription activator-like effector nucleases (TALENs). These are entitled site-directed nucleases (SDNs), representing the fundamental concept of using a DNA-cutting enzyme (nuclease) to create a specific DNA break at a particular location [204,205]. CRISPR/Cas systems are further divided into classes 1 and 2 based on effector molecules. The class 1 system has multiple effector molecules and is subdivided into three types, including I, III, IV, and 12 subtypes found in 90% of the CRISPR loci in bacteria and archaea targeting DNA and RNA. The class 2 system is characterized by a single effector molecule, with three types, including II, V, and VI, and nine subtypes, which represent 10% of the CRISPR loci targeting DNA and RNA and are found in bacteria. The most prevalent CRISPR/Cas systems utilized for gene editing are type II-A Cas9 from Streptococcus pyogenes and type V-A Cas12a (Cpf1) from Acidaminococcus sp. and Lachnospiraceae [206–208]. The CRISPR/Cas9-mediated removal of SlUDPGT52 resulted in improved drought tolerance because of increased reactive oxygen species (ROS) scavenging [209]. The efficacy of CRISPR/Cas9 in facilitating the introduction of de novo domestication of elite features from wild relatives to the cultivated tomato, as well as the reverse process, has been demonstrated. CRISPR/Cas9 technology has been utilized to manipulate and examine a range of attributes about tomatoes, including leaf, stem, and male sterility, parthenocarpy, fruit maturation, quality, nutrition, heat, drought, salinity stress, carbon–nitrogen metabolism, and herbicide resistance [210,211]. Genome editing technology has contributed to the heat-tolerant breeding of tomatoes via the identification of critical genes associated with acquired thermotolerance mechanisms, including SlIAA9, HsfA2, JA/COI1, HsfB1, and SlAGL6 [212]. It has improved tomato resistance by regulating genes such as lateral organ boundaries domain (SlLBD40) (Solyc02g085910), mitogen-activated protein kinase (SlMAPK3), and cytidine base editor (CBE) against biotic and abiotic stressors [213].

15. Development of Heat-Tolerant Tomato Varieties

There has been a global demand for the development of heat-tolerant varieties to effectively respond to both present and anticipated rises in heat stress. However, breeding for heat tolerance has encountered challenges related to the intricate nature of heat stress and plant reactions and the limited comprehension of the genetic underpinnings of heat tolerance characteristics [214]. The efficacy of heat-tolerant breeding is contingent on the proficient determination and description of constituent qualities that underlie heat resistance processes in the presence of heat stress, as well as the comprehensive understanding of their genetic structure throughout both the vegetative and reproductive phases [215,216]. Affordable and technologically sophisticated high-throughput genotyping is being applied; however, accurate phenotyping is a significant barrier to understanding the genetic basis of required but intricate traits, which slows down breeding programs [217,218]. Effective plant breeding programs have to prioritize the development of phenotyping techniques that are cost-effective, precise, reliable, less labor-intensive, reproducible, and easily applicable, targeting traits such as increased yield, resistance to biotic and abiotic stress, improved quality, photosensitivity, synchronous maturity, and detoxification ability. To develop tomato genotypes resistant to high temperatures, it is crucial to examine cultivated and wild tomato genetic resources thoroughly. To create heat-resistant tomato varieties that have high production and yield, it is essential to comprehend the genetic architecture of heat-tolerance traits to effectively regulate the increase in metabolites, osmoprotectant, photosynthetic activity efficiency, membrane stability, the number of flowers per inflorescence, inflorescence number, pollen number, female fertility, pollen viability, fruit set, fruit number, and fruit weight and the decrease in canopy temperature, style protrusion, and style length [219–221]. Previous breeding projects have not derived significant benefits from the sizeable range of wild tomatoes, mainly because of problems such as progeny sterility, self-incompatibility, and linkage drag [222]. For breaking linkage drag, various techniques, including chromosome segment substitution lines (CSSLs), advanced backcross quantitative trait loci (QTL) analysis, and backcross inbred lines (BILs), could be applied to generate lines that possess small fragments of donor parent chromosomes [223–225]. A practical approach in tomato breeding efforts to enhance resistance to abiotic and biotic stressors is incorporating native germplasm and wild relatives into existing varieties by introducing novel allelic combinations. Multiple tomato introgression lines have been developed by using wild cousins such as *Solanum pimpinellifolium*, *Solanum habrochaites*, and *Solanum pennellii*, which exhibit resistance to abiotic and biotic stressors [226–228].

16. Genetic Resource Development

Genetic resources or germplasm includes plants, parts of plants, and seeds, which are significant for breeding, research, and conservation. For example, seeds of an ancient heirloom tomato variety passed down to current time are just seeds produced by a gardener or company, but they are germplasm when part of a breeding program for new variety development, collected for the preservation of the genetic diversity, or preserved as genetically governed traits [229]. Lycopersicum tomato species are diploid (2n = 2x = 24) with similar chromosome numbers and structures [230] that produce perfect hermaphrodite flowers and have a complete range of mating systems from autogamous *L. cheesmanii* and *L. parviflorum* to obligately outcrossed self-incompatible biotypes of *L. chilense*, *L. hirsutum*, *L. peruvianum*, and *L. pennellii* [231], while the tendency of self-fertility with different levels of facultative outcrossing is present in *L. chmielewskii*, *L. esculentum*, *L. pimpinellifolium*, and the self-compatible biotypes of *L. hirsutum* and *L. pennellii* [232]. In the quest for specific genetic traits, contemporary and prospective researchers and crop breeders must have full access to landraces, diverse varieties, and relevant wild species. Multiple institutes have developed tomato genetic resources to cater to the needs of researchers and breeders studying heat tolerance and other agronomic features. About 62.8 thousand tomato accessions, both wild and domesticated varieties (*L. esculentum*), are present in gene banks across the globe and are ready to be used for genealogical research [233]. The Tomato Genetics Resource Center (TGRC) at the University of California (Davis) (http://tgrc.ucdavis.edu/, accessed on 11 December 2023), the United States of America, is a well-recognized and valuable repository of various germplasm and wild species. The World Vegetable Center (http://seed.worldveg.org, accessed on 11 December 2023), located in Taiwan, China, maintains a vast assortment of around 8835 tomato accessions, of which 6676 are readily available for procurement upon request. A wide variety of tomato genetic resources have been collected by the National Agriculture and Food Research Organization (NARO) gene bank (https://www.gene.affrc.go.jp, accessed on 22 January 2024) and the National Bio-Resource Project (NBRP) of tomato (https://tomato.nbrp.jp, accessed on 24 January 2024) in Japan. The National Bio-Resource Project (NBRP) maintains a collection of more than 10,000 Micro-Tom mutants that have been generated using the techniques of gamma-ray irradiation and ethyl methane sulfonate (EMS) mutagenesis [234,235]. The Micro-Tom plant is a model for investigating fruit production and its ability to withstand different abiotic and biotic challenges [236]. Data about Micro-Tom mutants can be retrieved from the TOMATOMA database, found at http://tomatoma.nbrp.jp/index.jsp (accessed on 8 February 2024) [237]. Several other genetic resources are available from which seeds or genetic material could be obtained, including the Solanaceae Genomics Network (SGN, http://solgenomics.net/, accessed on 8 February 2024), the United States Department of Agriculture (USDA) (https://www.usda.gov/, accessed on 14 February 2024), the Tomato Genetics Cooperative (TGC) (https://tgc.ifas.ufl.edu/, accessed on 17 February 2024) at the University of Florida, USA, the Ohio State Tomato Breeding and Genetics Program (OSTBGP) (https://tomato.cfaes.ohio-state.edu/, accessed on 17 February 2024), Vavilov Institute, Russia (VIR) (https://www.vir.nw.ru/en/, accessed on 3 March 2024), and Instituto de Investigaciones Fundamentales en Agricultura Tropical (INIFAT), Cuba [238,239]. These genetic resource reservoirs could be accessed and explored to obtain genes for tomato heat resistance improvement and other targeted breeding programs.

17. Conclusions and Future Aspects

The average global temperature has significantly increased because of global climate change, which has also put food security and agricultural output at risk [240]. Reduced photosynthesis, decelerated growth and development, and reduced nutrient uptake are just a few of the physiological and biochemical processes upset by heat stress in tomatoes, resulting in yield losses [220,241]. Over the next several years, it is anticipated that the damaging consequences of heat stress will get worse. Uncertainty surrounds the magnitude of the potential consequences associated with global warming. Changes can have both direct and indirect impacts on food production conditions. Direct changes can lead to significant changes in food production, resulting in increased mortality rates because of floods, storms, heat waves, and droughts. On the other hand, indirect effects may include unemployment in rural areas requiring specific climate conditions for crop growth, such as cultivating tomatoes in open fields [242,243].

Environmental change, particularly a rise in ambient temperatures, substantially affects plant growth, development, production, and yield, leading to a severe decline in crop yield and jeopardizing international food security. Increasing heat stress disrupts various physiological and biochemical systems in tomato plants. Tomato seed pollen viability and root development are significantly affected by elevated temperatures in different parts of the world. The emerging data indicate that reactive oxygen species (ROS) cause cellular oxidative damage but also serve as signaling molecules in the heat stress response (HSR), triggering adaptive responses. However, the exact mechanism underlying the interconnections among various signaling pathways linked with ROS has not yet been fully understood. Understanding the interaction between ROS and redox signals and identifying the precise redox pathways activated in different cell compartments is essential for adjusting HSR in response to varying HS intensity and duration levels. The molecular processes behind the pollen heat-stress response and thermotolerance remain largely unexplored. In the context of escalating global warming, there is an urgent need for molecular and genetic research to ascertain the genes responsible for conferring heat tolerance in tomatoes, thereby mitigating the detrimental effects of high temperatures. The primary objective of high-throughput phenotyping should be to investigate several aspects of plant physiology, including canopy temperature, pollen viability, photosynthetic efficiency, membrane thermostability, sugar content, and osmoprotectant activity, to obtain full inside knowledge.

In our opinion, to enhance the overall resilience of tomatoes against heat stress, it is crucial to elucidate the molecular and physiological mechanisms underlying the negative correlations among seed germination, plant growth, development, pollen viability, fruit sets, fruit size, fruit weight, other agronomic traits, and thermal stress. Collecting diverse tomato genetic resources, including various cultivars and wild species, would be valuable for future genetic engineering, particularly in developing heat-resistant tomato plants. Including wild tomato species in breeding programs incurs some drawbacks since introducing genes from wild relatives into advanced lines might alter the already established horticultural features owing to linkage drag. Transgenic technology has the potential to serve as an advantageous instrument for enhancing the heat stress resistance of tomatoes, especially when integrated with conventional techniques. Integrating marker-assisted breeding with high throughput phenotyping can significantly improve the breeding performance of tomatoes in terms of heat resilience. Understanding the genetic foundations of novel populations is of utmost importance, including approaches like chromosomal segment substitution lines (CSSLs), introgression lines (ILs), backcross inbred lines (BILs), and mutants for trait identification. Genome editing could identify the molecular mechanism of heat stress transcription factors and enhance heat tolerance features, like increasing the number of inflorescences and flowers per inflorescence.

Despite certain advancements in translational genomics, particularly with the backing of the gene-editing technology CRISPR/Cas9, some significant difficulties remain, for example, several features subject to quantitative regulation require several genes. Hence, it is imperative to manipulate several new genes to induce new desired phenotypes in modified tomato crops. Further challenges include the lack of effective delivery routes for gene editing reagents such as mRNA (sgRNA), DNA plasmid, and ribonucleoprotein (RNP), technical bottlenecks, and ethical concerns. Moreover, there is a lack of comprehensive genetic data regarding the necessary dietary components, and generating accurate alterations in DNA sequences is challenging. Nevertheless, several gene-editing techniques offer effective and precise gene editing of plants, including base editors, replicons, and targeted non-homologous insertions. The continuous progress in sequencing technology can be utilized to find reference genome sequences for previously unknown tomato wild cousins, which will serve as a great approach to exploit the genetic variability in these species. Genome editing facilitates the development of novel domestication tactics that selectively utilize tomato relatives. Establishing more vibrant collaboration between private plant breeding enterprises and public sector gene banks at regional, national, and worldwide levels is essential. It has significant benefits, particularly in enhancing the conservation and utilization of tomato genetic resources.

A holistic approach is required to comprehensively elucidate the causes of tomato susceptibility to heat stress and the development of heat-resistant varieties in the interfaces of continuously increasing global temperature. So, integrated strategies (Figure 7) based on sophisticated technologies involving high-throughput genotyping, genome editing, and multi-omics approaches like transcriptomics, genomics, metabolomics, proteomics, epigenomics, proteogenomics, lipidomics, interactomics, ionomics, phenomics, bioinformatics genetic engineering, genetic resources collection, preservation, and utilization would enable researchers and breeders to develop heat-tolerant tomato varieties with capabilities to combat increasing temperature stress for a long time.

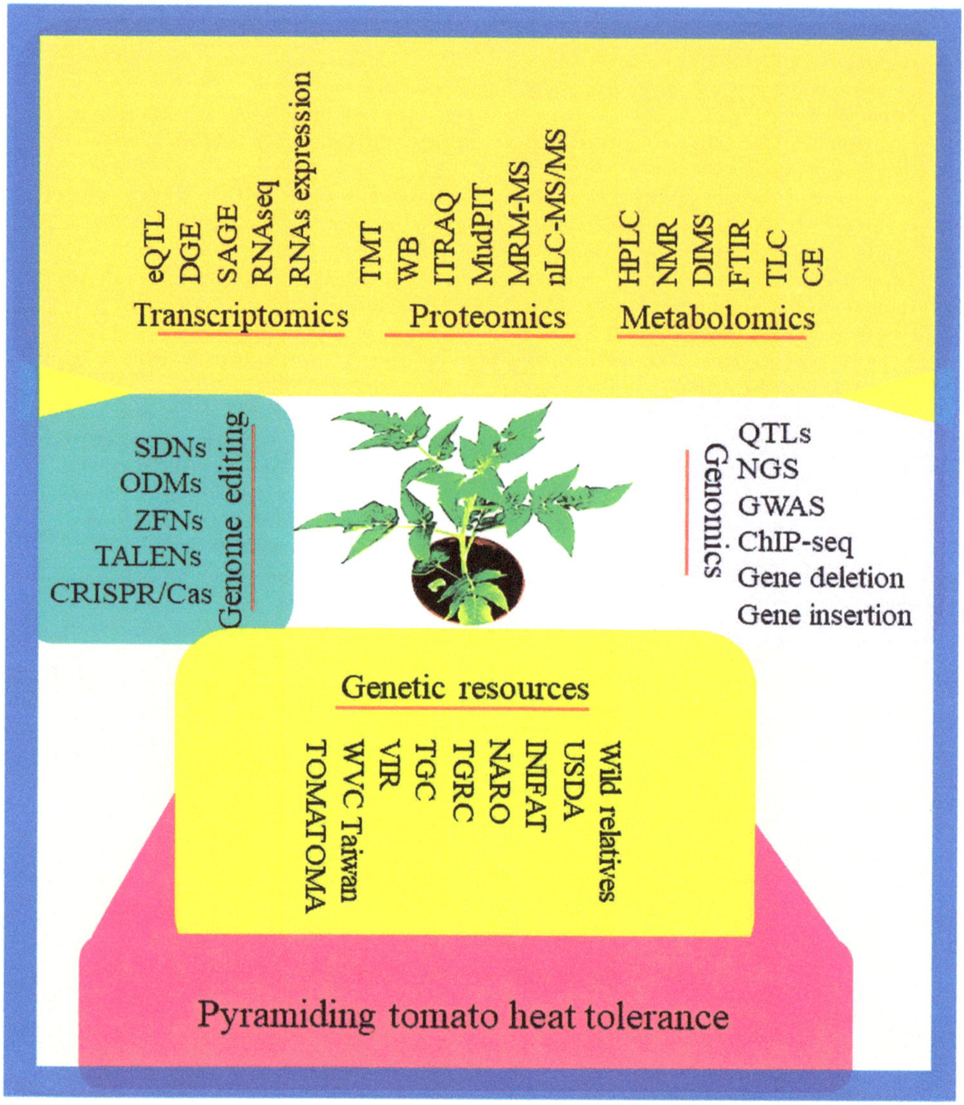

Figure 7. Improving heat tolerance in tomatoes through integrated approaches, including genomics, transcriptomics, proteomics, metabolomics, gene editing, and genetic resources.

Author Contributions: Conceptualization, Q.K. and Y.Z.; methodology monitoringand H.Y.; formal analysis, G.X., Q.K., H.Y., Z.L. and Y.Z.; investigation, Q.K., Y.Z., H.Y. and G.X.; visualization, Y.W., G.X., Y.Z., Z.L. and H.Y.; data curation, Y.W., G.X. and Q.K.; writing—original draft preparation, Q.K., Y.Z. and Z.L.; writing—review and editing, Y.W., G.X., Z.L. and H.Y.; funding acquisition, Y.Z. All authors have read and agreed to the published version of the manuscript.

Funding: This research was supported by the Zhejiang Provincial Natural Science Foundation of China under Grant No. LY22C150007, the Zhejiang Provincial Natural Science Foundation of China under Grant No. LY20C030002, the Key Research and Development Program of Lishui under Grant No. 2020ZDYF08, and Lishui University Initial Funding under Grant No. QD1503.

Institutional Review Board Statement: No experiments with humans or animals were performed in this study, so ethical clearance was not required.

Informed Consent Statement: All authors agreed to contribute to this manuscript.

Data Availability Statement: All the necessary data are included in this manuscript.

Conflicts of Interest: The authors declare that they have no conflicts of interest.

References

1. Alsamir, M.; Mahmood, T.; Trethowan, R.; Ahmad, N. An overview of heat stress in tomato (*Solanum lycopersicum* L.). *Saudi J. Biol. Sci.* **2021**, *28*, 1654–1663. [CrossRef] [PubMed]
2. Reimer, J.J.; Thiele, B.; Biermann, R.T.; Junker-Frohn, L.V.; Wiese-Klinkenberg, A.; Usadel, B.; Wormit, A. Tomato leaves under stress: A comparison of stress response to mild abiotic stress between a cultivated and a wild tomato species. *Plant Mol. Biol.* **2021**, *107*, 177–206. [CrossRef] [PubMed]
3. Gerszberg, A.; Hnatuszko-Konka, K. Tomato tolerance to abiotic stress: A review of most often engineered target sequences. *Plant Growth Regul.* **2017**, *83*, 175–198. [CrossRef]
4. Willcox, J.K.; Catignani, G.L.; Lazarus, S. Tomatoes and cardiovascular health. *Crit. Rev. Food Sci. Nutr.* **2003**, *43*, 1–18. [CrossRef] [PubMed]
5. Kumar, K.S.; Paswan, S.; Srivastava, S. Tomato-a natural medicine and its health benefits. *J. Pharmacogn. Phytochem.* **2012**, *1*, 33–43.
6. Toor, R.; Lister, C.; Savage, G. Antioxidant activities of New Zealand-grown tomatoes. *Int. J. Food Sci. Nutr.* **2005**, *56*, 597–605. [CrossRef]
7. Viuda-Martos, M.; Sanchez-Zapata, E.; Sayas-Barberá, E.; Sendra, E.; Pérez-Álvarez, J.; Fernández-López, J. Tomato and tomato byproducts. Human health benefits of lycopene and its application to meat products: A review. *Crit. Rev. Food Sci. Nutr.* **2014**, *54*, 1032–1049. [CrossRef]
8. Tomas, M.; Beekwilder, J.; Hall, R.D.; Sagdic, O.; Boyacioglu, D.; Capanoglu, E. Industrial processing versus home processing of tomato sauce: Effects on phenolics, flavonoids and in vitro bioaccessibility of antioxidants. *Food Chem.* **2017**, *220*, 51–58. [CrossRef]
9. Capobianco-Uriarte, M.d.l.M.; Aparicio, J.; De Pablo-Valenciano, J.; Casado-Belmonte, M.d.P. The European tomato market. An approach by export competitiveness maps. *PLoS ONE* **2021**, *16*, e0250867. [CrossRef]
10. Ben-Ari, G.; Lavi, U. Marker-assisted selection in plant breeding. In *Plant Biotechnology and Agriculture*; Elsevier: Amsterdam, The Netherlands, 2012; pp. 163–184.
11. Krishna, R.; Ansari, W.A.; Soumia, P.; Yadav, A.; Jaiswal, D.K.; Kumar, S.; Singh, A.K.; Singh, M.; Verma, J.P. Biotechnological Interventions in Tomato (*Solanum lycopersicum*) for Drought Stress Tolerance: Achievements and Future Prospects. *BioTech* **2022**, *11*, 48. [CrossRef]
12. Wahid, A.; Gelani, S.; Ashraf, M.; Foolad, M.R. Heat tolerance in plants: An overview. *Environ. Exp. Bot.* **2007**, *61*, 199–223. [CrossRef]
13. Bita, C.E.; Gerats, T. Plant tolerance to high temperature in a changing environment: Scientific fundamentals and production of heat stress-tolerant crops. *Front. Plant Sci.* **2013**, *4*, 273. [CrossRef] [PubMed]
14. Mishra, S.; Spaccarotella, K.; Gido, J.; Samanta, I.; Chowdhary, G. Effects of heat stress on plant-nutrient relations: An update on nutrient uptake, transport, and assimilation. *Int. J. Mol. Sci.* **2023**, *24*, 15670. [CrossRef] [PubMed]
15. Habib-ur-Rahman, M.; Ahmad, A.; Raza, A.; Hasnain, M.U.; Alharby, H.F.; Alzahrani, Y.M.; Bamagoos, A.A.; Hakeem, K.R.; Ahmad, S.; Nasim, W. Impact of climate change on agricultural production; Issues, challenges, and opportunities in Asia. *Front. Plant Sci.* **2022**, *13*, 925548. [CrossRef] [PubMed]
16. Costa, M.-C.D.; Farrant, J.M. Plant resistance to abiotic stresses. *Plants* **2019**, *8*, 553. [CrossRef] [PubMed]
17. Brown, C.; Alexander, P.; Arneth, A.; Holman, I.; Rounsevell, M. Achievement of Paris climate goals unlikely due to time lags in the land system. *Nat. Clim. Change* **2019**, *9*, 203–208. [CrossRef]
18. Akhoundnejad, Y.; Dasgan, H.Y. Physiological performance of some high temperature tolerant tomato genotypes. *Int. J. Sci. Technol. Res.* **2018**, *4*, 57–74.
19. Dasgan, H.; Akhoundnejad, Y. Determination of Tolerance Levels to High Temperature Stress of Different Local Tomato Genotypes, Examination of eir Morphological, Physiological and Agricultural Characteristics. 2013.
20. Awasthi, R.; Kaushal, N.; Vadez, V.; Turner, N.C.; Berger, J.; Siddique, K.H.; Nayyar, H. Individual and combined effects of transient drought and heat stress on carbon assimilation and seed filling in chickpea. *Funct. Plant Biol.* **2014**, *41*, 1148–1167. [CrossRef] [PubMed]
21. Barnabás, B.; Jäger, K.; Fehér, A. The effect of drought and heat stress on reproductive processes in cereals. *Plant Cell Environ.* **2008**, *31*, 11–38. [CrossRef]
22. El Haddad, N.; Rajendran, K.; Smouni, A.; Es-Safi, N.E.; Benbrahim, N.; Mentag, R.; Nayyar, H.; Maalouf, F.; Kumar, S. Screening the FIGS set of lentil (*Lens culinaris* Medikus) germplasm for tolerance to terminal heat and combined drought-heat stress. *Agronomy* **2020**, *10*, 1036. [CrossRef]

23. Machado, S.; Paulsen, G.M. Combined effects of drought and high temperature on water relations of wheat and sorghum. *Plant Soil* **2001**, *233*, 179–187. [CrossRef]
24. Teng, L.; Zhang, X.-P.; Qing, L.; Jin, L.; Chen, Y.-Q.; Peng, S. Yield penalty of maize (*Zea mays* L.) under heat stress in different growth stages: A review. *J. Integr. Agric.* **2022**, *21*, 2465–2476.
25. Fedyaeva, A.; Stepanov, A.; Lyubushkina, I.; Pobezhimova, T.; Rikhvanov, E. Heat shock induces production of reactive oxygen species and increases inner mitochondrial membrane potential in winter wheat cells. *Biochemistry* **2014**, *79*, 1202–1210. [CrossRef] [PubMed]
26. Gill, S.S.; Tuteja, N. Reactive oxygen species and antioxidant machinery in abiotic stress tolerance in crop plants. *Plant Physiol. Biochem.* **2010**, *48*, 909–930. [CrossRef] [PubMed]
27. Suzuki, N.; Koussevitzky, S.; Mittler, R.; Miller, G. ROS and redox signalling in the response of plants to abiotic stress. *Plant Cell Environ.* **2012**, *35*, 259–270. [CrossRef] [PubMed]
28. Zinn, K.E.; Tunc-Ozdemir, M.; Harper, J.F. Temperature stress and plant sexual reproduction: Uncovering the weakest links. *J. Exp. Bot.* **2010**, *61*, 1959–1968. [CrossRef] [PubMed]
29. Nafees, K.; Kumar, M.; Bose, B. Effect of different temperatures on germination and seedling growth of primed seeds of tomato. *Russ. J. Plant Physiol.* **2019**, *66*, 778–784. [CrossRef]
30. Vollenweider, P.; Günthardt-Goerg, M.S. Diagnosis of abiotic and biotic stress factors using the visible symptoms in foliage. *Environ. Pollut.* **2005**, *137*, 455–465. [CrossRef] [PubMed]
31. Tian, J.; Belanger, F.C.; Huang, B. Identification of heat stress-responsive genes in heat-adapted thermal Agrostis scabra by suppression subtractive hybridization. *J. Plant Physiol.* **2009**, *166*, 588–601. [CrossRef]
32. Tokić, M.; Leljak Levanić, D.; Ludwig-Müller, J.; Bauer, N. Growth and Molecular Responses of Tomato to Prolonged and Short-Term Heat Exposure. *Int. J. Mol. Sci.* **2023**, *24*, 4456. [CrossRef]
33. Qaisar, K.; Mumtaz, A.S.; Haris, K.; Jan, S.A.; Nazir, A.; Khan, S.A.; Noor, S.; Shah, S.H.; Ibrahim, M.I.; Muhammad, I. Exploring durable genetic resistance against leaf rust through phenotypic characterization and Lr34 linked STS marker in wheat germplasm. *Biosci. J.* **2016**, *32*, 986–998.
34. Giri, A.; Heckathorn, S.; Mishra, S.; Krause, C. Heat stress decreases levels of nutrient-uptake and-assimilation proteins in tomato roots. *Plants* **2017**, *6*, 6. [CrossRef] [PubMed]
35. Rajametov, S.N.; Yang, E.Y.; Jeong, H.B.; Cho, M.C.; Chae, S.Y.; Paudel, N. Heat treatment in two tomato cultivars: A study of the effect on physiological and growth recovery. *Horticulturae* **2021**, *7*, 119. [CrossRef]
36. Golam, F.; Prodhan, Z.H.; Nezhadahmadi, A.; Rahman, M. Heat tolerance in tomato. *Life Sci. J.* **2012**, *9*, 1936–1950.
37. Driedonks, N.; Wolters-Arts, M.; Huber, H.; de Boer, G.-J.; Vriezen, W.; Mariani, C.; Rieu, I. Exploring the natural variation for reproductive thermotolerance in wild tomato species. *Euphytica* **2018**, *214*, 67. [CrossRef]
38. Camejo, D.; Rodríguez, P.; Morales, M.A.; Dell'Amico, J.M.; Torrecillas, A.; Alarcón, J.J. High temperature effects on photosynthetic activity of two tomato cultivars with different heat susceptibility. *J. Plant Physiol.* **2005**, *162*, 281–289. [CrossRef] [PubMed]
39. Poudyal, D.; Rosenqvist, E.; Ottosen, C.-O. Phenotyping from lab to field–tomato lines screened for heat stress using Fv/Fm maintain high fruit yield during thermal stress in the field. *Funct. Plant Biol.* **2018**, *46*, 44–55. [CrossRef] [PubMed]
40. Mathur, S.; Agrawal, D.; Jajoo, A. Photosynthesis: Response to high temperature stress. *J. Photochem. Photobiol. B Biol.* **2014**, *137*, 116–126. [CrossRef] [PubMed]
41. Hu, S.; Ding, Y.; Zhu, C. Sensitivity and responses of chloroplasts to heat stress in plants. *Front. Plant Sci.* **2020**, *11*, 375. [CrossRef]
42. Zhang, Y.; Lai, X.; Yang, S.; Ren, H.; Yuan, J.; Jin, H.; Shi, C.; Lai, Z.; Xia, G. Functional analysis of tomato CHIP ubiquitin E3 ligase in heat tolerance. *Sci. Rep.* **2021**, *11*, 1713. [CrossRef]
43. Zahra, N.; Hafeez, M.B.; Ghaffar, A.; Kausar, A.; Al Zeidi, M.; Siddique, K.H.; Farooq, M. Plant photosynthesis under heat stress: Effects and management. *Environ. Exp. Bot.* **2023**, *206*, 105178. [CrossRef]
44. Gutteridge, S. The impact of a changing atmosphere on chloroplast function, photosynthesis, yield, and food security. *Essays Biochem.* **2018**, *62*, 1–11. [CrossRef] [PubMed]
45. Muhie, S.H. Optimization of photosynthesis for sustainable crop production. *CABI Agric. Biosci.* **2022**, *3*, 50. [CrossRef]
46. Ashraf, M.; Harris, P.J. Photosynthesis under stressful environments: An overview. *Photosynthetica* **2013**, *51*, 163–190. [CrossRef]
47. Kumar Tewari, A.; Charan Tripathy, B. Temperature-stress-induced impairment of chlorophyll biosynthetic reactions in cucumber and wheat. *Plant Physiol.* **1998**, *117*, 851–858. [CrossRef] [PubMed]
48. Baker, N.R.; Rosenqvist, E. Applications of chlorophyll fluorescence can improve crop production strategies: An examination of future possibilities. *J. Exp. Bot.* **2004**, *55*, 1607–1621. [CrossRef] [PubMed]
49. Janka, E.; Körner, O.; Rosenqvist, E.; Ottosen, C.-O. High temperature stress monitoring and detection using chlorophyll a fluorescence and infrared thermography in chrysanthemum (*Dendranthema grandiflora*). *Plant Physiol. Biochem.* **2013**, *67*, 87–94. [CrossRef] [PubMed]
50. Islam, M. Effect of temperature on photosynthesis, yield attributes and yield of tomato genotypes. *Int. J. Expt. Agric.* **2011**, *2*, 8–11.
51. Levy, A.; Rabinowitch, H.; Kedar, N. Morphological and physiological characters affecting flower drop and fruit set of tomatoes at high temperatures. *Euphytica* **1978**, *27*, 211–218. [CrossRef]
52. Saeed, A.; Hayat, K.; Khan, A.; Iqbal, S. Heat tolerance studies in tomato (*Lycopersicon esculentum* Mill.). *Int. J. Agric. Biol.* **2007**, *9*, 649–652.
53. Rick, C.M.; Dempsey, W.H. Position of the stigma in relation to fruit setting of the tomato. *Bot. Gaz.* **1969**, *130*, 180–186. [CrossRef]

54. Dane, F.; Hunter, A.G.; Chambliss, O.L. Fruit set, pollen fertility, and combining ability of selected tomato genotypes under high-temperature field conditions. *J. Am. Soc. Hortic. Sci.* **1991**, *116*, 906–910. [CrossRef]
55. Giorno, F.; Wolters-Arts, M.; Mariani, C.; Rieu, I. Ensuring reproduction at high temperatures: The heat stress response during anther and pollen development. *Plants* **2013**, *2*, 489–506. [CrossRef]
56. Speranza, A.; Calzoni, G.; Pacini, E. Occurrence of mono-or disaccharides and polysaccharide reserves in mature pollen grains. *Sex. Plant Reprod.* **1997**, *10*, 110–115. [CrossRef]
57. Pacini, E. Types and meaning of pollen carbohydrate reserves. *Sex. Plant Reprod.* **1996**, *9*, 362–366. [CrossRef]
58. Abdul-Baki, A.A.; Stommel, J.R. Pollen viability and fruit set of tomato genotypes under optimumand high-temperature regimes. *HortScience* **1995**, *30*, 115–117. [CrossRef]
59. El Ahmadi, A.B.; Stevens, M.A. Reproductive Responses of Heat-tolerant Tomatoes to High Temperatures1. *J. Am. Soc. Hortic. Sci.* **1979**, *104*, 686–691. [CrossRef]
60. Bhattarai, S.; Harvey, J.T.; Djidonou, D.; Leskovar, D.I. Exploring morpho-physiological variation for heat stress tolerance in tomato. *Plants* **2021**, *10*, 347. [CrossRef]
61. Cauchie, A.; Quinet, M. *Impacts of Heat and Water Stresses on the Cultivated Tomato Solanum Lycopersicum and Its Wild Relative Solanum Chilense*; Faculté des Bioingénieurs, Université Catholique de Louvain: Ottignies-Louvain-la-Neuve, Belgium, 2022.
62. Jadon, M. A Novel Method for Leaf Area Estimation based on Hough Transform. *J. Multim. Process. Technol.* **2018**, *9*, 33–44. [CrossRef]
63. Garnier, E.; Shipley, B.; Roumet, C.; Laurent, G. A standardized protocol for the determination of specific leaf area and leaf dry matter content. *Funct. Ecol.* **2001**, *15*, 688–695. [CrossRef]
64. Greer, D.H.; Weedon, M.M. Modelling photosynthetic responses to temperature of grapevine (Vitis vinifera cv. Semillon) leaves on vines grown in a hot climate. *Plant Cell Environ.* **2012**, *35*, 1050–1064. [CrossRef]
65. Young, L.W.; Wilen, R.W.; Bonham-Smith, P.C. High temperature stress of Brassica napus during flowering reduces micro-and megagametophyte fertility, induces fruit abortion, and disrupts seed production. *J. Exp. Bot.* **2004**, *55*, 485–495. [CrossRef]
66. Djanaguiraman, M.; Annie Sheeba, J.; Durga Devi, D.; Bangarusamy, U. Cotton leaf senescence can be delayed by nitrophenolate spray through enhanced antioxidant defence system. *J. Agron. Crop Sci.* **2009**, *195*, 213–224. [CrossRef]
67. Pressman, E.; Peet, M.M.; Pharr, D.M. The effect of heat stress on tomato pollen characteristics is associated with changes in carbohydrate concentration in the developing anthers. *Ann. Bot.* **2002**, *90*, 631–636. [CrossRef]
68. Peet, M.; Sato, S.; Gardner, R. Comparing heat stress effects on male-fertile and male-sterile tomatoes. *Plant Cell Environ.* **1998**, *21*, 225–231. [CrossRef]
69. Abdelmageed, A.H.; Gruda, N. Influence of heat shock pretreatment on growth and development of tomatoes under controlled heat stress conditions. *J. Appl. Bottny Food Qual.* **2007**, *81*, 26–28.
70. Huot, B.; Castroverde, C.D.M.; Velásquez, A.C.; Hubbard, E.; Pulman, J.A.; Yao, J.; Childs, K.L.; Tsuda, K.; Montgomery, B.L.; He, S.Y. Dual impact of elevated temperature on plant defence and bacterial virulence in Arabidopsis. *Nat. Commun.* **2017**, *8*, 1808. [CrossRef]
71. Djanaguiraman, M.; Perumal, R.; Jagadish, S.; Ciampitti, I.; Welti, R.; Prasad, P. Sensitivity of sorghum pollen and pistil to high-temperature stress. *Plant Cell Environ.* **2018**, *41*, 1065–1082. [CrossRef]
72. Khan, M.I.R.; Iqbal, N.; Masood, A.; Per, T.S.; Khan, N.A. Salicylic acid alleviates adverse effects of heat stress on photosynthesis through changes in proline production and ethylene formation. *Plant Signal. Behav.* **2013**, *8*, e26374. [CrossRef] [PubMed]
73. Fortunato, S.; Lasorella, C.; Dipierro, N.; Vita, F.; de Pinto, M.C. Redox signaling in plant heat stress response. *Antioxidants* **2023**, *12*, 605. [CrossRef] [PubMed]
74. Waszczak, C.; Carmody, M.; Kangasjärvi, J. Reactive oxygen species in plant signaling. *Annu. Rev. Plant Biol.* **2018**, *69*, 209–236. [CrossRef]
75. Hasanuzzaman, M. *Agronomic Crops: Volume 3: Stress Responses and Tolerance*; Springer Nature: Berlin/Heidelberg, Germany, 2020.
76. Hasanuzzaman, M.; Bhuyan, M.B.; Zulfiqar, F.; Raza, A.; Mohsin, S.M.; Mahmud, J.A.; Fujita, M.; Fotopoulos, V. Reactive oxygen species and antioxidant defense in plants under abiotic stress: Revisiting the crucial role of a universal defense regulator. *Antioxidants* **2020**, *9*, 681. [CrossRef]
77. Snezhkina, A.V.; Kudryavtseva, A.V.; Kardymon, O.L.; Savvateeva, M.V.; Melnikova, N.V.; Krasnov, G.S.; Dmitriev, A.A. ROS generation and antioxidant defense systems in normal and malignant cells. *Oxid. Med. Cell. Longev.* **2019**, *2019*, 6175804. [CrossRef]
78. Milkovic, L.; Cipak Gasparovic, A.; Cindric, M.; Mouthuy, P.-A.; Zarkovic, N. Short overview of ROS as cell function regulators and their implications in therapy concepts. *Cells* **2019**, *8*, 793. [CrossRef]
79. Belhadj Slimen, I.; Najar, T.; Ghram, A.; Dabbebi, H.; Ben Mrad, M.; Abdrabbah, M. Reactive oxygen species, heat stress and oxidative-induced mitochondrial damage. A review. *Int. J. Hyperth.* **2014**, *30*, 513–523. [CrossRef]
80. Turrens, J.F. Mitochondrial formation of reactive oxygen species. *J. Physiol.* **2003**, *552*, 335–344. [CrossRef]
81. Sharkey, T.D.; Zhang, R. High temperature effects on electron and proton circuits of photosynthesis. *J. Integr. Plant Biol.* **2010**, *52*, 712–722. [CrossRef]
82. Fernández-Crespo, E.; Liu-Xu, L.; Albert-Sidro, C.; Scalschi, L.; Llorens, E.; González-Hernández, A.I.; Crespo, O.; Gonzalez-Bosch, C.; Camañes, G.; García-Agustín, P. Exploiting Tomato Genotypes to Understand Heat Stress Tolerance. *Plants* **2022**, *11*, 3170. [CrossRef]

83. Hasanuzzaman, M.; Bhuyan, M.B.; Parvin, K.; Bhuiyan, T.F.; Anee, T.I.; Nahar, K.; Hossen, M.S.; Zulfiqar, F.; Alam, M.M.; Fujita, M. Regulation of ROS metabolism in plants under environmental stress: A review of recent experimental evidence. *Int. J. Mol. Sci.* **2020**, *21*, 8695. [CrossRef]
84. Rivero, R.; Ruiz, J.; Romero, L. Oxidative metabolism in tomato plants subjected to heat stress. *J. Hortic. Sci. Biotechnol.* **2004**, *79*, 560–564. [CrossRef]
85. Soengas, P.; Rodríguez, V.M.; Velasco, P.; Cartea, M.E. Effect of temperature stress on antioxidant defenses in Brassica oleracea. *ACS Omega* **2018**, *3*, 5237–5243. [CrossRef]
86. Pinhero, R.G.; Rao, M.V.; Paliyath, G.; Murr, D.P.; Fletcher, R.A. Changes in activities of antioxidant enzymes and their relationship to genetic and paclobutrazol-induced chilling tolerance of maize seedlings. *Plant Physiol.* **1997**, *114*, 695–704. [CrossRef]
87. Shu-Hsien, H.; Chih-Wen, Y.; Lin, C.H. Hydrogen peroxide functions as a stress signal in plants. *Bot. Bull. Acad. Sin.* **2005**, *46*, 1–10.
88. McClung, C.R.; Davis, S.J. Ambient thermometers in plants: From physiological outputs towards mechanisms of thermal sensing. *Curr. Biol.* **2010**, *20*, R1086–R1092. [CrossRef]
89. Osei-Bonsu, I.; Osei, M.; Agyare, R.; Adjebeng-Danquah, J.; Asare Bediako, K.; Gyau, J.; Adomako, J.; Ofori, P.; Prempeh, R.; Cho, M.-C. Assessing the heat stress tolerance potential of tomato lines under poly-house and open field conditions. *Cogent Food Agric.* **2022**, *8*, 2115665. [CrossRef]
90. Ahuja, I.; de Vos, R.C.; Bones, A.M.; Hall, R.D. Plant molecular stress responses face climate change. *Trends Plant Sci.* **2010**, *15*, 664–674. [CrossRef]
91. Nievola, C.C.; Carvalho, C.P.; Carvalho, V.; Rodrigues, E. Rapid responses of plants to temperature changes. *Temperature* **2017**, *4*, 371–405. [CrossRef]
92. Srivastava, S.; Pathak, A.D.; Gupta, P.S.; Shrivastava, A.K.; Srivastava, A.K. Hydrogen peroxide-scavenging enzymes impart tolerance to high temperature induced oxidative stress in sugarcane. *J. Environ. Biol.* **2012**, *33*, 657.
93. Fitter, A.H.; Hay, R.K. *Environmental Physiology of Plants*; Academic Press: Cambridge, MA, USA, 2012.
94. Chang, H.-C.; Tang, Y.-C.; Hayer-Hartl, M.; Hartl, F.U. SnapShot: Molecular chaperones, Part I. *Cell* **2007**, *128*, 212-e1. [CrossRef]
95. Garbuz, D. Regulation of heat shock gene expression in response to stress. *Mol. Biol.* **2017**, *51*, 352–367. [CrossRef]
96. Prasinos, C.; Krampis, K.; Samakovli, D.; Hatzopoulos, P. Tight regulation of expression of two Arabidopsis cytosolic Hsp90 genes during embryo development. *J. Exp. Bot.* **2005**, *56*, 633–644. [CrossRef]
97. Rasul, G.; Chaudhry, Q.; Mahmood, A.; Hyder, K. Effect of temperature rise on crop growth and productivity. *Pak. J. Meteorol.* **2011**, *8*, 53–62.
98. Zhou, R.; Kong, L.; Wu, Z.; Rosenqvist, E.; Wang, Y.; Zhao, L.; Zhao, T.; Ottosen, C.O. Physiological response of tomatoes at drought, heat and their combination followed by recovery. *Physiol. Plant.* **2019**, *165*, 144–154. [CrossRef]
99. Zhou, R.; Yu, X.; Ottosen, C.-O.; Rosenqvist, E.; Zhao, L.; Wang, Y.; Yu, W.; Zhao, T.; Wu, Z. Drought stress had a predominant effect over heat stress on three tomato cultivars subjected to combined stress. *BMC Plant Biol.* **2017**, *17*, 24. [CrossRef]
100. Gautier, H.; Diakou-Verdin, V.; Bénard, C.; Reich, M.; Buret, M.; Bourgaud, F.; Poëssel, J.L.; Caris-Veyrat, C.; Génard, M. How does tomato quality (sugar, acid, and nutritional quality) vary with ripening stage, temperature, and irradiance? *J. Agric. Food Chem.* **2008**, *56*, 1241–1250. [CrossRef]
101. Garg, N.; Cheema, D.S. Assessment of fruit quality attributes of tomato hybrids involving ripening mutants under high temperature conditions. *Sci. Hortic.* **2011**, *131*, 29–38. [CrossRef]
102. Singh, U.; Patel, P.K.; Singh, A.K.; Tiwari, V.; Kumar, R.; Rai, N.; Bahadur, A.; Tiwari, S.K.; Singh, M.; Singh, B. Screening of tomato genotypes underhigh temperature stress for reproductive traits. *Veg. Sci.* **2015**, *42*, 52–55.
103. Scarano, A.; Olivieri, F.; Gerardi, C.; Liso, M.; Chiesa, M.; Chieppa, M.; Frusciante, L.; Barone, A.; Santino, A.; Rigano, M.M. Selection of tomato landraces with high fruit yield and nutritional quality under elevated temperatures. *J. Sci. Food Agric.* **2020**, *100*, 2791–2799. [CrossRef]
104. Lokesha, A.; Shivashankara, K.; Laxman, R.; Geetha, G.; Shankar, A. Effect of high temperature on fruit quality parameters of contrasting tomato genotypes. *Int. J. Curr. Microbiol. Appl. Sci.* **2019**, *8*, 1019–1029. [CrossRef]
105. Kaur, N.; Gupta, A.K. Signal transduction pathways under abiotic stresses in plants. *Curr. Sci.* **2005**, 1771–1780.
106. Jagodzik, P.; Tajdel-Zielinska, M.; Ciesla, A.; Marczak, M.; Ludwikow, A. Mitogen-activated protein kinase cascades in plant hormone signaling. *Front. Plant Sci.* **2018**, *9*, 410346. [CrossRef]
107. Boudsocq, M.; Sheen, J. CDPKs in immune and stress signaling. *Trends Plant Sci.* **2013**, *18*, 30–40. [CrossRef]
108. Hemantaranjan, A.; Bhanu, A.N.; Singh, M.; Yadav, D.; Patel, P.; Singh, R.; Katiyar, D. Heat stress responses and thermotolerance. *Adv. Plants Agric. Res* **2014**, *1*, 12. [CrossRef]
109. Baniwal, S.K.; Bharti, K.; Chan, K.Y.; Fauth, M.; Ganguli, A.; Kotak, S.; Mishra, S.K.; Nover, L.; Port, M.; Scharf, K.-D. Heat stress response in plants: A complex game with chaperones and more than twenty heat stress transcription factors. *J. Biosci.* **2004**, *29*, 471–487. [CrossRef]
110. Laloum, T.; Martín, G.; Duque, P. Alternative splicing control of abiotic stress responses. *Trends Plant Sci.* **2018**, *23*, 140–150. [CrossRef]
111. Zhang, B.; Pan, X.; Cobb, G.P.; Anderson, T.A. Plant microRNA: A small regulatory molecule with big impact. *Dev. Biol.* **2006**, *289*, 3–16. [CrossRef]

112. Park, C.-J.; Seo, Y.-S. Heat shock proteins: A review of the molecular chaperones for plant immunity. *Plant Pathol. J.* **2015**, *31*, 323. [CrossRef]
113. Carra, S.; Alberti, S.; Benesch, J.L.; Boelens, W.; Buchner, J.; Carver, J.A.; Cecconi, C.; Ecroyd, H.; Gusev, N.; Hightower, L.E. Small heat shock proteins: Multifaceted proteins with important implications for life. *Cell Stress Chaperones* **2019**, *24*, 295–308. [CrossRef]
114. Mittler, R.; Finka, A.; Goloubinoff, P. How do plants feel the heat? *Trends Biochem. Sci.* **2012**, *37*, 118–125. [CrossRef]
115. Saidi, Y.; Finka, A.; Muriset, M.; Bromberg, Z.; Weiss, Y.G.; Maathuis, F.J.; Goloubinoff, P. The heat shock response in moss plants is regulated by specific calcium-permeable channels in the plasma membrane. *Plant Cell* **2009**, *21*, 2829–2843. [CrossRef]
116. Saidi, Y.; Finka, A.; Goloubinoff, P. Heat perception and signalling in plants: A tortuous path to thermotolerance. *New Phytol.* **2011**, *190*, 556–565. [CrossRef]
117. Finka, A.; Mattoo, R.U.; Goloubinoff, P. Meta-analysis of heat-and chemically upregulated chaperone genes in plant and human cells. *Cell Stress Chaperones* **2011**, *16*, 15–31. [CrossRef]
118. Larkindale, J.; Hall, J.D.; Knight, M.R.; Vierling, E. Heat stress phenotypes of Arabidopsis mutants implicate multiple signaling pathways in the acquisition of thermotolerance. *Plant Physiol.* **2005**, *138*, 882–897. [CrossRef]
119. Bokszczanin, K.L.; Fragkostefanakis, S.; Bostan, H.; Bovy, A.; Chaturvedi, P.; Chiusano, M.L.; Firon, N.; Iannacone, R.; Jegadeesan, S.; Klaczynskid, K. Perspectives on deciphering mechanisms underlying plant heat stress response and thermotolerance. *Front. Plant Sci.* **2013**, *4*, 55641. [CrossRef]
120. Clos, J.; Westwood, J.T.; Becker, P.B.; Wilson, S.; Lambert, K.; Wu, C. Molecular cloning and expression of a hexameric Drosophila heat shock factor subject to negative regulation. *Cell* **1990**, *63*, 1085–1097. [CrossRef]
121. Guertin, M.J.; Lis, J.T. Chromatin landscape dictates HSF binding to target DNA elements. *PLoS Genet.* **2010**, *6*, e1001114. [CrossRef]
122. Scharf, K.-D.; Berberich, T.; Ebersberger, I.; Nover, L. The plant heat stress transcription factor (Hsf) family: Structure, function and evolution. *Biochim. Et Biophys. Acta BBA Gene Regul. Mech.* **2012**, *1819*, 104–119. [CrossRef]
123. Xie, K.; Guo, J.; Wang, S.; Ye, W.; Sun, F.; Zhang, C.; Xi, Y. Genome-wide identification, classification, and expression analysis of heat shock transcription factor family in switchgrass (*Panicum virgatum* L.). *Plant Physiol. Biochem.* **2023**, *201*, 107848. [CrossRef]
124. Sakurai, H.; Enoki, Y. Novel aspects of heat shock factors: DNA recognition, chromatin modulation and gene expression. *FEBS J.* **2010**, *277*, 4140–4149. [CrossRef]
125. Döring, P.; Treuter, E.; Kistner, C.; Lyck, R.; Chen, A.; Nover, L. The role of AHA motifs in the activator function of tomato heat stress transcription factors HsfA1 and HsfA2. *Plant Cell* **2000**, *12*, 265–278. [CrossRef]
126. Görlich, D.; Kutay, U. Transport between the cell nucleus and the cytoplasm. *Annu. Rev. Cell Dev. Biol.* **1999**, *15*, 607–660. [CrossRef]
127. Zhou, M.; Zheng, S.; Liu, R.; Lu, J.; Lu, L.; Zhang, C.; Liu, Z.; Luo, C.; Zhang, L.; Yant, L. Genome-wide identification, phylogenetic and expression analysis of the heat shock transcription factor family in bread wheat (*Triticum aestivum* L.). *BMC Genom.* **2019**, *20*, 505. [CrossRef]
128. Nover, L.; Bharti, K.; Döring, P.; Mishra, S.K.; Ganguli, A.; Scharf, K.-D. Arabidopsis and the heat stress transcription factor world: How many heat stress transcription factors do we need? *Cell Stress Chaperones* **2001**, *6*, 177. [CrossRef]
129. Kotak, S.; Port, M.; Ganguli, A.; Bicker, F.; Von Koskull-Döring, P. Characterization of C-terminal domains of Arabidopsis heat stress transcription factors (Hsfs) and identification of a new signature combination of plant class A Hsfs with AHA and NES motifs essential for activator function and intracellular localization. *Plant J.* **2004**, *39*, 98–112. [CrossRef]
130. Zhuang, L.; Cao, W.; Wang, J.; Yu, J.; Yang, Z.; Huang, B. Characterization and functional analysis of FaHsfC1b from Festuca arundinacea conferring heat tolerance in Arabidopsis. *Int. J. Mol. Sci.* **2018**, *19*, 2702. [CrossRef]
131. Mishra, S.K.; Tripp, J.; Winkelhaus, S.; Tschiersch, B.; Theres, K.; Nover, L.; Scharf, K.-D. In the complex family of heat stress transcription factors, HsfA1 has a unique role as master regulator of thermotolerance in tomato. *Genes Dev.* **2002**, *16*, 1555–1567. [CrossRef]
132. Gong, B.; Yi, J.; Wu, J.; Sui, J.; Khan, M.A.; Wu, Z.; Zhong, X.; Seng, S.; He, J.; Yi, M. LlHSFA1, a novel heat stress transcription factor in lily (*Lilium longiflorum*), can interact with LlHSFA2 and enhance the thermotolerance of transgenic Arabidopsis thaliana. *Plant Cell Rep.* **2014**, *33*, 1519–1533. [CrossRef]
133. Zhu, B.; Ye, C.; Lü, H.; Chen, X.; Chai, G.; Chen, J.; Wang, C. Identification and characterization of a novel heat shock transcription factor gene, GmHsfA1, in soybeans (*Glycine max*). *J. Plant Res.* **2006**, *119*, 247–256. [CrossRef]
134. El-Shershaby, A.; Ullrich, S.; Simm, S.; Scharf, K.-D.; Schleiff, E.; Fragkostefanakis, S. Functional diversification of tomato HsfA1 factors is based on DNA binding domain properties. *Gene* **2019**, *714*, 143985. [CrossRef]
135. Fragkostefanakis, S.; Mesihovic, A.; Simm, S.; Paupière, M.J.; Hu, Y.; Paul, P.; Mishra, S.K.; Tschiersch, B.; Theres, K.; Bovy, A. HsfA2 controls the activity of developmentally and stress-regulated heat stress protection mechanisms in tomato male reproductive tissues. *Plant Physiol.* **2016**, *170*, 2461–2477. [CrossRef]
136. Fragkostefanakis, S.; Roeth, S.; Schleiff, E.; SCHARF, K.D. Prospects of engineering thermotolerance in crops through modulation of heat stress transcription factor and heat shock protein networks. *Plant Cell Environ.* **2015**, *38*, 1881–1895. [CrossRef]
137. Aldubai, A.A.; Alsadon, A.A.; Migdadi, H.H.; Alghamdi, S.S.; Al-Faifi, S.A.; Afzal, M. Response of tomato (*Solanum lycopersicum* L.) genotypes to heat stress using morphological and expression study. *Plants* **2022**, *11*, 615. [CrossRef]
138. Bharti, K.; Schmidt, E.; Lyck, R.; Heerklotz, D.; Bublak, D.; Scharf, K.D. Isolation and characterization of HsfA3, a new heat stress transcription factor of Lycopersicon peruvianum. *Plant J.* **2000**, *22*, 355–365. [CrossRef]

139. Baniwal, S.K.; Chan, K.Y.; Scharf, K.-D.; Nover, L. Role of heat stress transcription factor HsfA5 as specific repressor of HsfA4. *J. Biol. Chem.* **2007**, *282*, 3605–3613. [CrossRef]
140. von Koskull-Döring, P.; Scharf, K.-D.; Nover, L. The diversity of plant heat stress transcription factors. *Trends Plant Sci.* **2007**, *12*, 452–457. [CrossRef]
141. Rao, S.; Das, J.R.; Balyan, S.; Verma, R.; Mathur, S. Cultivar-biased regulation of HSFA7 and HSFB4a govern high-temperature tolerance in tomato. *Planta* **2022**, *255*, 31. [CrossRef]
142. Zhang, S.; Wang, S.; Lv, J.; Liu, Z.; Wang, Y.; Ma, N.; Meng, Q. SUMO E3 ligase SlSIZ1 facilitates heat tolerance in tomato. *Plant Cell Physiol.* **2018**, *59*, 58–71. [CrossRef]
143. Balyan, S.; Rao, S.; Jha, S.; Bansal, C.; Das, J.R.; Mathur, S. Characterization of novel regulators for heat stress tolerance in tomato from Indian sub-continent. *Plant Biotechnol. J.* **2020**, *18*, 2118–2132. [CrossRef]
144. Dai, X.; Shen, L. Advances and trends in omics technology development. *Front. Med.* **2022**, *9*, 911861. [CrossRef]
145. Muthuramalingam, P.; Jeyasri, R.; Rakkammal, K.; Satish, L.; Shamili, S.; Karthikeyan, A.; Valliammai, A.; Priya, A.; Selvaraj, A.; Gowri, P. Multi-Omics and integrative approach towards understanding salinity tolerance in rice: A review. *Biology* **2022**, *11*, 1022. [CrossRef]
146. Ashraf, U.; Mahmood, S.; Shahid, N.; Imran, M.; Siddique, M.; Abrar, M. Multi-omics approaches for strategic improvements of crops under changing climatic conditions. In *Principles and Practices of OMICS and Genome Editing for Crop Improvement*; Springer: Cham, Germany, 2022; pp. 57–92.
147. Yang, Y.; Saand, M.A.; Huang, L.; Abdelaal, W.B.; Sirohi, M.H. Applications of multi-omics technologies for crop improvement. *Front. Plant Sci.* **2021**, *12*, 563953. [CrossRef]
148. Faryad, A.; Aziz, F.; Tahir, J.; Kousar, M.; Qasim, M.; Shamim, A. Integration of OMICS technologies for crop improvement. *Protein Pept. Lett.* **2021**, *28*, 896–908. [CrossRef]
149. Thakkar, S.; Banerjee, A.; Goel, S.; Roy, S.; Bansal, K. Genomics-based approaches to improve abiotic stress tolerance in plants: Present status and future prospects. *Plant Perspect. Glob. Clim. Changes* **2022**, 195–219.
150. Sharma, N.; Siddappa, S.; Malhotra, N.; Thakur, K.; Salaria, N.; Sood, S.; Bhardwaj, V. Advances in potato functional genomics: Implications for crop improvement. *Plant Cell Tissue Organ Cult. PCTOC* **2022**, *148*, 447–464. [CrossRef]
151. Singh, R.; Kumar, K.; Bharadwaj, C.; Verma, P.K. Broadening the horizon of crop research: A decade of advancements in plant molecular genetics to divulge phenotype governing genes. *Planta* **2022**, *255*, 46. [CrossRef]
152. Zhu, F.; Ahchige, M.W.; Brotman, Y.; Alseekh, S.; Zsögön, A.; Fernie, A.R. Bringing more players into play: Leveraging stress in genome wide association studies. *J. Plant Physiol.* **2022**, *271*, 153657. [CrossRef]
153. Parihar, A.; Shiwani. Molecular breeding and marker-assisted selection for crop improvement. In *Plant Genomics for Sustainable Agriculture*; Springer: Cham, Germany, 2022; pp. 129–164.
154. Liang, Y.; Ma, F.; Zhang, R.; Li, W.; Dang, J.; Su, H.; Li, B.; Hu, T.; Zhang, M.; Liang, Y. Genome-wide identification and characterization of tomato 14-3-3 (SlTFT) genes and functional analysis of SlTFT6 under heat stress. *Physiol. Plant.* **2023**, *175*, e13888. [CrossRef]
155. Lou, D.; Liang, G.; Yu, D. OsSAPK2 confers abscisic acid sensitivity and tolerance to drought stress in rice. *Front. Plant Sci.* **2017**, *8*, 256670. [CrossRef]
156. Merrick, L.F.; Herr, A.W.; Sandhu, K.S.; Lozada, D.N.; Carter, A.H. Optimizing plant breeding programs for genomic selection. *Agronomy* **2022**, *12*, 714. [CrossRef]
157. Beyene, Y.; Gowda, M.; Pérez-Rodríguez, P.; Olsen, M.; Robbins, K.R.; Burgueño, J.; Prasanna, B.M.; Crossa, J. Application of genomic selection at the early stage of breeding pipeline in tropical maize. *Front. Plant Sci.* **2021**, *12*, 685488. [CrossRef]
158. Kress, W.J.; Soltis, D.E.; Kersey, P.J.; Wegrzyn, J.L.; Leebens-Mack, J.H.; Gostel, M.R.; Liu, X.; Soltis, P.S. Green plant genomes: What we know in an era of rapidly expanding opportunities. *Proc. Natl. Acad. Sci. USA* **2022**, *119*, e2115640118. [CrossRef]
159. Udriște, A.-A.; Iordachescu, M.; Ciceoi, R.; Bădulescu, L. Next-generation sequencing of local Romanian tomato varieties and bioinformatics analysis of the Ve locus. *Int. J. Mol. Sci.* **2022**, *23*, 9750. [CrossRef]
160. Roychowdhury, R.; Taoutaou, A.; Hakeem, K.R.; Gawwad, M.R.A.; Tah, J. Molecular marker-assisted technologies for crop improvement. In *Crop Improvement in the Era of Climate Change*; IK International Publishing House: New Delhi, India, 2014; pp. 241–258.
161. Sarkar, B.; Varalaxmi, Y.; Vanaja, M.; Kumar, N.R.; Prabhakar, M.; Jyothilakshmi, N.; Yadav, S.; Maheswari, M.; Singh, V. Genome-wide SNP discovery, identification of QTLs and candidate genes associated with morpho-physiological and yield related traits for drought tolerance in maize. *Res. Sq.* **2022**.
162. Kumar, A.; Sood, S.; Babu, B.K.; Gupta, S.M.; Rao, B.D. *The Finger Millet Genome*; Springer: Cham, Germany, 2022.
163. Fang, Y.; Qin, X.; Liao, Q.; Du, R.; Luo, X.; Zhou, Q.; Li, Z.; Chen, H.; Jin, W.; Yuan, Y. The genome of homosporous maidenhair fern sheds light on the euphyllophyte evolution and defences. *Nat. Plants* **2022**, *8*, 1024–1037. [CrossRef]
164. Thompson, S.D.; Prahalad, S.; Colbert, R.A. Integrative genomics. In *Textbook of Pediatric Rheumatology*; Elsevier: Amsterdam, The Netherlands, 2016; pp. 43–53.e43.
165. Amaro, A.; Petretto, A.; Angelini, G.; Pfeffer, U. Advancements in omics sciences. In *Translational Medicine*; Elsevier: Amsterdam, The Netherlands, 2016; pp. 67–108.

166. Khan, Q.; Chen, J.Y.; Zeng, X.P.; Qin, Y.; Guo, D.J.; Mahmood, A.; Yang, L.T.; Liang, Q.; Song, X.P.; Xing, Y.X. Transcriptomic exploration of a high sucrose mutant in comparison with the low sucrose mother genotype in sugarcane during sugar accumulating stage. *GCB Bioenergy* **2021**, *13*, 1448–1465. [CrossRef]
167. Conesa, A.; Madrigal, P.; Tarazona, S.; Gomez-Cabrero, D.; Cervera, A.; McPherson, A.; Szcześniak, M.W.; Gaffney, D.J.; Elo, L.L.; Zhang, X. A survey of best practices for RNA-seq data analysis. *Genome Biol.* **2016**, *17*, 13. [CrossRef]
168. Pertea, M.; Pertea, G.M.; Antonescu, C.M.; Chang, T.-C.; Mendell, J.T.; Salzberg, S.L. StringTie enables improved reconstruction of a transcriptome from RNA-seq reads. *Nat. Biotechnol.* **2015**, *33*, 290–295. [CrossRef]
169. Kwasniewski, M.; Daszkowska-Golec, A.; Janiak, A.; Chwialkowska, K.; Nowakowska, U.; Sablok, G.; Szarejko, I. Transcriptome analysis reveals the role of the root hairs as environmental sensors to maintain plant functions under water-deficiency conditions. *J. Exp. Bot.* **2016**, *67*, 1079–1094. [CrossRef]
170. Lowe, R.; Shirley, N.; Bleackley, M.; Dolan, S.; Shafee, T. Transcriptomics technologies. *PLoS Comput. Biol.* **2017**, *13*, e1005457. [CrossRef]
171. Mishra, P.; Singh, P.; Rai, A.; Abhishek, K.; Shanmugam, V.; Aamir, M.; Kumar, A.; Malik, M.Z.; Singh, S.K. Abiotic stress-mediated transcription regulation, chromatin dynamics, and gene expression in plants: Arabidopsis as a role model. In *Mitigation of Plant Abiotic Stress by Microorganisms*; Elsevier: Amsterdam, The Netherlands, 2022; pp. 321–345.
172. Frank, G.; Pressman, E.; Ophir, R.; Althan, L.; Shaked, R.; Freedman, M.; Shen, S.; Firon, N. Transcriptional profiling of maturing tomato (*Solanum lycopersicum* L.) microspores reveals the involvement of heat shock proteins, ROS scavengers, hormones, and sugars in the heat stress response. *J. Exp. Bot.* **2009**, *60*, 3891–3908. [CrossRef]
173. Sadder, M.T.; Alsadon, A.; Wahb-Allah, M. Transcriptomic analysis of tomato lines reveals putative stress-specific biomarkers. *Turk. J. Agric. For.* **2014**, *38*, 700–715. [CrossRef]
174. Zheng, Y.; Yang, Z.; Luo, J. Transcriptome analysis of sugar and acid metabolism in young tomato fruits under high temperature and nitrogen fertilizer influence. *Front. Plant Sci.* **2023**, *14*, 1197553. [CrossRef] [PubMed]
175. Gowda, G.N.; Zhang, S.; Gu, H.; Asiago, V.; Shanaiah, N.; Raftery, D. Metabolomics-based methods for early disease diagnostics. *Expert Rev. Mol. Diagn.* **2008**, *8*, 617–633. [CrossRef] [PubMed]
176. Turi, K.N.; Romick-Rosendale, L.; Ryckman, K.K.; Hartert, T.V. A review of metabolomics approaches and their application in identifying causal pathways of childhood asthma. *J. Allergy Clin. Immunol.* **2018**, *141*, 1191–1201. [CrossRef] [PubMed]
177. Scrivo, R.; Casadei, L.; Valerio, M.; Priori, R.; Valesini, G.; Manetti, C. Metabolomics approach in allergic and rheumatic diseases. *Curr. Allergy Asthma Rep.* **2014**, *14*, 445. [CrossRef] [PubMed]
178. Fiehn, O. Metabolomics—The link between genotypes and phenotypes. *Funct. Genom.* **2002**, *48*, 155–171.
179. Verslues, P.E.; Bailey-Serres, J.; Brodersen, C.; Buckley, T.N.; Conti, L.; Christmann, A.; Dinneny, J.R.; Grill, E.; Hayes, S.; Heckman, R.W. Burning questions for a warming and changing world: 15 unknowns in plant abiotic stress. *Plant Cell* **2023**, *35*, 67–108. [CrossRef] [PubMed]
180. Mashabela, M.D.; Masamba, P.; Kappo, A.P. Metabolomics and chemoinformatics in agricultural biotechnology research: Complementary probes in unravelling new metabolites for crop improvement. *Biology* **2022**, *11*, 1156. [CrossRef] [PubMed]
181. Putri, S.P.; Yamamoto, S.; Tsugawa, H.; Fukusaki, E. Current metabolomics: Technological advances. *J. Biosci. Bioeng.* **2013**, *116*, 9–16. [CrossRef] [PubMed]
182. Ghatak, A.; Chaturvedi, P.; Weckwerth, W. Metabolomics in plant stress physiology. In *Plant Genetics and Molecular Biology*; Springer: Cham, Germany, 2018; pp. 187–236.
183. Zheng, Y.-J.; Yang, Z.-Q.; Wei, T.-T.; Zhao, H.-L. Response of tomato sugar and acid metabolism and fruit quality under different high temperature and relative humidity conditions. *Phyton-Int. J. Exp. Bot.* **2022**, *91*, 2033–2054. [CrossRef]
184. Dhatt, B.K.; Abshire, N.; Paul, P.; Hasanthika, K.; Sandhu, J.; Zhang, Q.; Obata, T.; Walia, H. Metabolic dynamics of developing rice seeds under high night-time temperature stress. *Front. Plant Sci.* **2019**, *10*, 484751. [CrossRef] [PubMed]
185. Paupière, M.J.; Müller, F.; Li, H.; Rieu, I.; Tikunov, Y.M.; Visser, R.G.; Bovy, A.G. Untargeted metabolomic analysis of tomato pollen development and heat stress response. *Plant Reprod.* **2017**, *30*, 81–94. [CrossRef] [PubMed]
186. Luengwilai, K.; Saltveit, M.; Beckles, D.M. Metabolite content of harvested Micro-Tom tomato (*Solanum lycopersicum* L.) fruit is altered by chilling and protective heat-shock treatments as shown by GC–MS metabolic profiling. *Postharvest Biol. Technol.* **2012**, *63*, 116–122. [CrossRef]
187. Wilhelm, M.; Schlegl, J.; Hahne, H.; Gholami, A.M.; Lieberenz, M.; Savitski, M.M.; Ziegler, E.; Butzmann, L.; Gessulat, S.; Marx, H. Mass-spectrometry-based draft of the human proteome. *Nature* **2014**, *509*, 582–587. [CrossRef] [PubMed]
188. Vercauteren, F.G.; Bergeron, J.J.; Vandesande, F.; Arckens, L.; Quirion, R. Proteomic approaches in brain research and neuropharmacology. *Eur. J. Pharmacol.* **2004**, *500*, 385–398. [CrossRef]
189. Meyer, J.G. Qualitative and quantitative shotgun proteomics data analysis from data-dependent acquisition mass spectrometry. In *Shotgun Proteomics: Methods and Protocols*; Springer: Cham, Germany, 2021; pp. 297–308.
190. Ong, S.-E.; Blagoev, B.; Kratchmarova, I.; Kristensen, D.B.; Steen, H.; Pandey, A.; Mann, M. Stable isotope labeling by amino acids in cell culture, SILAC, as a simple and accurate approach to expression proteomics. *Mol. Cell. Proteom.* **2002**, *1*, 376–386. [CrossRef] [PubMed]
191. Ong, S.-E.; Mann, M. A practical recipe for stable isotope labeling by amino acids in cell culture (SILAC). *Nat. Protoc.* **2006**, *1*, 2650–2660. [CrossRef] [PubMed]

192. Nedelkov, D.; Nelson, R.W.; Florens, L.; Washburn, M.P. Proteomic analysis by multidimensional protein identification technology. In *New and Emerging Proteomic Techniques*; Springer: Cham, Germany, 2006; pp. 159–175.
193. Ross, P.L.; Huang, Y.N.; Marchese, J.N.; Williamson, B.; Parker, K.; Hattan, S.; Khainovski, N.; Pillai, S.; Dey, S.; Daniels, S. Multiplexed protein quantitation in Saccharomyces cerevisiae using amine-reactive isobaric tagging reagents. *Mol. Cell. Proteom.* **2004**, *3*, 1154–1169. [CrossRef] [PubMed]
194. Yakkioui, Y.; Temel, Y.; Chevet, E.; Negroni, L. Integrated and quantitative proteomics of human tumors. In *Methods in Enzymology*; Elsevier: Amsterdam, The Netherlands, 2017; Volume 586, pp. 229–246.
195. Mahmood, T.; Yang, P.-C. Western blot: Technique, theory, and trouble shooting. *North Am. J. Med. Sci.* **2012**, *4*, 429.
196. Rai, A.K.; Satija, N.K. Importance of targeted therapies in acute myeloid leukemia. In *Translational Biotechnology*; Elsevier: Amsterdam, The Netherlands, 2021; pp. 107–133.
197. Zecha, J.; Satpathy, S.; Kanashova, T.; Avanessian, S.C.; Kane, M.H.; Clauser, K.R.; Mertins, P.; Carr, S.A.; Kuster, B. TMT labeling for the masses: A robust and cost-efficient, in-solution labeling approach*[S]. *Mol. Cell. Proteom.* **2019**, *18*, 1468–1478. [CrossRef]
198. Khan, Q.; Qin, Y.; Guo, D.-J.; Lu, Z.; Xie, X.-Q.; Yang, L.-T.; Liang, Q.; Song, X.-P.; Xing, Y.-X.; Li, Y.-R. Proteome based comparative investigation of a high sucrose sugarcane mutant in contrast to the low sucrose mother variety by using TMT quantitative proteomics. *Sugar Tech* **2022**, *24*, 1246–1259. [CrossRef]
199. Jegadeesan, S.; Chaturvedi, P.; Ghatak, A.; Pressman, E.; Meir, S.; Faigenboim, A.; Rutley, N.; Beery, A.; Harel, A.; Weckwerth, W. Proteomics of heat-stress and ethylene-mediated thermotolerance mechanisms in tomato pollen grains. *Front. Plant Sci.* **2018**, *9*, 407077. [CrossRef] [PubMed]
200. Parrine, D.; Wu, B.-S.; Muhammad, B.; Rivera, K.; Pappin, D.; Zhao, X.; Lefsrud, M. Proteome modifications on tomato under extreme high light induced-stress. *Proteome Sci.* **2018**, *16*, 20. [CrossRef] [PubMed]
201. Li, H.; Qin, Y.; Wu, X.; O'Hair, J.; Potts, J.; Zhou, S.; Yang, Y.; Fish, T.; Thannhauser, T.W. Identification of heat-induced proteomes in meiotic pollen mother cells of tomato'Maxifort'using single-cell-type tandem mass tag (TMT) proteomics. *Veg. Res.* **2022**, *2*, 2.
202. Cardi, T.; Batelli, G.; Nicolia, A. Opportunities for genome editing in vegetable crops. *Emerg. Top. Life Sci.* **2017**, *1*, 193–207. [PubMed]
203. Kumar, S.; Rymarquis, L.A.; Ezura, H.; Nekrasov, V. CRISPR-Cas in agriculture: Opportunities and challenges. *Front. Plant Sci.* **2021**, *12*, 672329. [CrossRef] [PubMed]
204. Gaj, T.; Gersbach, C.A.; Barbas, C.F. ZFN, TALEN, and CRISPR/Cas-based methods for genome engineering. *Trends Biotechnol.* **2013**, *31*, 397–405. [CrossRef] [PubMed]
205. Podevin, N.; Davies, H.V.; Hartung, F.; Nogué, F.; Casacuberta, J.M. Site-directed nucleases: A paradigm shift in predictable, knowledge-based plant breeding. *Trends Biotechnol.* **2013**, *31*, 375–383. [CrossRef] [PubMed]
206. Makarova, K.S.; Wolf, Y.I.; Iranzo, J.; Shmakov, S.A.; Alkhnbashi, O.S.; Brouns, S.J.; Charpentier, E.; Cheng, D.; Haft, D.H.; Horvath, P. Evolutionary classification of CRISPR–Cas systems: A burst of class 2 and derived variants. *Nat. Rev. Microbiol.* **2020**, *18*, 67–83. [CrossRef]
207. Makarova, K.S.; Koonin, E.V. Annotation and classification of CRISPR-Cas systems. *CRISPR Methods Protoc.* **2015**, *1311*, 47–75.
208. Zetsche, B.; Gootenberg, J.S.; Abudayyeh, O.O.; Slaymaker, I.M.; Makarova, K.S.; Essletzbichler, P.; Volz, S.E.; Joung, J.; Van Der Oost, J.; Regev, A. Cpf1 is a single RNA-guided endonuclease of a class 2 CRISPR-Cas system. *Cell* **2015**, *163*, 759–771. [CrossRef]
209. Qiu, L.; Chen, X.; Hou, H.; Fan, Y.; Wang, L.; Zeng, H.; Chen, X.; Ding, Y.; Hu, X.; Yan, Q. Genome-wide characterization of the tomato UDP-glycosyltransferase gene family and functional identification of SlUDPGT52 in drought tolerance. *Hortic. Adv.* **2023**, *1*, 14. [CrossRef]
210. Tiwari, J.K.; Singh, A.K.; Behera, T.K. CRISPR/Cas genome editing in tomato improvement: Advances and applications. *Front. Plant Sci.* **2023**, *14*, 1121209. [CrossRef]
211. Chandrasekaran, M.; Boopathi, T.; Paramasivan, M. A status-quo review on CRISPR-Cas9 gene editing applications in tomato. *Int. J. Biol. Macromol.* **2021**, *190*, 120–129. [CrossRef] [PubMed]
212. Hoshikawa, K.; Pham, D.; Ezura, H.; Schafleitner, R.; Nakashima, K. Genetic and molecular mechanisms conferring heat stress tolerance in tomato plants. *Front. Plant Sci.* **2021**, *12*, 786688. [CrossRef]
213. Yu, W.; Wang, L.; Zhao, R.; Sheng, J.; Zhang, S.; Li, R.; Shen, L. Knockout of SlMAPK3 enhances tolerance to heat stress involving ROS homeostasis in tomato plants. *BMC Plant Biol.* **2019**, *19*, 354. [CrossRef] [PubMed]
214. De la Peña, R.C.; Ebert, A.W.; Gniffke, P.A.; Hanson, P.; Symonds, R.C. Genetic adjustment to changing climates: Vegetables. In *Crop Adaptation to Climate Change*; John Wiley & Sons: Hoboken, NJ, USA, 2011; pp. 396–410.
215. Driedonks, N.; Rieu, I.; Vriezen, W.H. Breeding for plant heat tolerance at vegetative and reproductive stages. *Plant Reprod.* **2016**, *29*, 67–79. [CrossRef] [PubMed]
216. Khan, Q.; Kashif, M.; Shah, S.J. Comprehensive analysis of the mechanism underlying plastic microbiome and plants interaction, with future perspectives. *J. Soil Plant Environ.* **2022**, *1*, 31–43. [CrossRef]
217. Tanger, P.; Klassen, S.; Mojica, J.P.; Lovell, J.T.; Moyers, B.T.; Baraoidan, M.; Naredo, M.E.B.; McNally, K.L.; Poland, J.; Bush, D.R. Field-based high throughput phenotyping rapidly identifies genomic regions controlling yield components in rice. *Sci. Rep.* **2017**, *7*, 42839. [CrossRef]
218. Araus, J.L.; Kefauver, S.C. Breeding to adapt agriculture to climate change: Affordable phenotyping solutions. *Curr. Opin. Plant Biol.* **2018**, *45*, 237–247. [CrossRef]

219. Driedonks, N.J. From Flower to Fruit in the Heat-Reproductive Thermotolerance in Tomato and Its Wild Relatives. Ph.D. Thesis, Radboud University Nijmegen, Nijmegen, The Netherlands, 2018.
220. Xu, J.; Driedonks, N.; Rutten, M.J.; Vriezen, W.H.; de Boer, G.-J.; Rieu, I. Mapping quantitative trait loci for heat tolerance of reproductive traits in tomato (*Solanum lycopersicum*). *Mol. Breed.* **2017**, *37*, 58. [CrossRef] [PubMed]
221. Hazra, P.; Samsul, H.; Sikder, D.; Peter, K. Breeding tomato (*Lycopersicon esculentum* Mill) resistant to high temperature stress. *Int. J. Plant Breed.* **2007**, *1*, 31–40.
222. Zhang, H.; Mittal, N.; Leamy, L.J.; Barazani, O.; Song, B.H. Back into the wild—Apply untapped genetic diversity of wild relatives for crop improvement. *Evol. Appl.* **2017**, *10*, 5–24. [CrossRef] [PubMed]
223. Tanksley, S.; Nelson, J. Advanced backcross QTL analysis: A method for the simultaneous discovery and transfer of valuable QTLs from unadapted germplasm into elite breeding lines. *Theor. Appl. Genet.* **1996**, *92*, 191–203. [CrossRef] [PubMed]
224. Ali, M.L.; Sanchez, P.L.; Yu, S.-b.; Lorieux, M.; Eizenga, G.C. Chromosome segment substitution lines: A powerful tool for the introgression of valuable genes from Oryza wild species into cultivated rice (*O. sativa*). *Rice* **2010**, *3*, 218–234. [CrossRef]
225. Bessho-Uehara, K.; Furuta, T.; Masuda, K.; Yamada, S.; Angeles-Shim, R.B.; Ashikari, M.; Takashi, T. Construction of rice chromosome segment substitution lines harboring Oryza barthii genome and evaluation of yield-related traits. *Breed. Sci.* **2017**, *67*, 408–415. [CrossRef]
226. Barrantes, W.; López-Casado, G.; García-Martínez, S.; Alonso, A.; Ruiz, J.J.; Fernández-Muñoz, R.; Granell, A.; Monforte, A.J. Exploring new alleles involved in tomato fruit quality in an introgression line library of Solanum pimpinellifolium. *Front. Plant Sci.* **2016**, *7*, 191030. [CrossRef] [PubMed]
227. Celik, I.; Gurbuz, N.; Uncu, A.T.; Frary, A.; Doganlar, S. Genome-wide SNP discovery and QTL mapping for fruit quality traits in inbred backcross lines (IBLs) of Solanum pimpinellifolium using genotyping by sequencing. *BMC Genom.* **2017**, *18*, 1. [CrossRef] [PubMed]
228. Haggard, J.E.; Johnson, E.B.; St. Clair, D.A. Multiple QTL for horticultural traits and quantitative resistance to Phytophthora infestans linked on Solanum habrochaites chromosome 11. *G3 Genes Genomes Genet.* **2015**, *5*, 219–233. [CrossRef]
229. Tripodi, P.; D'Alessandro, A.; Francese, G. An integrated genomic and biochemical approach to investigate the potentiality of heirloom tomatoes: Breeding resources for food quality and sustainable agriculture. *Front. Plant Sci.* **2023**, *13*, 1031776. [CrossRef]
230. Rick, C.M.; Butler, L. Cytogenetics of the tomato. *Adv. Genet.* **1956**, *8*, 267–382.
231. Rick, C.M. Genetic resources in Lycopersicon. *Tomato Biotechnol. Proc.* **1987**, 17–26.
232. Rick, C.M. The potential of exotic germplasm for tomato improvement. In *Plant Improvement and Somatic Cell Genetics*; Vasil, I.K., Scowcroft, W.R., Frey, K.J., Eds.; CABI: Wallingford, UK, 1982.
233. Rossi, M.; Goggin, F.L.; Milligan, S.B.; Kaloshian, I.; Ullman, D.E.; Williamson, V.M. The nematode resistance gene Mi of tomato confers resistance against the potato aphid. *Proc. Natl. Acad. Sci. USA* **1998**, *95*, 9750–9754. [CrossRef] [PubMed]
234. Matsukura, C.; Aoki, K.; Fukuda, N.; Mizoguchi, T.; Asamizu, E.; Saito, T.; Shibata, D.; Ezura, H. Comprehensive resources for tomato functional genomics based on the miniature model tomato Micro-Tom. *Curr. Genom.* **2008**, *9*, 436–443. [CrossRef] [PubMed]
235. Watanabe, S.; Mizoguchi, T.; Aoki, K.; Kubo, Y.; Mori, H.; Imanishi, S.; Yamazaki, Y.; Shibata, D.; Ezura, H. Ethylmethanesulfonate (EMS) mutagenesis of Solanum lycopersicum cv. Micro-Tom for large-scale mutant screens. *Plant Biotechnol.* **2007**, *24*, 33–38. [CrossRef]
236. Ezura, H. Toward in silico design and engineering of solanaceae and cucurbitaceae crops. In *Functional Genomics and Biotechnology in Solanaceae and Cucurbitaceae Crops*; Springer: Cham, Germany, 2016; pp. 251–258.
237. Shikata, M.; Hoshikawa, K.; Ariizumi, T.; Fukuda, N.; Yamazaki, Y.; Ezura, H. TOMATOMA update: Phenotypic and metabolite information in the micro-tom mutant resource. *Plant Cell Physiol.* **2016**, *57*, e11. [CrossRef]
238. Foolad, M.R. Genome mapping and molecular breeding of tomato. *Int. J. Plant Genom.* **2007**, *2007*, 64358. [CrossRef] [PubMed]
239. Kalloo, G. *Genetic Improvement of Tomato*; Springer Science & Business Media: Cham, Germany, 2012; Volume 14.
240. Suliman, A.A.; Elkhawaga, F.A.; Zargar, M.; Bayat, M.; Pakina, E.; Abdelkader, M. Boosting Resilience and Efficiency of Tomato Fields to Heat Stress Tolerance Using Cytokinin (6-Benzylaminopurine). *Horticulturae* **2024**, *10*, 170. [CrossRef]
241. Cammarano, D.; Jamshidi, S.; Hoogenboom, G.; Ruane, A.C.; Niyogi, D.; Ronga, D. Processing tomato production is expected to decrease by 2050 due to the projected increase in temperature. *Nat. Food* **2022**, *3*, 437–444. [CrossRef]
242. Silva, R.; Kumar, L.; Shabani, F.; Picanço, M. Assessing the impact of global warming on worldwide open field tomato cultivation through CSIRO-Mk3· 0 global climate model. *J. Agric. Sci.* **2017**, *155*, 407–420. [CrossRef]
243. Dutta, T.K.; Phani, V. The pervasive impact of global climate change on plant-nematode interaction continuum. *Front. Plant Sci.* **2023**, *14*, 1143889. [CrossRef] [PubMed]

MDPI AG
Grosspeteranlage 5
4052 Basel
Switzerland
Tel.: +41 61 683 77 34

Metabolites Editorial Office
E-mail: metabolites@mdpi.com
www.mdpi.com/journal/metabolites

www.ingramcontent.com/pod-product-compliance
Lightning Source LLC
Chambersburg PA
CBHW040316270926
51850CB00007BC/1499
9783725868506